PERCEPTION
OF DISPLAYED
INFORMATION

OPTICAL PHYSICS AND ENGINEERING

Series Editor: William L. Wolfe

Optical Sciences Center
University of Arizona
Tucson, Arizona

1968:

M. A. Bramson
Infrared Radiation: A Handbook for Applications

1969:

Sol Nudelman and S. S. Mitra, Editors
Optical Properties of Solids

1970:

S. S. Mitra and Sol Nudelman, Editors
Far-Infrared Properties of Solids

1971:

Lucien M. Biberman and Sol Nudelman, Editors
Photoelectronic Imaging Devices
    Volume 1: Physical Processes and Methods of Analysis
    Volume 2: Devices and Their Evaluation

1972:

A. M. Ratner
Spectral, Spatial, and Temporal Properties of Lasers

1973:

Lucien M. Biberman, Editor
Perception of Displayed Information

W. B. Allan
Fibre Optics: Theory and Practice

Albert Rose
Vision: Human and Electronic

CONTRIBUTORS

Lucien M. Biberman
Department of Electrical Engineering
University of Rhode Island
Kingston, Rhode Island
and
Institute for Defense Analyses
400 Army-Navy Drive
Arlington, Virginia

Richard R. Legault
Associate Director
Environmental Research Institute of Michigan
P. O. Box 618
Ann Arbor, Michigan

Frederick A. Rosell
Westinghouse Electric Corporation
Defense and Electronics Systems Division
Baltimore, Maryland

Otto H. Schade, Sr.
Consultant
32 Francisco Avenue
W. Caldwell, New Jersey

Alvin D. Schnitzler
Institute for Defense Analyses
400 Army-Navy Drive
Arlington, Virginia

Harry L. Snyder
Director, Human Factors Laboratory
Department of Industrial Engineering
   and Operations Research
Virginia Polytechnic Institute
   and State University
Blacksburg, Virginia

Robert H. Willson
Westinghouse Electric Corporation
Defense and Electronics Systems Division
Baltimore, Maryland

PERCEPTION OF DISPLAYED INFORMATION

Edited by

Lucien M. Biberman

Institute for Defense Analyses
Science and Technology Division
Arlington, Virginia

and

Department of Electrical Engineering
University of Rhode Island
Kingston, Rhode Island

PLENUM PRESS · NEW YORK-LONDON · 1973

Library of Congress Catalog Card Number 72-97695
ISBN 978-1-4684-2030-2   ISBN 978-1-4684-2028-9 (eBook)
DOI 10.1007/978-1-4684-2028-9
Softcover reprint of the hardcover 1st edition 1973

PREFACE

As this book took form, its contents furnished the material for a graduate course at the University of Rhode Island. Toward the end of that course, the class reviewed the literature on display characteristics and design. The universal criticism voiced in those reviews was that there was lots of hardware information but no criteria upon which one could base a sound design. Though one could learn all about the size and brightness of various displays, one could not form any judgment about how effectively the display transferred information to an observer.

As I reviewed our nearly completed text, an announcement crossed my desk stating that one of the professional societies in a seminar was to consider if one should not attempt to formulate a theory concerning *information transfer from displays to an observer*. That was the first title chosen for our book, before our publisher told us that "that was a paragraph, not a title." The group of contributors to this book have labored long in the conviction that there was a real need to develop and present a consolidated theory based upon the work of a number of pioneers, including Barnes and Czerny, de Vries, Rose, Coltman and Anderson, Schade, Johnson, van Meeteren, and others, who established the various parts of a substantial theoretical and experimental background that seemed ripe for consolidation. Starting in the middle-to-late sixties, efforts at that consolidation were undertaken by the authors of this book, first individually, then collectively, with the contributors getting together, then separating for more intensive individual studies.

This is the second cooperative effort for most of us. In the first book, *Photoelectronic Imaging Devices*, I had written the summary chapter of Volume 2 when Rosell submitted his version to me, and that better one went to press. In this present effort, I had prepared a rough draft of a chapter entitled "Closing the Loop" in which I attempted to pull together all the work of the distinguished men acknowledged above.

However, Rosell again sent in a section for his chapter that nicely closed the loop and furnished the psychophysical experimental evidence to nail it all down; thus, again it is his material that goes to press.

We now can say that we can, and in fact can teach others to, compute such things as the signal-to-noise ratio that is necessary for one to distinguish or identify a Chevrolet "Impala" from, say, a Chevrolet "Nova" when its image fills, say, one percent of a display. This sort of calculation, and indeed the very idea that such an erudite theory as that of Rose or the empirical data of Johnny Johnson could ever really lead to understanding the physical parameters necessary for the various visual task requirements on displays, was hotly and scornfully contested and rejected in correspondence from some of my friends in psychology.

However, this material has been taught briefly to a senior course in introductory photoelectronics and in more detail in a graduate course in special topics in electrical engineering at the University of Rhode Island. The problems include the calculation of the signal-to-noise ratio required to enable the counting of the stripes or the stars or the number of points on a star of a flag image that fills some fraction of the display area, at various levels of confidence. The answers to the problems showed some spread, but careful review showed that the students sometimes approximated things like stars in a wide variety of geometric models that made calculations simpler but different. When that was compensated for, the answers were uniform and correct in both the senior and graduate classes. What is more, the problems were based upon data previously experimentally verified. Thus at the end of this academic year, we can feel satisfied that we indeed have a method that is sound in theory and practice and that the theory and method of calculation can be learned quickly.

In an attempt to unify the contributions from a number of contributors, I have prepared an introduction to each chapter that introduces the topic, relates it to the material that has gone before, and points out how it forms the foundation for the material yet to be presented. In this manner, as editor, I have attempted not only to cover the important topics spanning over seven decades in time, but also to present work that was done within three months of the time we went into the final publication process. It is my hope that in doing so we have produced a text that can be used equally well by practicing electrooptical system designers and serious senior or first-year graduate students in electrical engineering, psychology, or human engineering who are interested in communications in the broad context of communicating with people through imagery on electrooptical displays.

LUCIEN M. BIBERMAN

March 1973
Kingston, R.I.

ACKNOWLEDGMENTS

This work became possible when the men who wrote it and the organizations supporting them joined in a common cause, this book.

We particularly wish to point out that the formal work began when the Institute for Defense Analyses management decided to accept it as part of its central research program. This form of support was heightened by the encouragement and advice offered by George H. Heilmeier and Edwin N. Myers of the Office of the Director of Defense Research and Engineering.

We are indebted to Frank McCann of the Air Force Avionics Laboratory, Wright-Patterson Air Force Base for his support to Westinghouse under Contract No. F-33615-70C-1461 of the analysis and research leading to Chapter 5.

We especially wish to thank Frank Scott and Robert Hufnagel and their associates at Perkin–Elmer for their published and unpublished material we have quoted.

We also wish to acknowledge the help of Ronald Erickson, Hershel Self, and G. C. Brock in providing us with the reports of their research.

We acknowledge with thanks the permission of the Society of Motion Picture and Television Engineers, the Society of Photographic Scientists and Engineers, and the Optical Society of America to extract and reproduce major amounts of basic archival material quoted here.

The material for the book would have been difficult to assemble without the help of the Institute for Defense Analyses library staff and the reference and interlibrary loan office of the library at the University of Rhode Island.

We would also like to acknowledge the effort of the Institute for Defense Analyses illustration staff for their help in providing the majority of the final line drawings and the advice and assistance of Beverley

F. Roberts and Richard A. Cheney in reviewing and assembling our manuscript.

The final manuscript itself was prepared principally by Ms. Barbara Ruby of Westinghouse, Baltimore, Ms. Anne Keheley and Ms. Nora A. Gleason of the Institute for Defense Analyses, and Ms. Joan Lamoureux of the University of Rhode Island. Ms. Lamoureux's patience with repeated changes and reorganizations of much material deserves and gets my most sincere thanks.

Finally, I wish to thank my wife Anne and my daughter Judith for their patience in listening to and reading parts of the book where elements of style and clarity were in question.

Lucien M. Biberman

CONTENTS

Chapter 4

Analysis of Noise-Required Contrast and Modulation in Image-Detecting and Display Systems

Alvin D. Schnitzler

Chapter 5

Recent Psychophysical Experiments and the Display Signal-to-Noise Ratio Concept

F. A. Rosell and R. H. Willson

Chapter 6

Image Reproduction by a Line Raster Process

Otto H. Schade, Sr.

Chapter 7

The Aliasing Problems in Two-Dimensional Sampled Imagery

Richard Legault

Chapter 8

A Summary

Lucien M. Biberman

INTRODUCTION

Lucien M. Biberman

As we have often said, it is a truism that a good picture is better than a bad picture, but it has not been abundantly clear, especially to the designers of most electrooptical imaging systems, what criteria must be used to decide if the picture is good or bad. The lens designers and the airborne camera designers have done considerable research in order to know beforehand that the optics they propose to build will permit the purchaser of their equipment to resolve the specified test chart that he expects to see with their equipment. From this need to meet contractual specifications for camera and lens performance grew a sharper understanding of lens quality. Though the questions about image quality, signal, and noise in photographs are still actively argued, the effects of image quality are now sufficiently well known that one can specify hard physical parameters and expect a definite range of performance from photointerpreters using imagery produced by equipment meeting definite specifications.

There are two major sets of factors governing the performance of man and his image-forming devices. One set of factors relates to the physics of light, optics, the solid state, materials, and the engineering approaches to the design of photoelectronic devices. The second set is related to the less well-known factors of psychophysics and vision and the interrelations among visual tasks, the quality of the image, the time available, and other subjective matters related to the observer and his task.

Schade points out that:

> "The *psychophysical factors* came first in photography (and TV) ranking 'print-quality' according to 'graininess,' 'gray scale,' and 'resolving power' long before the *objective equivalents* were understood correctly; simply because the psychophysical factors are determined by direct observation. Even very old data on the eye are still valid because the eye has not changed for a thousand years.
>
> "The correlation with objective factors, however, has required much experimental work with images of precisely known objective factors (signal, noise, MTF) which have been found to determine what we can see: resolving power, gray scale at any contrast, illumination, and gray scale value."

Little by little, work has been carried out establishing the link between the subjective measures and the objective measures of image quality. With the increasing belief that such reconciliation was about to be accomplished work, investigators began to probe (1) the relationship between subjective measures of quality and human judgment of quality; (2) the relationship between human judgment of quality and the objective measures of quality; (3) the relationship between subjective measures of quality and observer performance; and finally (4) the relationship between the objective measure of quality and observer performance.

Let us examine that tortuous listing of relationships. Limiting resolution* is a subjective measure: (1) One can examine the relationship between limiting resolution and people's opinions of photographic quality; (2) one can measure limiting resolution and how well a person can identify objects; (3) one can measure modulation transfer function and/or the graininess of film, as Scott and Hufnagel did, and correlate that with personal opinions of image quality on a sound statistical basis; and finally (4) one can measure such quantitative factors as acutance, the modulation transfer function, or grain noise (or some function of all three) and relate that to observer performance.

Finally, it may even be possible to derive from first principles a theory or set of theories that relate the physical characteristics of a picture or other form of image with the response of a human observer. The great achievement is the derivation of the theory and the experimental, or should I say clinical, verification. Since the end test in the experiment

---

* The topic of resolution, the various definitions, and related concepts are treated in Chapter 2, Section 2.2.1.

is human response, there is always more room for experimental error than the measurement of, say, acutance. We do not claim the ultimate in understanding of these relationships today, but it will, I believe, become clear that if one chooses some physical measure like the signal-to-noise ratio in an image element as a function of its spatial frequency, one can predict some aspects of perception about as well as one can measure them. To me, this is exciting and is the reason for our enthusiasm with the work of Rosell in both theory and clinical experiment.

Image quality is not a singular prerequisite for good observer performance. The image must be large enough and bright enough to be seen. Some will say it must also have enough contrast, and I shall not argue, but if I were to specify the signal-to-noise ratio as previously mentioned, I should have covered all I needed to say. This topic is a tortuous one. The relating of Blackwell's classical experiment with those of Rose, Schade, Coltman and Anderson, and more recently Rosell and Willson is a difficult task that is treated by Legault (1971) and is further extended by Schnitzler in Chapter 4 and Rosell in Chapter 5.

The problem of image size in detection tasks has long been understood. Steedman and Baker (1960) analyzed this problem long ago but have been ignored, since their factual and important data complicate the problem of system design, installation, or procurement. More of the early airborne real-time systems were deficient due to display size than for almost any other reason. Image size and the perception of objects is treated in Chapter 2, Section 2.4, and extensively in Chapters 4 and 5.

John Johnson of Night Vision Laboratories did research on this which he reported in 1958 (Johnson, 1958) and which has been and is being used as the present "bible."* Johnson's original paper was initially poorly presented in that important diagrams and tables are separated from the explanatory text (see Fig. 1.1 redrawn from the original). Thus where Johnson states that, "For a target to be recognizable, there must be system 'resolution' sufficient to place 4.0 ± 0.8 line pairs across the critical dimension of that target," the table and the diagram show "resolution across minimum dimension" and do not repeat "line pairs," which is the universal standard of resolution terminology except in the

---

* Schnitzler in Chapter 4 and Rosell in Chapter 5 incorporate Johnson's criteria into the results of theoretical studies by Rose and others to produce the concepts allowing one to calculate the signal-to-noise ratio required to detect, recognize, or identify objects against clear and cluttered backgrounds.

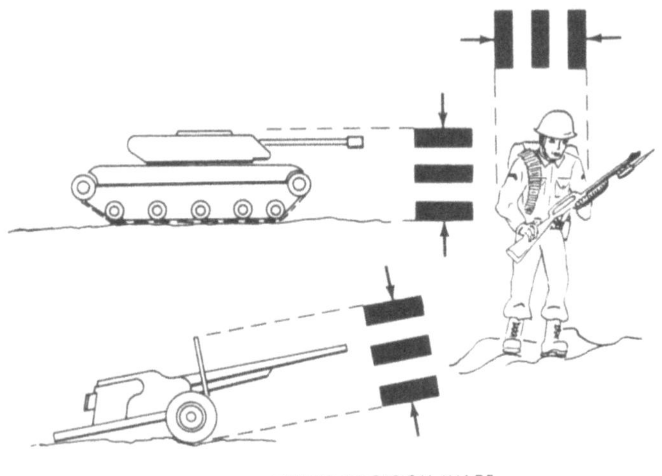

| TARGET | RESOLUTION PER MINIMUM DIMENSION IN LINE PAIRS | | | |
BROADSIDE VIEW	DETECTION	ORIENTATION	RECOGNITION	IDENTIFICATION
TRUCK	0.90	1.25	4.5	8.0
M-48 TANK	0.75	1.20	3.5	7.0
STALIN TANK	0.75	1.20	3.3	6.0
CENTURION TANK	0.75	1.20	3.5	6.0
HALF-TRACK	1.00	1.50	4.0	5.0
JEEP	1.20	1.50	4.5	5.5
COMMAND CAR	1.20	1.50	4.3	5.5
SOLDIER (STANDING)	1.50	1.80	3.8	8.0
105 HOWITZER	1.00	1.50	4.8	6.0
AVERAGE	1.0 ± 0.25	1.4 ± 0.35	4.0 ± 0.8	6.4 ± 1.5

Fig. 1.1. Required resolution for detection, orientation, recognition, and
identification.

television industry, which talks about "TV lines" (one "line pair" equals
two "TV lines"). Further, Johnson's paper assumes a knowledgeable
readership, and does not explain the implicit relationships between
resolution, contrast, and their dependence upon the signal-to-noise ratio.

As a result, designers commonly misuse Johnson's data, quoting
"lines on target" instead of line pairs in the minimum critical dimension.
The "line pair" versus "TV lines" confusion often results in systems being

underdesigned or underspecified by a factor of two. The "lines on target" versus "line pairs per minimum dimension" not only includes the line pair error but includes the length-to-width ratio of typical targets as well. These two errors result in typical underdesign factors of perhaps four or more.

The relationship between performance of an observer coupled to an image-forming system is stated (in Chapter 3) to be related to the difference between the modulation transfer function of the equipment and the demand modulation function (DMF) of the observer.

The DMF is defined by F. Scott (1966). It is the plot, of the modulation as a function of spatial frequency, needed to produce just-resolvable three-bar patterns in a specified film with specified development. This definition of the DMF is often called the aerial image modulation (AIM) curve. We use DMF to mean the modulation required to produce just-resolvable imagery on displays. The typical charts used are shown in Fig. 2-4. The DMF is not very well known except for a very few isolated tasks like the recognition of USAF standard three-bar test target patterns.

The detailed information of the demand modulation function for resolving the USAF three-bar target charts was obtained for a wide variety of films developed in various ways. The need was to judge just how much modulation was needed at the film plane to produce just-resolvable three-bar patterns in developed film to meet contract acceptance tests.

No such commercial motives justified comparable studies of people recognition, aircraft detection, or terrain identification. Nor has the problem of background character and its effect on observer demand functions been investigated to any significant extent.

Experience in tests where targets are immersed in varying degrees of clutter, like finding a golf ball in the "rough," indicates that the searcher must approach to such a short distance that detection and recognition occur simultaneously. In fact, under some conditions, detection, recognition, and identification occur almost simultaneously. Clutter, like the raster, is a form of annoying interference. Though clutter is difficult to analyze, rasters have been thoroughly analyzed, as in Chapters 6 and 7.

The proper design of devices starts, therefore, with an understanding of the difficulty of the visual task and thus of the value of the required signal-to-noise versus spatial frequency. The MTF and noise of the system or, in other words, the S/N versus spatial frequency, must then satisfy the demand curve to ensure, before detail design begins, that a real and useful device can, at least theoretically, be produced. It is quite probable that had this process been carried out in the past, many electro-optical system designs for specific requirements would have been recognized as inadequate before hardware came into existence. As it is, there is some good data on the demand function, but with reasonably few exceptions, this is still largely related to the detection of the USAF three-bar chart. However, one *can* infer much from the performance obtained with three-bar target charts. For example, by adding two elemental squares to the target, one forms the letter capital E. From such simple investigations one can determine that one requires five resolved lines in both horizontal and vertical directions to resolve alphanumeric symbols, *providing there is no seriously interfering raster effect.*

Schade has pointed out that for many practical applications the three-bar target oriented both vertically *and* horizontally gives a consistent indication of system quality not duplicated in television test charts. Rapidly converging wedge patterns could also be useful. Very slowly converging wedges or long, parallel lines do not give similar results.

The form and performance of imaging devices are closely related to the application; that is, to the character of the scene and its spectral composition, contrast, and radiance and to the difficulty and degree of detail in the visual task to be performed and the time available for its performance. The amount of time the observer has to make his observation, together with the detail required in his observation, is usually a factor overlooked in discussing the capability of these sensors. This factor is treated in Chapter 2, Section 2.4.

Since visual systems are devised to transmit information of intelligence to an observer through the visual sensory system, i.e., the eye and the brain of a human observer, it is important to understand what governs the transmission of information to an observer. In the case of visual aids employing electrooptical displays it becomes important to understand how the display affects the transmission of pictorial information and what governs the transmission of information from a display to an observer.

In 1969, in search of a new domestic television set, I read ratings, talked to knowledgeable colleagues, and went out to shop for a television receiver. In the course of visiting a number of stores, I was strongly impressed by the actions and apparent criteria of many other shoppers. A surprisingly large number of prospective buyers got up close to the picture tube and examined the raster structure. Most seemed very pleased with pictures in which the "lines" were sharp and clean. I intruded to ask a number of buyers so engaged how they were making their choice since they obviously were making a close technical (if subjective) comparison. Flattered at my recognition of them as men of discrimination and sound judgment, they invariably pointed out to me that they were looking for a set that was in good adjustment and sharp focus—"just look at how sharp the lines are, real good detail in *that* set."

Unfortunately, many technical personnel in electrooptical device design do not seem to be aware that there is any inconsistency in such reasoning. Few seem to exhibit an appreciation of the factors that govern image quality let alone quantitative measures for judging information transfer.

Rectangular objects parallel to the raster lines tend to lose their horizontal boundaries (but not the vertical) due to such boundaries looking like just another raster line and to the aliasing effects produced by the process of sampling the imagery in one dimension.

The above remarks apply primarily to television formatted displays shown on cathode ray tubes.

The advent of digitally addressed displays, such as those employing a matrix of light-emitting diodes, plasma elements, liquid crystal displays, and two-dimensionally addressed channel plate amplifier displays, makes *two*-dimensional rasters a potentially serious problem in the very near future unless the problem is tackled early in that class of display. The aliasing effects now can become limiting in both dimensions.

Since aliasing may not be a familiar effect, we shall briefly describe it here. Lavin (1971) describes aliasing as the generation of a spurious spectrum that is always a problem in frequency-domain analysis. In sampled-data systems like television, it results from sampling information at rates below the Nyquist criteria.

The usual form of aliasing is due to the generation of sums and differences between the frequencies present in the original scene and the sampling (or raster) frequency when improper filtering before and after sampling is employed. The most commonly noted effects are similar to the effect of a bas-relief, edge enhancement, or the barber-pole effect in a color television receiver in which colors rapidly cascade up or down across a plaid cloth pattern or necktie.

Aliasing in image-forming systems produces real and spurious imagery in the output of a system that is related to the real imagery in the input. Thus a picket fence may show as a fence with more pickets in the image than the real fence in the scene. (See Fig. 6.7, Chapter 6.) Schade in Chapter 6 and Legault in Chapter 7 consider aliasing in detail.

The broad extent over which this lack of understanding applied within an industry engaged in the production and merchandising of photographic and electronic images is perhaps best illustrated by a letter of Schade (1964b) summarizing his work to the Board of Editors of the Society of Motion Picture and Television Engineers. That letter should be required reading for all seriously concerned with electrooptically formed imagery.

There has been a series of approaches by others, by those interested in theory, by those using experiment, and by those experimentally evaluating the theories. A description of some of their work follows in Chapter 2.

Though this book consistently speaks about electrooptical displays, it really addresses the problems of live presentation on cathode ray tubes and does not address the problems of laser displays, scan converters, projection systems, etc.

We try to infer much from image evaluation studies. Unfortunately, most of those studies are done on photographic images and, in most of the TV or line scan studies, are restricted to stationary images. The value of threshold SNR_D (displayed signal-to-noise ratio) is higher for photographic images than for live displays because on the latter, the eye integrates a number of frames and thus operates on the noise in a more favorable manner than for a photograph, in which the noise is "frozen" into the image.

Because of restrictions of space and time, we do not consider the electronics or the economics of better displays though we do point out, for example, that spot wobble without excellent vertical stability is not effective. The vertical stability rather than the spot wobble is the expensive part.

Similarly we do not discuss interlacing and its effects. Schade points out that conclusions drawn from experiments with imaging systems that do not store an image should not be applied to systems that employ tubes with charge storage targets.

Cameras with such storage elements produce signals that are dependent on the time interval between successive line scans of the same line area. This is not the case for nonstorage devices like image dissector tubes. Equal signals in successive scans of a uniform field are not obtained from storage surfaces when random interlacing is employed, except when the interlacing fails because the beam is too broad, astigmatic, or unstable.

Schade further points out that whenever he finds a very high-resolution camera or display to be anisotropic, he suspects instability in the raster or astigmatism in the electrooptics. He notes that most electrooptical systems *are* astigmatic, especially off axis.

Though these last few remarks are more concerned with cameras, perhaps, than displays, the result, to the man studying TV imagery, is real, and troublesome. We hope these brief comments may help clarify some topics we did not really attempt to cover.

IMAGE QUALITY

Lucien M. Biberman

2.1. EDITOR'S INTRODUCTION

Some of the earliest research on television systems was concerned with the level of picture quality in terms of resolution, flicker, and shades of gray that would be needed for commercial broadcast purposes for entertainment or education. For these objectives easy viewing of large images with relatively little detail was needed. From these requirements came the 525-line, 30-cycle, 60-field picture as we know it in the U. S.

The work of Baldwin (1940) established much of the basis upon which image quality decisions were made for television in the U. S. His work should be considered required reading.

Though Strehl put forth a metric for image quality in 1902, image quality was not intensively studied for another several decades.

Since the 1940's interest has increasingly developed in the definition and formulation of mathematical models for image quality. Some of that work will be reviewed later in this chapter, while some of the most recent work will be treated by Snyder in Chapter 3.

It was not until the commercial broadcasting industry began to broadcast motion picture films in the 1940's that their engineers realized they had serious problems since they could only reproduce very poor pictures from high-quality films. The motion picture industry by that time had developed fine cameras and fine films capable of showing a great amount of detail on the projection screen, but all attempts to transmit such imagery via television resulted in severe image quality loss.

Schade and the other major contributors whose work was devoted to understanding image quality established quantitative specifications for equipment to produce given levels of image quality and solved the problem.

Though there were others before and after him who contributed significantly to the understanding of image quality in both motion pictures and television, it is my opinion that much of the credit for the present state of understanding belongs to Schade.

It must be noted that image reproduction, whether by photography, printing, or television, involves a one- or two-dimensional sampling of the imagery in the reproduction process. In photography the sampling is random through the medium of very small, developed grains of silver halide. In high-quality photography, the grains are sufficiently fine, i.e., small, that the grains visually fuse, that is, they appear to the naked eye to form a continuous image.

Before the end of the nineteenth century, printing of pictures was essentially a line reproduction process and etchings or wood cuts formed the usual printer's plate. In the late nineteenth century Horgan (1913) put into commercial use the process by which the various shades of gray or halftones were reproduced by a number of uniformly spaced black dots on a white paper. The relative area of the black dots to the white unprinted area determined the level of the gray scale.

As long as the basic spacing of the dot "screen" was small, the dots, like the grains of silver in a photograph, fused visually and the observer saw a continuous tone picture. The quality of such reproduction depends upon the fineness of the "screen." Usual newsprint uses a coarse screen that gives a reasonably continuous halftone appearance when viewed at or beyond the normal reading distances. If the reader wishes to look for more detail and brings the picture closer to his eye, he sees not more detail but the individual dots instead.

The better the printing process, the finer the screen and the closer one can examine imagery before one notices the dot structure.

It is clear that the process of sampling by dots in a two-dimensional sampling or by line scans in a one-dimensional sampling adversely affects image quality. Thus we should like to point out that the image quality criteria discussed so extensively in recent literature applies to continuous tone imagery, and that this quality of the continuous tone image may be seriously degraded if a sampling process is employed in which the sampling frequency is not large compared to the frequencies of interest in the imagery. ·

In this chapter we will discuss image quality of continuous tone images and then will review the various theoretical and experimental papers put forward concerning the effect on image visibility of one- and two-dimensional rasters on image quality, even though these topics are treated quantitatively by Schade in Chapter 6 and Legault in Chapter 7.

2.2. THE QUALITY OF CONTINUOUS TONE IMAGES

2.2.1. Concepts and Definitions

Since the first man attempted pictorial reproduction of things there have been judgments made of the quality of the reproduction. The most common one still uttered is usually in words akin to "that's a good (poor) likeness." With the more recent scientific development of optics, criteria such as resolving power and limiting resolution have come into acceptance.

Schade points out in a 1972 personal communication that these are insufficient, and that:

> "This broad term 'quality' should include all three basic param-
> eters (gray scale, MTF, noise) *and also* uniformity, isotropy, geo-
> metric distortion which are of prime importance for obtaining the
> total information *of which the format is capable*. The raster introduces
> another parameter because of both 'aliasing' and interference with
> vision. Others are brightness of the display and flicker and of course,
> image size and magnification, and observation time.
>
> "The 'perfect' display is perhaps in a practical sense a piece of the
> real world shown on the screen which looks the same as the real world
> to the unaided eye, or, for a different spectral band (IR) 'appears' to
> be a piece of the real world, sharp and clear and without distortion.
> In the latter case we have no subjective quality equivalent because we
> cannot see the infrared world directly. Here we must establish an
> objective standard."

Often in references we come across resolution, resolving power, acuity, acutance, etc. We shall review definitions and present below the widely accepted, if not rigorously defined, variants of these terms.

Resolving power is a term initially introduced by Lord Rayleigh to describe the quality of optical devices and components. He arbitrarily defined two equal luminous point sources to be resolved when the center of the diffraction pattern of the first fell into the first dark ring of the diffraction pattern of the second. Resolving power is normally used to describe the performance of a device when used with two point sources, and is expressed in terms of an angular separation between the two sources, usually in microradians or seconds of arc. When point sources A and B are resolved by the Rayleigh (1881) criteria, the profile of luminance through the images is as shown in Fig. 2.1. Thus the definition of resolution is sometimes said to be that angular separation that results in a crossover at 81% of the peak of the two central disks. Later workers have dropped the former concept and talked only about the ratio of peak to crossover, and, still later, in terms of contrast. An often-quoted statement is that resolution occurs when the contrast of the peak to crossover is 2 or 5%. Still fuzzier definitions say the eye resolves objects when their contrast exceeds $\geq 2\%$. This of course is a function of several things, such as brightness and shape of the object, but these factors are rarely specified in the usual quotation of "resolution."

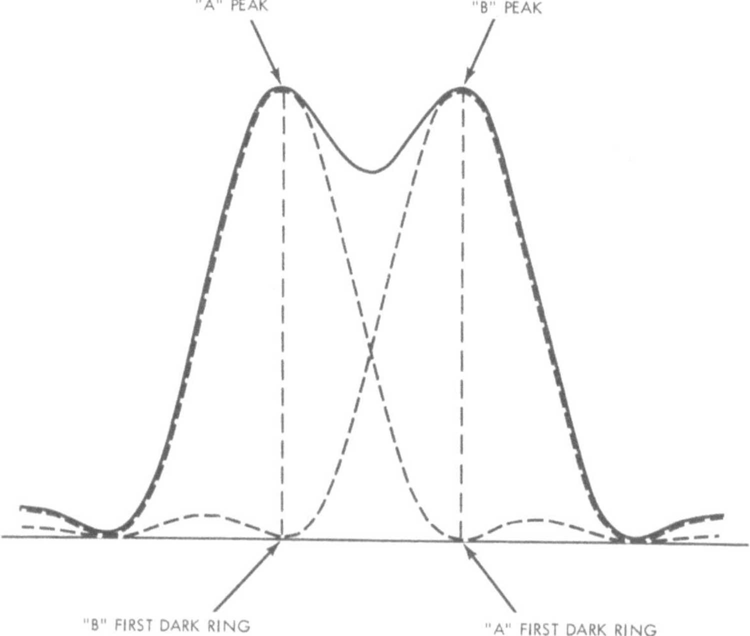

Fig. 2.1. Rayleigh criterion of resolution. Images of sources A and B are just resolved.

The foregoing discussion, however, may leave much to be desired by the reader who demands a more rigorous treatment than fits within the scope of this book. For them I recommend the book by Papoulis (1970) and especially his treatment of resolution on p. 448 *et seq.* Although he eloquently disposes of the subject in a few short paragraphs, many readers will need to read the first 447 pages to appreciate pp. 448–451. For those interested in a rigorous treatment along the more usual lines found in physical optics texts, we recommend the material in Born and Wolf (1965) under 7.6.3 entitled "Application of the Fabry–Perot Interferometer to the Study of the Fine Structure of Spectral Lines" and under 8.6.2, "Resolving Power of Image-Forming Systems."

Many eminent workers, including Schade, refer to resolving power in line pairs per millimeter in the image plane. This, of course, is a simple shift from angular separation to spatial frequency.

Slocum, in an internal report of Hughes Aircraft Co., has presented a good summary of commonly used resolution criteria, of which there are many. This data was also presented later in an open publication version (Slocum *et al.*, 1967). Although we prefer the Rayleigh criterion, we believe the reader should be aware of all the commonly used criteria and their interrelations. The following has been abridged and edited* from Slocum's original paper.

Several techniques exist for measuring and specifying display resolution. These can result in widely different resolution numbers for the same device. The cascading of several devices in series such as a scan converter tube, video amplifier, and cathode ray tube creates additonal complexities in specifying or predicting a total system resolution expecially when the resolution of each individual device is specified differently. Therefore a standard for comparing and combining the respective resolutions of several devices would be useful. This resolution standard may be arbitrarily selected but should be meaningful in application to sensor displays and should be capable of convenient and consistent measurement. The most frequently used techniques for specifying display device resolution are: shrinking raster, limiting television response, and spatial frequency response or modulation transfer function (MTF).

---

* Reproduced in its present form with the kind permission of G. K. Slocum.

Since each of the above commonly reported forms of resolution is usually defined in terms of the spot size of an assumed Gaussian electron beam giving rise to a Gaussian distribution of illuminance on a display phosphor, we shall briefly summarize the characteristics of that distribution. For more rigorous detail we refer the reader to any standard work on statistics or systems analysis.

The relative current density as a function of the radius of a Gaussian electron beam can be shown to be

$$\frac{J}{J_0}(r) = \exp\left[-\left(\frac{r}{r_0}\right)^2\right]$$

If J_0 is the peak current density at radius $r = 0$, then r_0 is the radius at which $J/J_0 = 1/e = 0.3679$.

The relative distribution of current density can also be shown as

$$\frac{J}{J_0} = \exp\left[-0.5\left(\frac{r}{\sigma}\right)^2\right]$$

where $\sigma = r_0/\sqrt{2}$.

Thus at a radius of σ the current density is $J_0 e^{-0.5} = 0.6065 J_0$. For that reason σ is often defined as the radius $r_0/\sqrt{2}$ or the radius at which the current density or luminance is 0.6065 that of peak.

Shrinking Raster Resolution. Shrinking raster resolution is determined by writing a raster of equally spaced lines on the display and reducing or "shrinking" the raster line spacing until the lines are just on the verge of blending together to form an indistinguishable blur. An experienced observer normally determines this flat-field condition at about 2–5% peak-to-peak luminance variation. Since the energy distribution in a CRT spot is frequently very nearly Gaussian, the flat-field response factor occurs at a line spacing of approximately 2σ, where σ is the spot radius at the 60.65% amplitude of the spot intensity distribution.

Television Resolution (TV Limiting Response). A television wedge pattern may be used to measure spot size by determining the point at which the lines of the well-illuminated, high-contrast wedge are just resolved. The number of TV lines per unit distance is then the number of black and white lines at the point of limiting resolution. The wedge pattern is equivalent to a square wave modulation function, and therefore the TV resolution is often referred to as the limiting square wave response. (One needs to be careful to remember that, in television parlance, one cycle of the square wave produces a black and white interval and is considered as two television lines.) Assuming a Gaussian spot distribution, the limiting square wave response occurs at a television line spacing of 1.18σ. Thus there are approximately 1.7 times as many limiting television lines per unit distance as shrinking raster lines for a display with the same spot size.

Modulation Transfer Function (MTF). Schade, in his sine wave response technique, analyzes the display resolution by the use of a sine wave test signal, rather than the square wave signals employed in a TV test pattern or the photographic bar patterns commonly employed in the optical field, i.e., the input sine wave pattern is 100% modulated; the sine wave response test produces a curve of response called the modulation transfer function. The International Commission on Optics (ICO) defines the modulation transfer function as the modulus of the Fourier transform of the line spread function. The earlier definition used above follows directly from the ICO definition. An MTF is shown in Fig. 2.2. When several devices are cascaded such as a scan converter and CRT, the MTF's of the individual devices are multiplied together to provide the total system MTF. This capability for computing the system MTF from individual device MTF's is a major advantage of using the MTF resolution measurement. Another advantage of the MTF technique is the graphic capability it provides for the determination of the visual acuity limit, described below, of a given display system. The MTF response can be related to the other resolution measurements (shrinking raster and television) when the spot distribution is known. If a Gaussian spot shape is assumed, for example, and a sine wave test signal were set on the display at a half-cycle spacing corresponding to the shrinking raster resolution line spacing, the resultant observable modulation on the display would be approximately 29%. Table 2.1 has been calculated for a Gaussian spot distribution and can be used to convert from one resolution measurement to another. In real displays distributions may deviate considerably from Gaussian and large errors can occur. See Chapter 6.

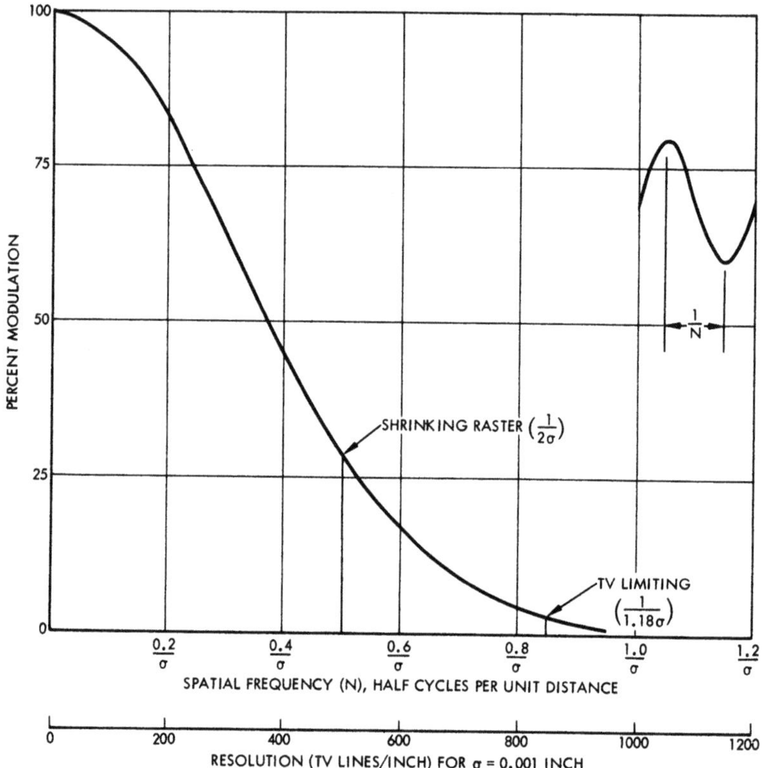

Fig. 2.2. Relative modulation transfer function [after Slocum (1967)].

Sensor Resolution. Sometimes sensor resolution is imprecisely defined by the sensor 3 dB response occurring at a line spacing of 1.67σ or a frequency $n = 0.6/\sigma$ and is 1.2 times the shrinking raster resolution.

2.2.1.1. *Visual Acuity of the Observer*

The detail discrimination threshold of the human eye, i.e., visual acuity, has been investigated exhaustively. Visual acuity is the reciprocal of the angle subtended by the minimum size standard test object that can be resolved 50% of the time by a human observer. The angle resolved by a normal eye is approximately 1 min of arc; normal acuity is that reciprocal, i.e., one. Under poor light and/or contrast acuity will degrade down to values of less than 0.1. Various types of acuity, such as minimum detectable, minimum separable, vernier, and stereo, have been defined.

TABLE 2.1

Conversion Table for Various Measures of Display Resolution*

From		To							
		TV limiting	10% MTF	TV_{50}	Shrinking raster	50% Amplitude	50% MTF	Optical	Equivalent passband
TV limiting	1.18σ	—	0.80	0.71	0.59	0.50	0.44	0.42	0.33
10% MTF	1.47σ	1.25	—	0.88	0.74	0.62	0.55	0.52	0.42
TV_{50} (3 dB)	1.67σ	1.4	1.14	—	0.84	0.71	0.63	0.59	0.47
Shrinking raster	2.00σ	1.7	1.36	1.2	—	0.85	0.75	0.71	0.56
50% Amplitude	2.35σ	2.0	1.6	1.4	1.17	—	0.88	0.83	0.66
50% MTF	2.67σ	2.26	1.8	1.6	1.33	1.14	—	0.94	0.75
Optical (l/e)	2.83σ	2.4	1.9	1.7	1.4	1.2	1.06	—	0.80
Equivalent passband (N_e)	3.54σ	3.0	2.4	2.1	1.77	1.5	1.33	1.25	—

* After Slocum.

Minimum separable visual acuity applies in the case of shape recognition in which, generally, closely spaced image details must be discerned. It is known to vary as a function of adaptation level, image brightness, contrast, exposure time, image motion, vibration, spectral characteristics, angular position of the target relative to the line of sight, etc. Visual acuity is usually defined in terms of arbitrary regular test patterns with generally sharp edges, although some studies have been conducted with sine wave patterns.

Discrimination of imagery detail differs from visual acuity measurements in that it requires detection of discontinuities characterized by diffuse edges and irregular intensity distributions. The published acuity data are statistics representing specified performance levels (usually 50% detection probability). Thus they provide information in a probabilistic rather than in a deterministic sense. Therefore in any specific instance visual performance may fall far short of or exceed predictions based on published data. In general, standard visual acuity data must

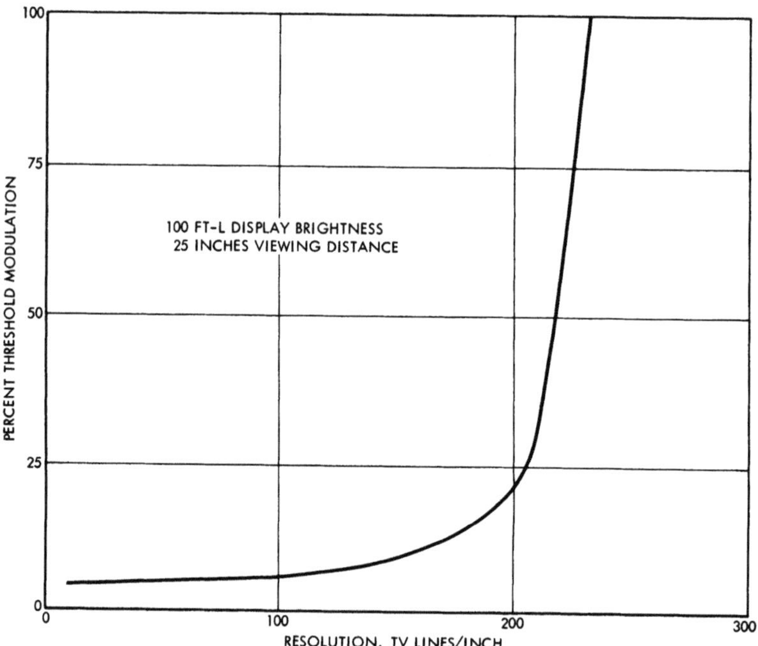

Fig. 2.3. Modulation for visual threshold [after Patel (1966)].

be modified by field factors related to the conditions under which actual (not laboratory) performance is to be evaluated to obtain realistic operator performance estimates under operational conditions. Unmodified data can be used to establish average expected limits of performance under ideal conditions.

A minimum visual threshold modulation curve (Patel, 1966) is plotted in Fig. 2.3. These data are for a sine wave test pattern with an average brightness of 100 ft-L viewed at 25 in. These data will vary as a function of the geometry of the pattern, i.e., how large its amplitude is and its size relative to the total field presented. (See discussion in Chapter 5.) This curve neglects image motion, exposure time, wavelength, and vibration effects. The visual threshold modulation curve sets the lower limit on useful system modulation. To be visually discernible an image detail must exceed the threshold modulation of Fig. 2.3. The maximum usable limiting resolution of a sensor/display system (for a specified viewing distance) is indicated by the point at which the modulation transfer function (MTF) of the system crosses the corresponding DMF curve. The concept of noise equivalent modulation and its calculation is presented by Schnitzler in Chapter 4.

Lavin (1971) says:

> "Limiting resolution is one of the simplest of all performance qualities to measure; the technique involves the use of a 'standard' resolving power chart such as those shown in [Fig. 2.4].
>
> "Unfortunately the values obtained for limiting resolution or resolving power are strongly influenced by the geometry of the test chart used and by factors such as the length and number of the test-chart lines, or whether the chart is prepared as black lines on a white background, black lines on a gray background, etc.
>
> "Whenever the resolving power is specified, the geometry of the resolution chart employed and its modulation must also be stated. Limiting-resolution-measurement variations in excess of 25% have been observed with changes only in the number and length of lines of the resolution chart. The same system and the same observer were employed for each test. Even greater changes are noted when low-modulation resolution charts are employed having different geometries."

Usage varies and my best advice is to read with care and to determine, in context, what is meant by resolution.

In human vision, as previously defined, the reciprocal of resolving power expressed in minutes of arc is called visual *acuity*. Thus under

Fig. 2.4. USAF, NBS, Igor Limansky, and EIA resolution charts. [Note: the usual Igor Limansky resolution chart consists of 9 individual charts arranged in a 3 × 3 array.]

some conditions the eye can resolve 1 min of arc and is said to have an acuity of one. Under less than optimum conditions of seeing acuity may range down below 0.1.

Acutance, on the other hand, is a measure of the sharpness of an edge expressed in terms of the mean square of the gradient of radiant or luminous flux or density (in a photographic image) with distance from the edge. If one plots density (the log of the reciprocal of the transmission) versus distance for a sharp edge on a photographic plate over a range of density ΔD and measures the position X_i and the density D_i at n positions, then the mean-squared gradient is

$$\overline{G_x{}^2} = \left[\sum_{i=1}^{n} (\Delta D_i / \Delta x_i)^2 \right] \Big/ n$$

and acutance is defined as

$$A = \overline{G_x{}^2} / D'$$

where D' is the total range in density over which the measurements are made.

For comparative measurements of acutance a choice of equal values of D' is necessary, since unfortunately acutance will vary somewhat as D' varies. There is as yet no single choice of D' that satisfies both the theoretical and practical sensibilities of those concerned with image quality.

Acutance and resolving power are often closely related but this need not be the case.

Acutance has been used by many as a measure of image quality. Higgins and Wolfe (1955) found that the expression $A[1 - \exp(-kR^2)]$, where k is a constant and R is resolving power, and subjective impressions of image sharpness were well correlated.

In an address in 1967, G. C. Brock,* in discussing acutance, said:

> "The concept of acutance probably suffered, at least in aerial photography, from the date of its introduction. Functionally, it had the merit of exposing the folly of testing systems by their high-contrast resolving power, but the worst of such anomalies were avoided by the obvious approach of using low-contrast resolving power as advocated

---

* Unpublished address to the Soc. Photo. Sci. and Eng., quoted by permission of the author.

by Selwyn and Howlett. Today we may say that the great merit of low contrast resolving power testing is that it discriminates against low modulation regions of the transfer function that contribute little to microcontrast on images of any shape. But at that time there was prejudice against low contrast testing, and acutance attracted well deserved interest. However, it requires microdensitometry, which was less commonly available then, than now, and the general idea of doing extensive physical measurements was less acceptable. As the climate for more sophisticated testing improved, interest in acutance was swamped in the rising Fourier flood, and attention was deflected from the real merits of measuring something directly related to the quality of images as we see them.

"Acutance obviously has no place in lens testing and system analysis and does not offer the display of quality versus size that is so desirable in aerial photography. Nevertheless, its emphasis on edge quality is very appropriate, since edges are common in the highly detailed scenes we examine. In some form it may yet find a place for comparative evaluation of aerial negatives, in circumstances where reduction of the edge trace to an MTF is unnecessary or unjustified."

2.2.2. Factors Related to Geometrical and Physiological Optics

Hardy and Perrin (1932) in their classic text summarized the effect of the aperture size of a lens on the character of an image of a distant point source. They show that if the aperture is very large compared to the wavelength of incident monochromatic light, the wavefront approximates a plane wave of infinite extent. The lens thus produces a simple point reconstruction of the infinitely distant source. If, however, the aperture is less than half a wavelength of the incident monochromatic light, all portions of the distant image may be considered to be of similar amplitude and phase, and thus will propagate in all directions.

Between these two extremes lies the realm of the apertures of real devices. For apertures larger than half a wavelength and smaller than an infinite number of wavelengths the aperture produces a diffraction pattern. A distant point source imaged by an aberrationless lens will produce a diffraction pattern, the central disk of which contains 84% of the total flux appearing in the image. This diffraction pattern was analyzed by Airy in 1834, resulting in the following:

$$E = Kp^4\left[1 - \frac{1}{2}m^2 + \frac{1}{3}\left(\frac{m^2}{2!}\right)^2 - \frac{1}{4}\left(\frac{m^3}{3!}\right)^2 + \frac{1}{5}\left(\frac{m^4}{4!}\right)^2 + \cdots\right]^2$$

where $K = $ const and $m = (\pi p/\lambda) \sin \alpha$, with p the radius of the aperture, λ the wavelength, and E the irradiance at any point in the image lying at an angle α off the optical axis.

If one constructs a table of m versus E, one finds $E = 1$ when $m = 0$ and this is called the central disk or the Airy disk. When $m = 0.61\pi$, $E = 0$ and this is called the first dark ring or zone. When $m = 0.81\pi$, $E = 0.174$ and this region is called the first bright ring. The second and third dark rings occur at 1.116π and 1.619π. The third maximum, or second bright ring, occurs when $m = 1.333\pi$ and the value of radiance in that ring is only 0.0041 of that of the central disk.

All the foregoing applies to distortionless apertures. When a real lens or mirror is used some aberrations are introduced. This results in a shift of radiant flux out of the central disk into the surrounding bright and dark rings.

In the case of two point sources separated by small angular distances, modest aberrations have little effect. In the case of broad-area sources with much low-contrast detail, the image is made up of many Airy disks and attendant bright and dark circles overlapping to form an image of degraded contrast, especially for fine details of initially low source contrast.

Strehl in 1902 attempted to describe image quality in terms of a shift of flux out of the central disk into the surrounding bright rings. His work was originally published in German and is hard to obtain. Linfoot (1964) describes the Strehl "Definitionshelligkeit," which is the ratio of the monochromatic flux in the Airy disk of a real system with aberrations to that in an ideal, aberration-free system.

In the ideal system the flux in the central bright core is 0.84 while $1 - 0.84$ is the flux in the surrounding rings; in real systems the terms are $0.84V$ and $1 - 0.84V$, where V is the Strehl ratio.

To a real extent the Strehl criterion of image quality is useful today for broad-area, low-contrast imagery such as aerial photography.

In practice the Strehl ratio can be measured simply only on isolated point sources. On extended images the ratio cannot be extracted as previously defined since a large number of overlapping cores and rings from adjacent points coalesce in an area image. Thus one must use another approach. This process is well described by Linfoot in his text on optical image evaluation (Linfoot, 1964) in Chapter V, Section 1, "Strehl Definition and Correlation Quality."

Actually the Strehl measure may be related in modern terminology to the two-dimensional modulation transfer function integrated over all frequencies in both coordinates:

$$\int\limits_{-\infty}^{\infty}\!\!\int T(v_{x,y})\, dv_x\, dv_y$$

The application of the Strehl criterion is discussed as a measure of image quality in Section 2.2.3.2.

More recently a number of workers have considered image quality and the informational value of imagery from the standpoint of communications theory. One of the most recent studies is by Schnitzler in Chapter 4, while one of the first was by Linfoot in his series of papers spanning the 1950's. His work not only pinpoints the areas of concern, but his writing is so clear as to make its reading a pleasure.

We quote below some of the points made by him in his 1958 paper (Linfoot, 1958).* In synopsis he says the following.

> "The problem of optical image evaluation assumes a more tractable form when treated as a problem in communication theory. The complex amplitude distribution over the entry pupil of an optical system can be regarded as an intercepted part of a message sent out by the object and the formation of an image in the focal plane of the system as a decoding of this intercepted message which presents information about the object in a more convenient form.
>
> "An optical system can properly be said to be of high quality only if the amount of information contained in its images approaches the maximum possible with the prescribed aperture and receiving surface, and it is an agreeable consequence of the special properties of centered optical systems of the usual type that those which are efficient according to this criterion also form images which are sharp and clear in the usual sense of the words. Even so, fine detail may be metamorphosed and it is worth considering whether a departure from maximum image-fidelity may not some times allow a useful gain in the amount of information recorded."

Linfoot goes on to show that for some classes of applications one can process images to improve informational content, perhaps at the

---

* Reproduced in abridged form from *Physica* by permission of the North-Holland Publishing Co., Amsterdam.

expense of fidelity. He summarizes one of his arguments as follows:

> "If the arithmetical recoding of optical images were a standard practice today, instead of a prospect for the future opened up by the advent of the fast computing machines, we would go on to add that informationally optimized designs were always to be preferred. When, however, the process of recoding (under the name of image interpretation) has to be carried out in the nervous systems of human beings, the choice between, say, a fidelity-maximized design and an informationally optimized one ought to be made on the basis of the experimentally determined needs and capabilities of the human interpreter."

It is clear that if an astronomer wished to locate a star by means of a plate he might well settle for an image that was a darkened section of emulsion the integral of which could give an estimate of the star's radiant intensity while all else was reduced to a clarified emulsion. Since there are many stars, perhaps he would settle for all stars of less than a given magnitude appearing as black circular images while all else was reduced to a clear negative. This clearly is not faithful image reproduction but it certainly maximizes the signal-to-noise ratio of some particular portion of the imagery.

On the other hand, portrait photographers speak of "soft lenses" and even go as far as to introduce sheer fabrics between lens and plate to diffuse and further soften imagery for aesthetic effects. What then is the criterion for image quality, or are there several? Linfoot proposes three: (1) structure content, (2) correlation quality, (3) image fidelity.

Following Linfoot's lead we may examine image fidelity as the mean square difference between the radiance distribution $\mathscr{R}_0$ in a perfect replication of the object and that, $\mathscr{R}_I$, in the image within an area a of the image plane.

The mean square difference within a is thus

$$(1/a) \iint_a [\mathscr{R}_0(x, y) - \mathscr{R}_I(x, y)]^2 \, dx \, dy$$

One can normalize the function by dividing through by

$$(1/a) \iint_a [\mathscr{R}_0(x, y)]^2 \, dx \, dy$$

This results in Linfoot's definition of *fidelity defect*:

$$\left\{ \iint\limits_a [(\mathscr{R}_0 - \mathscr{R}_I)^2 \, dx \, dy] \right\} \Big/ \left(\iint\limits_a \mathscr{R}_0{}^2 \, dx \, dy \right)$$

which in a well-designed and adjusted device shows numerical values between zero and one.

Similarly, Linfoot defines *image fidelity* as one minus the fidelity defect, or

$$\Phi_a = 1 - \left\{ \left[\iint\limits_a (\mathscr{R}_0 - \mathscr{R}_I)^2 \, dx \, dy \right] \Big/ \left[\iint\limits_a (\mathscr{R}_0)^2 \, dx \, dy \right] \right\}$$

Thus Φ_a normally lies between zero and one.

The second Linfoot criterion is *structural content* of an image:

$$T_a = \left(\iint\limits_a \mathscr{R}_I{}^2 \, dx \, dy \right) \Big/ \left(\iint\limits_a \mathscr{R}_0{}^2 \, dx \, dy \right)$$

T_a can be thought of as a measure of the reproduction of fine object detail rather than the fidelity of the position of that detail within an area a.

The third Linfoot criterion, *correlation quality*, follows from conventional practice and is defined as

$$Q_a = \left(\iint\limits_a \mathscr{R}_0 \mathscr{R}_I \, dx \, dy \right) \Big/ \left(\iint\limits_a \mathscr{R}_0{}^2 \, dx \, dy \right)$$

Linfoot goes on to show that the three quality factors, quality of fidelity, structural content, and correlation quality, are not independent but are related by the equation

$$\Phi_a + T_a = 2Q_a$$

Linfoot (1958) shows that considerable information is lost due to fundamental processes of information transfer from object to image. His illustration concerns the transmission by either emitted or reflected incoherent radiation of information concerning the spatial radiance distribution of the emitting or reflecting surface.

The radiation from the surface carries a full "message" concerning the configuration of the surface but only a very small fraction of that "total message" is usually intercepted by the image-forming optical system. If D is the diameter of the receiving optics and λ is the wavelength of the transmitted information, then the received "message" contains a great deal of information about the spatial distributions of radiance down to angular spatial frequencies approaching $2D/\lambda$, a parameter of the system sometimes called the "diffraction limit" of resolution. This limitation is due to the geometry of the receiving pupil or optical aperture and applies to an optical system without aberrations or imperfections.

Real systems of course possess imperfections in their optical surfaces and thus do not yield $2D/\lambda$ as a limiting frequency. Further, they add distortions and such other effects as veiling glare, which, though not decreasing the information in the image, add a significant amount of noise and make it more difficult if not impossible to extract the information that is present. One may perhaps loosely compare these processes to a change in state of a material which increases its entropy, thus decreasing the mount of energy available for useful purposes.

The publications by Linfoot (1966) and Fellgett and Linfoot (1955) are highly recommended to the reader interested in acquiring an analytical grasp of image evaluation from the more fundamental viewpoint of the informational content of images.

We shall not dwell further on Linfoot's analysis here since he concerns himself with levels of informational value per unit solid angle or per unit area far in excess of that we expect to discuss in the informational transfer from electrooptical displays, the limitations of which are usually caused by aberrations and electronic effects and thus are not really concerned with the effects discussed for images formed by diffraction-limited systems or even those remotely approaching the diffraction limit.

More to our interest is an understanding of the limitations of the eye on information transfer and the resolution limits as a function of both illuminance and contrast. For such purposes the work of Blackwell (1946a, b) is of great importance.

Related papers of Rose and of Schade, Coltman, and Coltman and Anderson, and later the experiments of Rosell and Willson, are discussed in Chapters 4 and 5 and form a basis of understanding the seeing of spatial frequencies quite coarse compared to those considered by Linfoot.

Good optical systems can resolve perhaps 1000 line pairs per millimeter, or perhaps a microradian. A healthy young human eye can resolve about $\frac{1}{3}$ mrad. Thus the unaided eye is often the limiting element in a complex optical system that ends up employing the human eye as the sensing element.

Schade (1971), for example, has demonstrated high-resolution camera tubes capable of resolving, say, 10,000 TV lines per picture height. Unfortunately, the display of such imagery can only be done effectively if perhaps 1% of the total image is displayed to the observer at any one time.

Thus we move on to consider and reference appropriate papers concerned with the levels of image quality one might expect to use effectively on a display that can be utilized effectively by a single observer, as contrasted to huge "situation displays" to be maintained, updated, and viewed by many people.

In attempting to understand the display–eye interface, it is desirable to review some of the early work leading to the present still incomplete understanding of the eye and its functions. Some work has been done on living rods and cones removed from animal eyes, and similar work has been done on human eyes, though most of the latter work has been accomplished either by peering into the living eye through a still intact cornea or by interrogating an observer about the sensations of vision or the objects perceived by him in carefully controlled experiments. Perhaps one of the most extensive of the latter was carried out by Blackwell and reported by him in 1946. In this paper Blackwell (1946a) reported*:

> "This paper pertains to the determination of the contrast threshold of the normal human observer under a wide variety of experimental conditions. The typical experimental procedure consisted in projecting a spot of light on a white screen some sixty feet from a group of observers who individually reported whether the stimulus had been seen. A large number of such presentations, made with varying brightness of the stimulus, provided data from which, by statistical analysis, the contrast threshold could be determined. Experiments of this sort were

---

* Reproduced in highly abridged form by permission of the author' and the Optical Society of America.

repeated with stimuli of varying sizes and with values of screen brightness covering a range from full daylight to slightly less than the darkest night. In all, more than two million responses to the test stimulus were recorded, some four hundred and fifty thousand of which have been statistically analyzed and reported herein.

"Five appropriate stimulus contrasts were selected on the basis of preliminary observations. Both during preliminary observations and during the regular session, each of the stimuli was presented in random sequence an equal number of times. The five stimuli were detectable by the observers with varying probability. The largest stimulus contrast was usually detected with a probability of 95 percent and the smallest stimulus contrast, with a probability of 10 percent. Three additional stimuli were selected so that an adequate function relating probability of detection and stimulus contrast was obtained. Threshold contrast was detected with a probability of 50 percent, due allowance having been made for chance success.

"The observers were young women, aged 19–26 years, whose visual acuity in each eye and in both eyes was approximately 20/20 without refractive correction.

"Observing occupied approximately half their time, the remainder being devoted to the statistical analysis of their data. *A priori* objections to this procedure were soon overcome by the obvious stability of individual experimental results.

"Observers were never considered 'trained' until they had made approximately 6400 observations under varied experimental conditions. In general, the experimental data of this report were obtained with observers who had been occupied from six months to a year in preliminary experiments. Consequently, they were veterans of from 35,000–75,000 observations when the experiments reported here were begun. The exceptional experience of the young women in the task of observing resulted in unusual sensitivity and gratifying stability of response.

"Adaptation brightnesses were investigated varying from zero to 100 footlamberts. Circular stimuli varied in diameter from 121.0 to 3.60 minutes of arc. Threshold contrasts were obtained for each of a group of nine observers for each of seventy-seven experimental sessions, consisting of 320 stimulus presentations each. A total of approximately 220,000 observations was made, therefore, under experimental conditions differing only in adaptation brightness, stimulus area, and stimulus contrast."

Contrast C was defined by Blackwell as $C = (B_s - B_o)/B_o$ for stimuli brighter than the observation screen, and $C = (B_0 - B_s)/B_0$ for stimuli darker than the observation screen, where B_0 is the brightness of the observation screen (background) and B_s the brightness of the stimulus. Values of C range from 0 to $+\infty$ for stimuli brighter than the screen and from 0 to $+1$ for stimuli darker than the screen. In the experiments of this section only stimuli brighter than the observation screen were investigated.

Contrast has been and is defined in many different ways, some carrying different names, i.e., contrast ratio as used in photography.

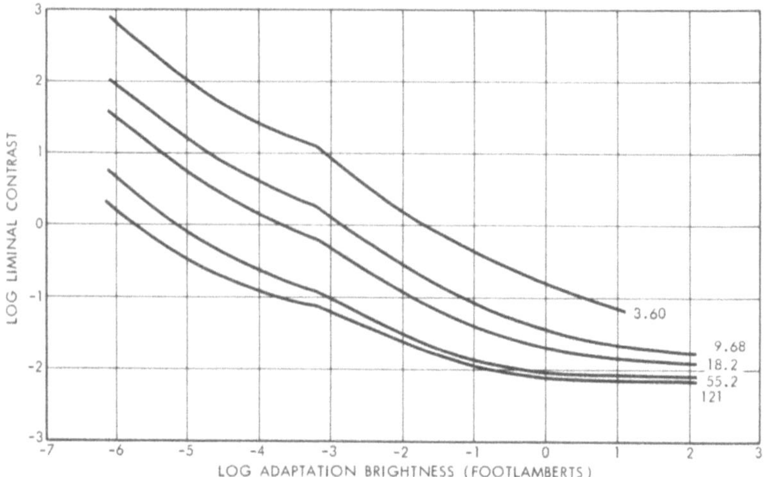

Fig. 2.5a. The arithmetical mean of threshold contrasts, computed from individual probability curves, plotted as a function of adaptation brightness for five stimulus area.

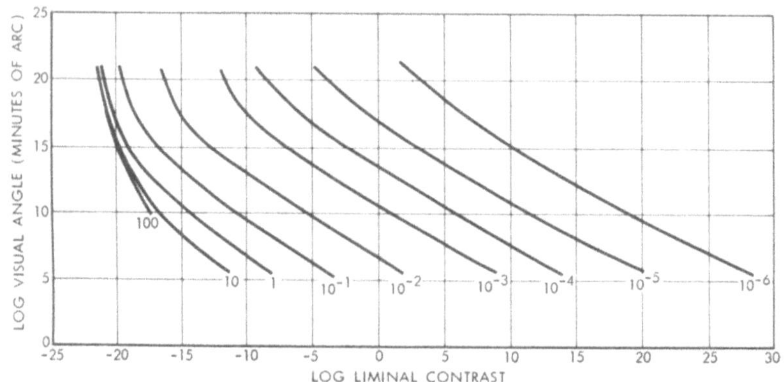

Fig. 2.5b. Interpolations from Fig. 2.5a. Each curve represents the relation between threshold contrast and stimulus area for a given adaptation brightness.

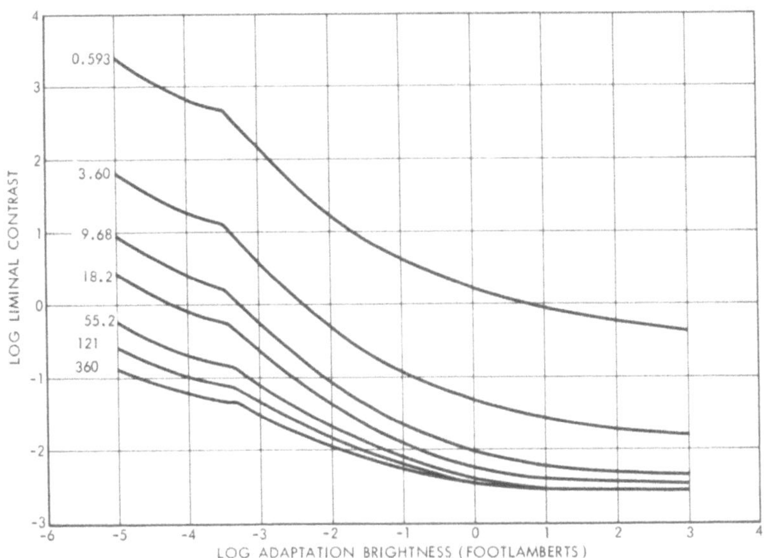

Fig. 2.5c. The arithmetical mean of threshold contrasts computed from individual probability curves, plotted as a function of adaptation brightness for seven stimulus areas.

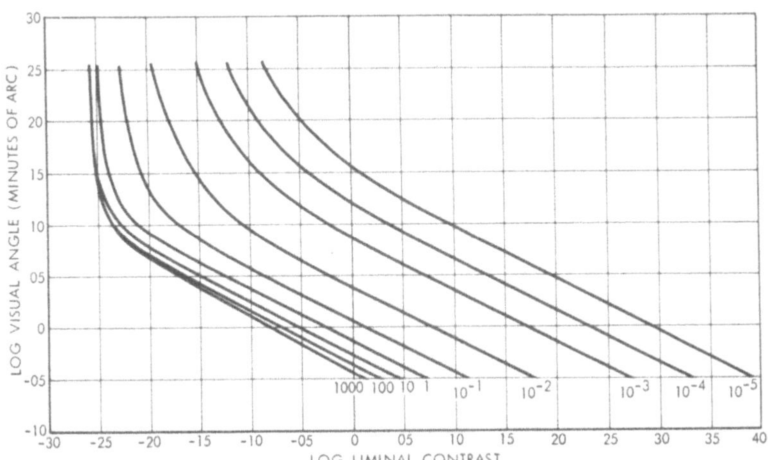

Fig. 2.5d. Interpolations from Fig. 2.5c. Each curve represents the relation between threshold contrast and stimulus area for a given adaptation brightness.

Usually contrast is taken to mean the ratio

$$\frac{\text{(brightness of the brighter)} - \text{(brightness of the darker)}}{\text{brightness of the brighter}}$$

This book uses several of the definitions, each where appropriate. In analyzing Blackwell's data we use the Blackwell definition. However, though we define with care, we caution the reader to be alert.

Blackwell's results are typified by the four sets of curves given in Figs. 2.5a–2.5d.

Few experiments, especially psychophysical experiments, have ever gathered so much data with so much care. With this abundant data as a foundation, it would appear that one could simply predict the performance of human observers obtaining information through an electro-optical system such as a television chain.

Unfortunately, this is not the case, since the displayed image of an electrooptical device can be very noisy. The fluctuations in display luminance usually do not result from the *inherent level* of luminance at the display, i.e., photon statistics, but rather from the noise in the pre-amplifiers or the beam current and other noise sources throughout the whole electrooptical chain, which results in the display luminance being modulated by such noise generators in addition to the usually much smaller levels associated with the Poisson distribution of the photons emitted by an otherwise noiseless display at the same level of luminance.

Legault (1971) compares the results of similar experiments by Rose (1942, 1948b), Coltman (1954), and Coltman and Anderson (1960), three classic experiments which gave results somewhat different than those of Blackwell. Legault goes on to examine the work of Rosell (1968) and Beurle and Hills (1968) and uses that data to show the strong dependence of the effective sampling aperture of the human eye upon the object being sampled or detected. Legault applies such data and is able to reconcile the results of Blackwell, Rose, and Coltman.

More recent results of Rosell presented in Chapter 5 reconcile even more data previously thought to be contradictory.

Recently the use of a number of transfer functions has been intensively investigated in relation to image quality. Of these the spread function, the modulation transfer function, and the optical transfer function have come into most prominent usage. Many investigators, especially those whose training was primarily rooted in experimental psychology, denied that the MTF or information theoretic concepts applied to the eye since the eye was a complex adaptive device and thus could not be so described. A contrary point of view is put forward well by Wolfe (1962)*:

> "A picture can be produced by either a painter or a photographer. The painter's tool is a brush. If he wishes to show fine detail, he uses a fine brush, but if he considers the broad outlines of the objects to be more important, he uses a broad brush. The photographer's tool, however, is the composite spread function $fp(x)$ of his photographic system, which consists of the combination of all its elemental spread functions, e.g., that of the camera lens, the negative film, the enlarger lens, the positive film, and, in the case of systems having television-type links, the pickup tube, etc. Like the painter, he can control the size of his brush—his spread function—and, to some degree, its shape.
>
> "The stimulus received by the higher receptor center in the brain of an observer looking at the picture is determined by the spread function of the physical system $fp(x)$ combined in some manner with that of the visual system $fv(x)$ to give $fpv(x)$. It is assumed in this paper that $fv(x)$ may be convoluted with $fp(x)$.
>
> "The visual spread function $fv(x)$ likewise comprises several elemental spread functions, e.g., that of the eye lens, the ocular media, the retina, and the higher receptor centers. The visual spread function represents not merely the physical spread function of the eye, but rather the spread function of the entire visual mechanism, objective and subjective.
>
> "A major aspect of this subjective response is definition or detail reproduction. In other words, if an observer looks at two pictures, A and B, of identical subject matter, and if $fpv(x)$ is narrower for A than for B, then the observer, on the average, will state that A has better definition than B. For a quantitative study, both the subjective and the physical aspects must be specified numerically."

In a series of psychophysical experiments Wolfe measured the half-width of the human visual spread function. Table 2.2 compares Wolfe's data with those of several other investigators, most of whose work was done within a ten-year span.

It is clear from Table 2.2 that the eye is a troublesome subject on which to conduct precise experiments, but it is also clear that at a field luminance of 70 ft-L the value of the measured spread function σv (the

---

* Reproduced in highly abridged form by permission of the author and the Optical Society of America.

TABLE 2.2

Summary of Determinations of Size of the Human Visual Spread Function

Author(s)	Date*	Technique	Field luminance, ft-L	Half-width, σv, μm
Wolfe	1962	Picture viewing: paired comparison	70	3–6
Wolfe	1962	Picture viewing: triads	70	>8
Selwyn	1948	Threshold	—	3.4
Schade	1948a	Threshold	7	9
Schade	1956	Threshold	70	5
Ludvigh	1953	Composite blur disk	50	16
Flamant	1955	Physical photometry	—	10
Stultz and Zweig	1959	Graininess observations	<10	14
Lowry and DePalma	1961	Mach phenomenon	20	<3

* Unless otherwise indicated, date of archival publication.

half-width at the retina) is contained between the values of $\sim$3–8 μm.

In is interesting to note that if one takes 5 μm as a value for σv, then the half-power of the projected angle is 5 μm/17 mm, where 17 mm is the focal length of the cornea. But this value is approximately 1 min of arc, the value of limiting resolution for a 20/20 light-adapted eye.

There are many related pieces of research that do not seem to tie into the work just reviewed but which actually must be considered repeated, extended, or just adopted in the design of communications systems involving visual imagery. Such papers include those listed below, with abbreviated titles, and in full in the reference list:

1. Bowen *et al.* (1960). Optimum Symbols for Radar Displays.
2. Rabedeau and Bates (1965). Image Quality Requirements in Optical Systems for Reproducing Typewritten Documents.
3. Howell and Kraft (1959). Size, Blur, and Contrast [. . .] Affecting the Legibility of [. . .] Symbols [. . .].
4. Crook *et al.* (1950). Effect of Amplitude of Apparent Vibration, [. . .] on Numeral Reading.

5. Smith, R. L., *et al.* (1966). Effects of Display Magnification, [...], Cues, [...].
6. Greening and Wyman (1970). Experimental Evaluation of a Visual Detection Model.

2.2.3. Experimental Programs for Studying the Informative Content of Images

2.2.3.1. *The Experimental Procedures*

In a search for physical measures of image quality, Frank Scott and Robert Hufnagel collaborated in a series of experiments first reported in a pair of papers presented before the Optical Society of America in October 1965.* Their experiments consisted in the subjective ranking of several series of prints carefully produced from a common negative of known MTF and sensitometric parameters. The series of prints was degraded in image quality first by producing prints without grain but with successively decreasing values of a Gaussian-shaped MTF characteristic. Some of the prints were then produced but with known amounts of grain superimposed. Last, a series of prints was produced with and without grain, with either a Gaussian-shaped MTF curve or with an MTF with a "bump" in an otherwise smooth characteristic. These prints were then evaluated in series of trials with and without time constraints by a number of observers carefully selected and instructed in a procedure detailed by Scott (1968). The following material is a condensation of their papers given in 1965.

The scene they selected for study is an aerial view of an industrial complex containing many buildings, cars, trucks, railway cars, and supply loading and storage areas. The scale of the photograph is approximately 3300:1 and its size approximately 4×4 in. The sensitometric characteristics of this photograph were known and its modulation transfer function was determined from microdensitometric edge gradient analysis.

The specific image structure characteristics desired in the photographs representing the stimuli in the experiments were obtained by a modified contact-printing procedure shown in Fig. 2.6. In this arrangement the original photograph, being a positive transparency, is physically

---

* We quote extensively from those unpublished papers in the following pages with the kind permission of the authors.

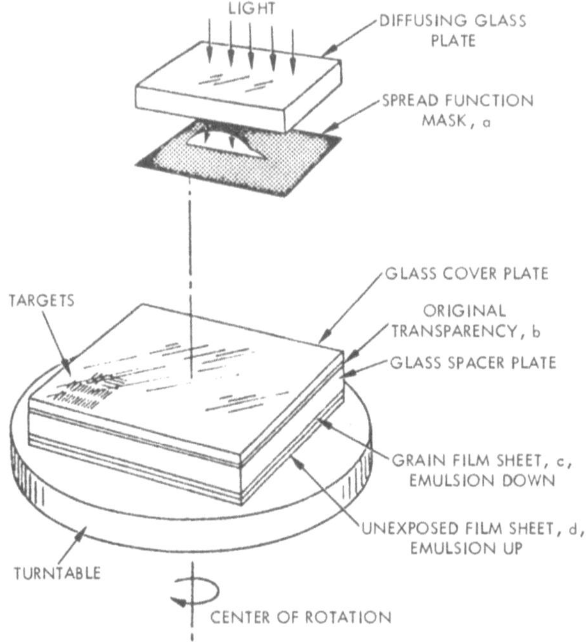

Fig. 2.6. Modified contact-printing procedure. On the basis of the geometry, the relationship of the point spread function $A(x)_p$ to the mask point function $A(x)_m$ is

$$A(x)_p = A(x)_m \frac{B/n_b + C/n_c}{D}$$

where B is the thickness of the glass spacer plate, C is the thickness of the grain film sheet, D is the distance between the mask and the transparency, n_b is the refractive index of the glass spacer plate, and n_c is the refractive index of the grain film sheet.

separated from the unexposed fine-grain film by a glass spacer. The exposures were made using a spatially varying intensity source, the intensity variation being circularly symmetric about its center, which represented the spread function which, when convolved with the spread function of the original photograph, produced the desired spread function on the printing film. The shape of the resulting spread function is determined primarily by the shape of the intensity distribution of the light source and the shape of the spread function of the original transparency. The size of the resulting spread function was controlled by adjusting the size of dimensions A, B, and C. Since variable-transmittance masks having a specific transmittance distribution are difficult to make, they used variable-area masks and rotated the original photograph–spacer–printing film sandwich about its center. The variable-area point-spread-function masks were prepared by plotting the specific function on polar coordinate graph paper.

Graininess was simulated by placing, in intimate contact with the printing film, a sheet of film bearing the image of photographic grains. Since the photographs were intended to be viewed with the naked eye, the granularity to be perceptible was relatively coarse. Images of grain structure having a relatively coarse granular structure were made by contact printing ground glass onto litho film. Various levels of graininess were obtained by varying the grind of the ground glass and by varying the printing exposure. From conventional granularity measurement procedures it was determined that the grain sheets so produced had characteristics identical with real film.

The negative pictures produced by this modified contact-printing technique were then contact printed onto photographic paper. The area in one corner of the photograph was allocated to targets which were used to monitor and measure the image structure characteristics obtained. The target array included a low-contrast, three-bar target, variable-transmittance sine wave patterns, and an area of constant density for granularity measurements. The sine wave patterns were not covered by the grain sheets; thus the microdensitometer trace of the sine wave pattern image was essentially noise-free, yielding enhanced measurement accuracy. With the precise manipulation of the image structure characteristics obtainable with this process and with carefully performed, precision measurements of the image structure, the resulting photographs were exactly describable in objective terms and all photographs produced were identical except for the image structure characteristics studied. Thus the photographs were considered ideally suited to experiments of this type.

The First Experiment. The set of photographs for the first experiment was designed to determine the interaction between granularity and point spread function size or degree of blur. Hence the photographs in the first set were identical to one another except in the granu-

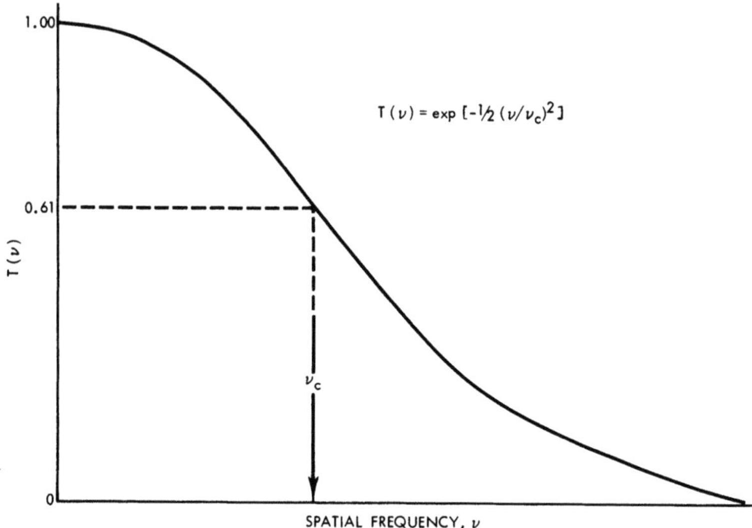

Fig. 2.7. Definition of characteristic spatial frequency ν_c.

larity level and the size of the spread function. The shape of the spread function was held constant and was of the Gaussian form. Being Gaussian-shaped, the transfer functions are completely describable by one parameter. They selected that spatial frequency corresponding to a given modulation transfer factor. This is similar to the standard deviation of a normal distribution. They called it the characteristic frequency ν_c (Fig. 2.7), discussed later. The granularity level of each photograph was designated by its rms granularity value for a 24-μm-diameter scanning aperture.

For their first experiment a series of 19 pictures was prepared with carefully controlled parameters of grain and MTF. The characteristics of the first series are shown in Fig. 2.8a.

In the first experiment 71 different subjects participated, of which 45 were practicing professional photointerpreters while the remaining 26 subjects were not photointerpreters but were familiar with viewing aerial photographs. Each subject was given identical instructions. These were, specifically:

> "In judging the quality of these pictures [Fig. 2.8b] do *not* consider the pictorial or pleasing aspects of the pictures. Rather, rank these pictures in the order in which they convey to you the information

in the scene from an intelligence or information gathering point of view. Do not limit your judgment on the basis of only one object. Examine many parts of the image. There are no resolution targets in the image and you may use a loupe or any magnification desired. Rank these pictures in order by placing them on the table in a row from best to poorest, with the best on your left. If and when you consider two or more pictures to be equal, place these one above another forming a column."

The ranking of each photograph by each observer was given a rank number. The photograph ranked best was given a rank number of one, the photograph second best, two, and so forth. If two photographs were ranked as being identical, those photographs were given a rank number which is the average of the two.

For the whole group of 71 subjects the rms rank spread for individual readings for those pictures with grain was 1.5 rank units, while for the photographs having no grain the rms spread was 3 rank units. This spread is, of course, significantly reduced by averaging.

The Second Experiment. The second experiment was designed to determine the effect of the shape of the modulation transfer function both in the presence and absence of grain on the subjective rank. Accordingly, the second set of photographs had two transfer functions: one was the Gaussian shape used in the first set of pictures and the shape

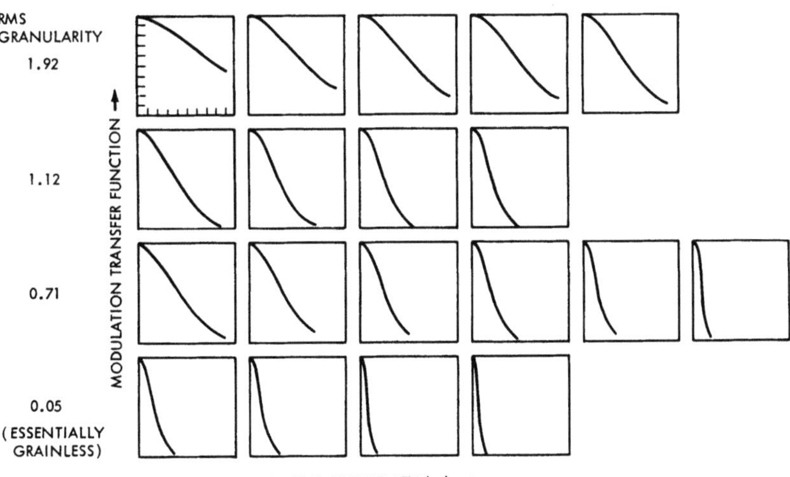

Fig. 2.8a. Parameters used in first experiment.

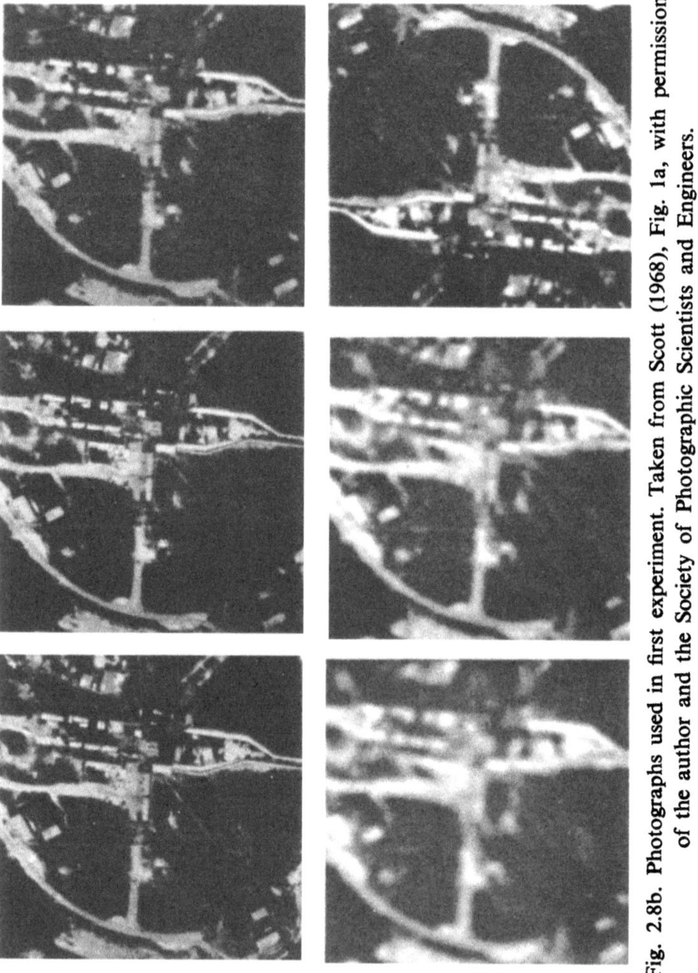

Fig. 2.8b. Photographs used in first experiment. Taken from Scott (1968), Fig. 1a, with permission of the author and the Society of Photographic Scientists and Engineers.

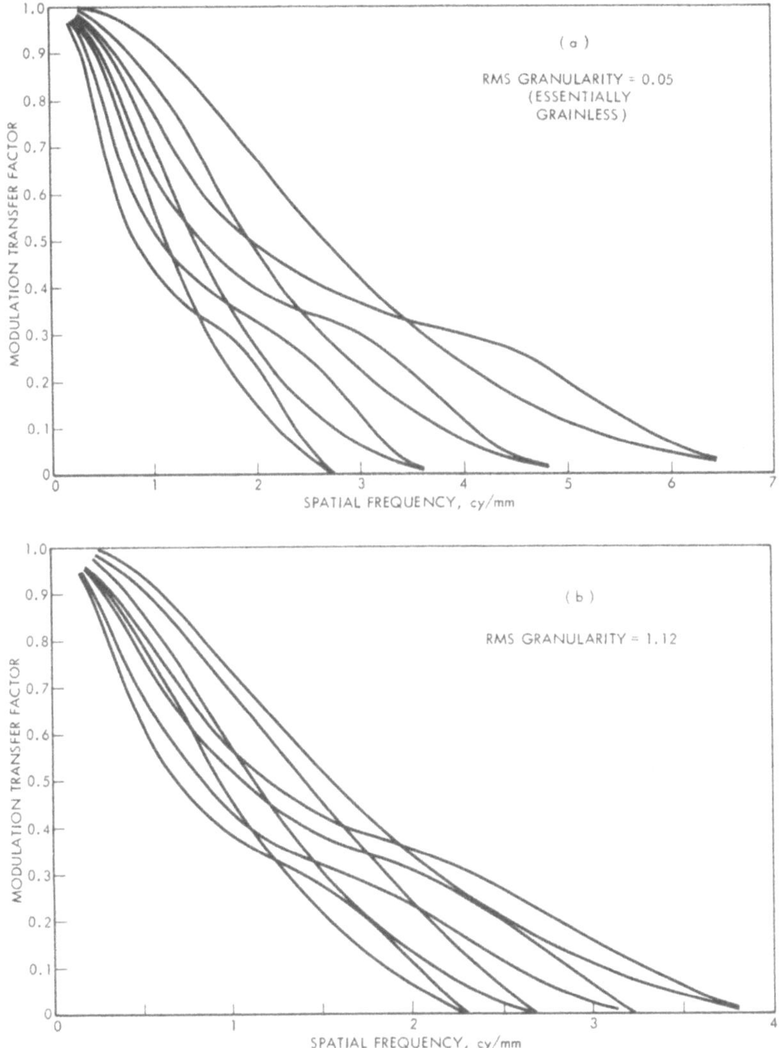

Fig. 2.9. Parameters of second experiment.

of the other transfer function had a bump in it as shown in Fig. 2.9.
Thus pictures having these two transfer function shapes with and without
grain were made. The size of the spread function corresponding to the
bump transfer function is designated by that spatial frequency for which

the modulation transfer factor is a given value, which, as mentioned earlier, we call the characteristic frequency v_c. The instructions given the subjects and the procedure employed in tabulating the responses were identical to those employed in the first experiment. After tabulation of the responses the experiment was repeated with another group of professional photointerpreters so that the results of the two groups of subjects could be compared to one another. There was no significant difference.

2.2.3.2. *Analysis of the Experimental Results*

Hufnagel proceeded to use the data from the first experiment to evaluate a number of factors of merit, or, as he prefers to call it, "the summary measure of image quality."

He assumed that a summary measure *was not invalid* if a plot of the objective and subjective rankings fall on a rising straight line, i.e., if the measure is larger, the observer should also rank it higher. Weaver (1968) gives good insight to this choice of criterion. By "not invalid" Hufnagel makes the point that there is much to complicate such studies and the summary measure may not necessarily be good for conditions extended much beyond those of these experiments.

If, within the experimental error, all the points do not lie on a single monotonic line, then the candidate summary measure has failed and he rejects it as a possible true summary measure. Hufnagel continues:

> "Acceptance through a single test does not imply that a measure is a valid one. To be a summary measure the candidate must pass *all* possible tests. We cannot perform all possible tests, and thus can never prove that some measure is universally valid. However, if a measure passes each of a finite number of diverse tests, we may assume that it is at least approximately valid."

Hufnagel chose the four measures

$$(1) \quad \iint_{-\infty}^{\infty} T(v_{x,y})\, dv_x\, dv_y, \qquad\qquad (2) \quad \int_{-\infty}^{\infty} T(v)\, dv$$

$$(3) \quad \iint_{-\infty}^{\infty} |\, T(v_x, v_y)\,|^2\, dv_x\, dv_y, \qquad (4) \quad \int_{-\infty}^{\infty} |\, T(v)\,|^2\, dv$$

in which $T(v)$ is the transfer function of the image.

The first summary measure is the two-dimensional measure proposed by Strehl in 1902. The second expression is derived from the first and is called by some the "line Strehl" definition. The third expression is N_e proposed by Schade (1964b). The last expression is a measure of the sharpness of an edge and is related to acutance.

Hufnagel goes on to discuss his results as follows:

> "Rather than discuss all of the candidate measures of image quality tested, we shall show just a few, from the simple to the complex, which are illustrative of the candidates tried and of the results obtained. We chose four common measures as proposed by various workers. Let me remind you that $T(v)$ is the transfer function of the photo, and that v is spatial frequency. These measures of image quality contain neither grain nor contrast as a parameter. Thus, we cannot expect them to work with grainy photographs. Our test of these measures, therefore, must be only with grainless photos. [Figure 2.10] shows a typical result from a grainless test series, where the Strehl definition is the candidate summary measure. All the points do not fall on a single line and thus

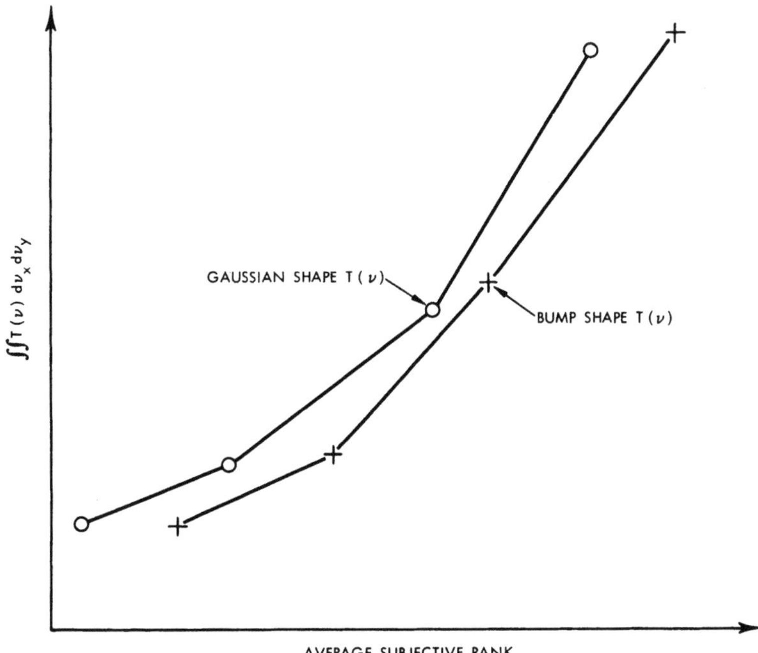

Fig. 2.10. Trial summary measure versus subjective rank.

this quantity does not meet the test of a summary measure. Of the four quantities tested the best results were obtained with the Strehl definition. Thus far we have found no simple function which works.

"Since it was decided that to be 'not invalid' all points must lie on a straight line, it is clear that none of the four measures is *really* valid but the closest is the Strehl criterion.

"In the absence of a good mathematical description we can perhaps best describe the results of the ranking test in a graphical way. In [Fig. 2.11] we have plotted the Gaussian and 'bump' transfer functions in such a way that they describe photos which are identical in subjective quality. The important thing to note in this figure is that the two transfer functions are nearly equal at high spatial frequencies, but that they are very dissimilar at low frequencies.

"Since the photos are of equal subjective quality, we may therefore conclude that the region of transfer function dissimilarity is not significantly weighted in the summary measure. Restated in a converse, but positive sense: *For grainless photographs* the $T(v)$ region of most importance corresponds to the higher spatial frequencies and/or the lower $T(v)$ modulus values. This equal quality *relative* relationship between the two shaped transfer functions held regardless of the actual spatial frequencies involved. This is to be expected since the observers

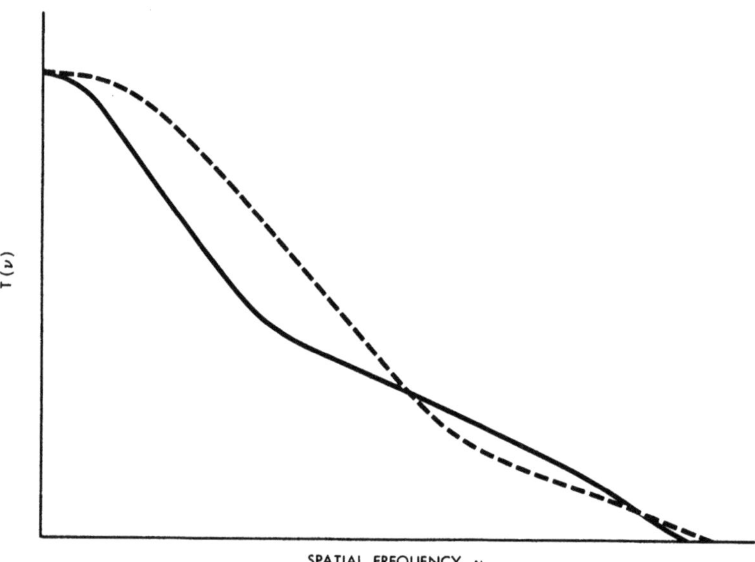

Fig. 2.11. Transfer functions with the same subjective image quality (for grainless photos).

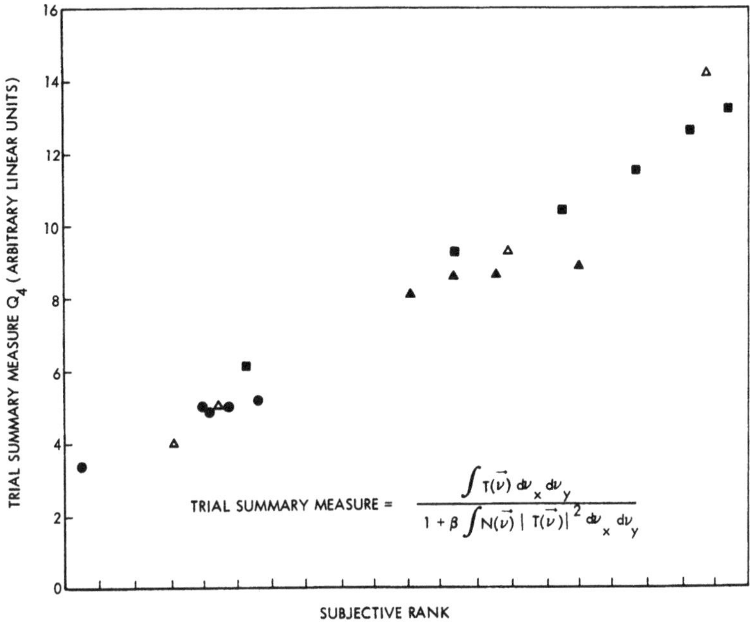

Fig. 2.12. Test of summary measure with blur-grain series subjective ranking data.

were allowed in the experiment to adjust the magnification employed and hence the spatial frequencies of the photos did not matter.

"So far we have talked only about grainless photographs. Now let us describe some of our efforts to account for grain. We have tried a number of simple combinations of the transfer function and grain parameters. The combination shown in [Fig. 2.12] worked best. We called this quantity, Q_4^2, a grain modified Strehl definition:

$$Q_4^2 = \frac{\int T(\nu)\, dv_x\, dv_y}{1 + \beta \int N(\nu)\, |\, T(\nu)\,|^2\, dv_x\, dv_y}$$

$N(\nu)$ is the grain 'noise' Wiener spectral density. The integral in the denominator represents grain noise remaining after transmission through a filter which matches the transfer function. Beta is a constant adjusted to yield the best fit to the experimental data. Here the fit is almost within the experimental error. It does not work quite this well for the other photograph series, but the fit is not too bad.

"We might be content to stop here were it not for some other discoveries.

"In the course of our work we noticed some interesting effects concerning the preferred magnification used by the subjects in viewing the pictures. To discuss this effect we must consider the modulation transfer function for the human visual system as shown in [Fig. 2.13]. The solid line part of the curve is taken from the work by Bryngdahl with the dotted extrapolation based on the results of several other workers. The important thing to note in this figure is the peak in the modulation transfer function. This peak is related to the well-known Mach band effect observed in visual phenomena. The spatial frequency scale corresponds to unit magnification, but the observer is free to vary the scales simply by varying his viewing distance or magnification. In [Fig. 2.14] we have plotted equal subjective quality transfer functions on the bottom and on the top visual system transfer functions with the spatial frequency scale adjusted to correspond to the experimentally observed preferred magnifications employed by the viewers. The plots on the left correspond to grainless photographs: the plots on the right-hand side correspond to photographs with moderate grain.

"To interpret these results, we note first for the grainless case that the observer appears to adjust his eye transfer function so that the rise in its magnitude with frequency just matches the corresponding decrease in the photograph transfer function. Perhaps the viewer attempts to maintain the transfer function near unity for as high a spatial frequency as possible. We note further that the peak of the eye transfer function occurs in the spatial frequency region where the photograph transfer function is weighted most heavily in forming the subjective quality."

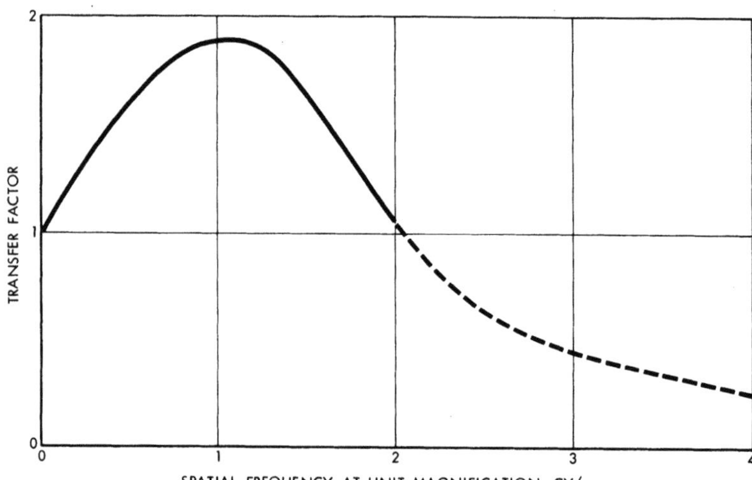

Fig. 2.13. Eye modulation transfer function.

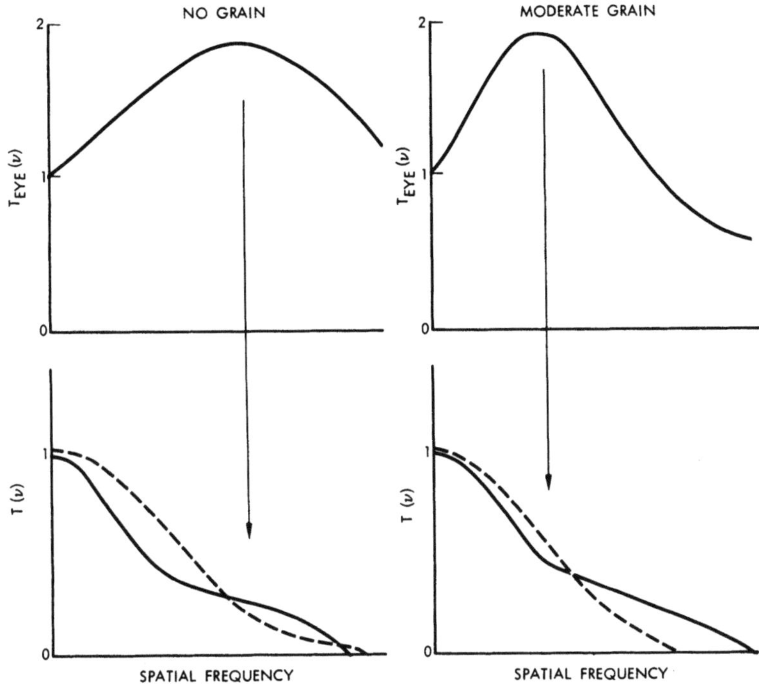

Fig. 2.14. Subjective quality transfer functions.

Hufnagel continues:

"Now let us examine the right-hand side of [Fig. 2.14] to see what happens when the grain is added to the photograph. First of all, we see that the observer reduces his preferred magnification apparently in an attempt to reduce the effect of the grain. He cannot reduce his magnification too far, however, without seriously impairing his ability to see the fine detail in the photograph. Thus, the magnification must be adjusted by the observer to yield some kind of optimized apparent image quality. Note that again the peak in the visual transfer function occurs in the region where the two equal quality photo transfer functions are most alike.

"These results lead us to a new class of candidate summary measures. We define one such candidate summary measure

$$Q_3 = \left. \frac{\int_{-\infty}^{+\infty} T_{\text{photo}}^2(\nu) T_{\text{eye}}^2(\nu/m)\, d\nu}{1 + \alpha \int\int_{-\infty}^{+\infty} N(\nu) T_{\text{eye}}^2(\nu/m)\, d\nu_x\, d\nu_y} \right]$$

Evaluated at optimum m

Q_3 = Grain and eye modified line detectability

"Here Q_3 may be interpreted as being a measure of line detectability as modified by the effect of the grain and the eye. As before $N(v)$ is the grain noise spectrum and α is a constant adjusted for a best empirical fit. For a given $T(v)$ and $N(v)$ there exists a magnification which maximizes Q_3. In [Fig. 2.15] we have plotted these maximized Q_3 values against the subjective rank data. The fit is quite good, and in fact may even be better than that obtained in [Fig. 2.12]. Not only does Q_3 agree with the subjective rank data, but it also predicts with reasonable accuracy the preferred optimum magnification. However, we must conclude that Q_3 and Q_4 give about equally good results, and that on the basis of experimental data to date we have no means of preferring one over the other. Thus we have not one but two 'acceptable' measures of image quality which are based on quite different concepts.

"Let me conclude this section of the paper by saying that we have tried about 20 different types of candidate measures and several variants of each. None have proved perfect: several were quite good but conceptually quite different. It is clear that we need more experimental results to weed out the unacceptable summary measures, and to test others yet to be developed.

"Our conclusions to date may be summarized as follows:

"In the absence of grain, and at normal contrast levels the modulation transfer function in the region $0.1 \leq T(k) \leq 0.4$, appears to be

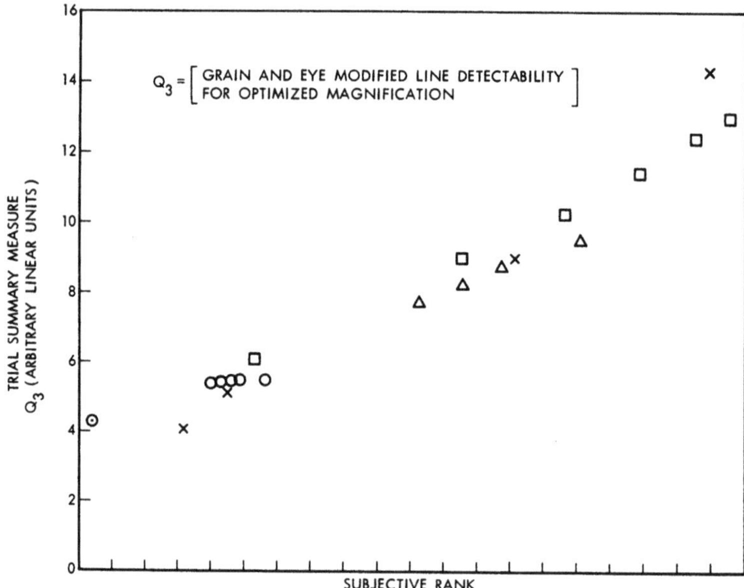

Fig. 2.15. Test of summary measure with blur-grain series subjective ranking data.

the most important in determining subjective quality. In the presence of moderate grain this region of importance rises to include $T(k)$ at larger modulus values. In the presence of extreme grain the modulus region of interest approaches unity and thus becomes independent of the transfer function. In this limit the quality depends only on grain, image contrast, and certain secondary effects, such as nonlinearity in the tonal response. The summary measure may well contain the eye transfer function. But if it does, it should correctly predict the preferred magnification.

"We are encouraged to find that one can obtain an orderly result from experimental data potentially so chaotic as these. Perhaps there is such a thing as a 'Summary Measure.'

"We have uncovered a potentially large number of mathematical expressions which all agree with the presently existing experimental results. It is clear from this experience that many more accurate and well controlled experiments need be done. Without these one must be content to live with the many 'right answers' which have been suggested here and in the past by others.

"We are planning to repeat some of the experiments described here with a new independent set of photographs. In addition we will examine the effect of additional variables such as contrast. But a study of contrast will most likely bring us face-to-face with problems concerning nonlinear behavior in the response of the visual system.

"The problems ahead appear overwhelming and one might even think unconquerable—were it not for the limited successes already achieved."

The work reported above was redone with some variations and some extensions of the techniques. F. Scott (1968) reported the somewhat more extensive results in 1968, with the following additional conclusions*:

"Observers can just distinguish approximately 5% differences in the size of spread functions of grainless photographs of identical scenes; for grainless photographs of different scenes, observers distinguished the spread function size with 10% sensitivity. The procedures used in the experiments showed that comparative subjective image-quality judgments can be made with reasonably short viewing time periods without detriment of performance; and that the training of professional photointerpreters does not contribute significantly to the accuracy and precision of subjective image-quality judgments.

"Finally, it was demonstrated that subjective image-quality judgments made by professional photointerpreters agree well with photo-interpreter performance."

* Quoted with permission of the author and the original publishers, the Society of Photographic Scientists and Engineers.

The F. Scott (1968) paper is complete with extensive statistics and bears out well the data of the first experiment. We chose to report and extract the unpublished papers on the first set of experiments because of the flavor it had of then-current research being reported by two enthusiastic investigators just breathing the life of creative imaginative research into their reporting of the first (and thus most exciting) series of these experiments. However, we recommend the reading of the more formal paper by F. Scott (1968).

In discussions of these papers with Schade he pointed out that the above experiments investigated *some aspects* of image quality. At this point the reader should ask, Is this summary figure of merit better than others such as resolving power, MTFA, or SNR_D? Schade then asks if the measures are more significant than those of J. Johnson (1958), Baldwin (1940), Strehl [see Linfoot (1964)], or Rayleigh (1881).

At this stage of the game one must answer both yes and no.

Why then do we quote so extensively from these papers? The work is a fine example of the procedures used in early searches for such measures of image quality. For some purposes existing measures may be enough; for others much is still to be desired.

Work continues, but the reader is reminded that there is a valuable literature going back to Lord Rayleigh, and that the Strehl function, or the work of Baldwin in searching for quality levels for television standards, should not be lightly passed over.

In his search for a summary measure as discussed above, Hufnagel used the system transfer function $T(v)$ or MTF, but did not apply the concept of the demand function or DMF introduced by F. Scott (1966) somewhat later. The concept of combining both the MTF and the DMF into a single quantity, the threshold quality factor, or TQF, was later introduced by Charman and Olin (1965) and has been used by others in a series of experiments that are described in some detail in Chapter 4 by Snyder, who like many others calls the Charman and Olin "TQF" by the name MTFA, for the area between the MTF curve and the demand curve. The advance in concept provided by the MTFA is that it introduces the threshold or DMF factor of Scott into a measure of image quality. It serves the function well for general-purpose varied imagery, *but* since the MTF is a decreasing function of spatial frequency and DMF is a rising function of frequency, the fit to targets of a particular frequency rather than the average response to a host of things in a scene leaves much to be desired. The conspicuity, or the degree to which a target stands out from its surround, is indeed more related to its signal-to-noise ratio and the demand function of the eye for an object of the size of that particular target than it is to the MTFA of a generalized scene.

As is pointed out in the next three chapters, if we wish specifically applicable results on seeing targets against backgrounds, we had better use specific data to get our specific answers. The methods, successes, and failures lead strongly to a point today where we can say the following.

1. The conspicuity of an object is related to its S/N level above that required by the eye for an object of its size.

2. The probability of detection, recognition, or identification is a function of the S/N of the image, the functions being different for each type of task.

3. Where the Johnson criterion indicates that one needs, say, four line pairs of resolution across the target object to achieve a given visual task, we shall show that when the signal-to-noise ratio in the image reaches a level where its threshold at that frequency crosses the DMF curve, the object becomes liminally (50% of the time) visible. Factors such as competing objects of interest, clutter, and raster effects in scanned and sampled imagery degrade such performance, but in a reasonably predictable manner, so compensation can be applied.

This is the thrust of our book. This is the point we develop until we can make it stick. We treat the earlier material and that which follows as an evolutionary trail toward that goal.

2.3. LINE-SCANNED IMAGERY

2.3.1. Definitions, Confusions, and General Problems

Some difficult subjects are made much more difficult by the language used; for example, words with multiple meanings and connotations are used without specifying which meaning is being used.

Johnson (1958) in his early experiments spoke of the required number of resolvable line pairs across the critical dimension of a target for detection or recognition, etc. In these experimental determinations Johnson set up resolution test charts adjacent to the object to be recognized, identified, etc. He then could determine the number of line pairs per linear dimension associated with these visual tasks. (See Fig. 1.1, Chapter 1.)

In the later parlance of television systems "line pairs" was dropped in favor of "TV lines" (one line pair being equivalent to two TV lines) and as jargon developed the term became "lines." Unfortunately in television usage the raster in the vertical direction is also specified as "TV lines per picture height," "lines per picture height," or just "lines." A common designation for a TV system or a display system is "525-line" or an "875-line" display, which refers to the sweep generator parameters that would generate 525 or 875 scans per picture height if the scanning efficiency were 100%, i.e., there was no dead time due to retrace or "flyback." Actually there is a dead time equivalent to 35 lines, so that a 525-line system actually presents 490 lines per vertical dimension of the picture.

The horizontal dimension of most common displays is 4/3 the vertical dimension. When one measures a commonly reported parameter of a display, the horizontal limiting resolution, one normally feeds a large video signal to the display and measures the spacing at which one can no longer consistently tell adjacent resolution elements apart. This spatial frequency is the frequency corresponding to the limiting resolution.

Limiting resolution is based upon the Rayleigh criterion and it is a good one that has stood the test of decades as a good and easily identifiable parameter. It just does not carry much information in describing the quality of an image-forming system.

Now the difficulty that arises is that horizontal resolution is also often specified in "lines per picture height" or just in lines, picture height hopefully being implied, though sometimes lines per picture width is really meant. Limiting resolution (in the horizontal direction) will vary with the bandwidth and the video signal-to-noise ratio driving the display. Thus we have the concept of "lines per picture height" (meaning of course horizontal limiting resolution) varying with video signal-to-noise ratio, bandwidth, and other parameters, while the "lines per picture height," forming the raster, are quite independent of the video signal-to-noise ratio, etc.

Both, however, are casually referred to as "lines," "lines per picture height," or "TV lines per picture height," and usually without much of a clue as to whether horizontal resolution or vertical (raster lines) is being discussed.

The important parameter, as we show later, is the number of "lines" that indicate informational value, and thus is defined as the resolution or number of lines per unit length or per picture height at some value of contrast and brightness. This informational value can also be specified as the signal-to-noise ratio at some spatial frequency, but the terminology has been so nonspecific that image quality has sometimes been studied as a function of the number of lines making up the raster rather than the number of lines that determine resolution.

Unfortunately a significant number of investigators and designers concerned with observer performance have become preoccupied with the number of lines a television set produces *in its raster*. It is unfortunate because the number of lines in a raster of a television display with *no* input is identical to the number of lines on the same set with an input of high signal-to-noise ratio resulting in a clear, clean, and bright picture. The zero input, of course, produces only a clear and clean pattern of horizontal lines with no picture and no information.

In spite of that clearly obvious point, systems and components still are described as producing 500 or 700 or 1000 lines.

In fact it is quite true that in many cases signal levels are small and limited in frequency and noise may be small but very broad in its bandwidth; thus the picture quality of a 500-line system will be superior to that of, say, a 1000-line system because the 500-line raster calls for only one-fourth the bandwidth of the 1000-line raster and thus has twice the signal-to-noise ratio at the useful (lower) frequencies while eliminating high-frequency noise.

Thus if the factors of significance such as signal-to-noise ratio are ignored in favor of such parameters as raster lines per picture height, one will get results like those obtained in past studies by men who did not comprehend the communication theory involved in television chains. They reported poor performance in simulated experiments on satellite-borne television systems when the systems were operated with rasters of many lines and good results when the rasters consisted of fewer lines, although they had expected quite the opposite.

One *should* expect quite the opposite, but *only* if the signal-to-noise ratio is sufficiently large at the broad bandwidths corresponding to a high-line-number raster, say 1000 lines. With such large values of S/N the

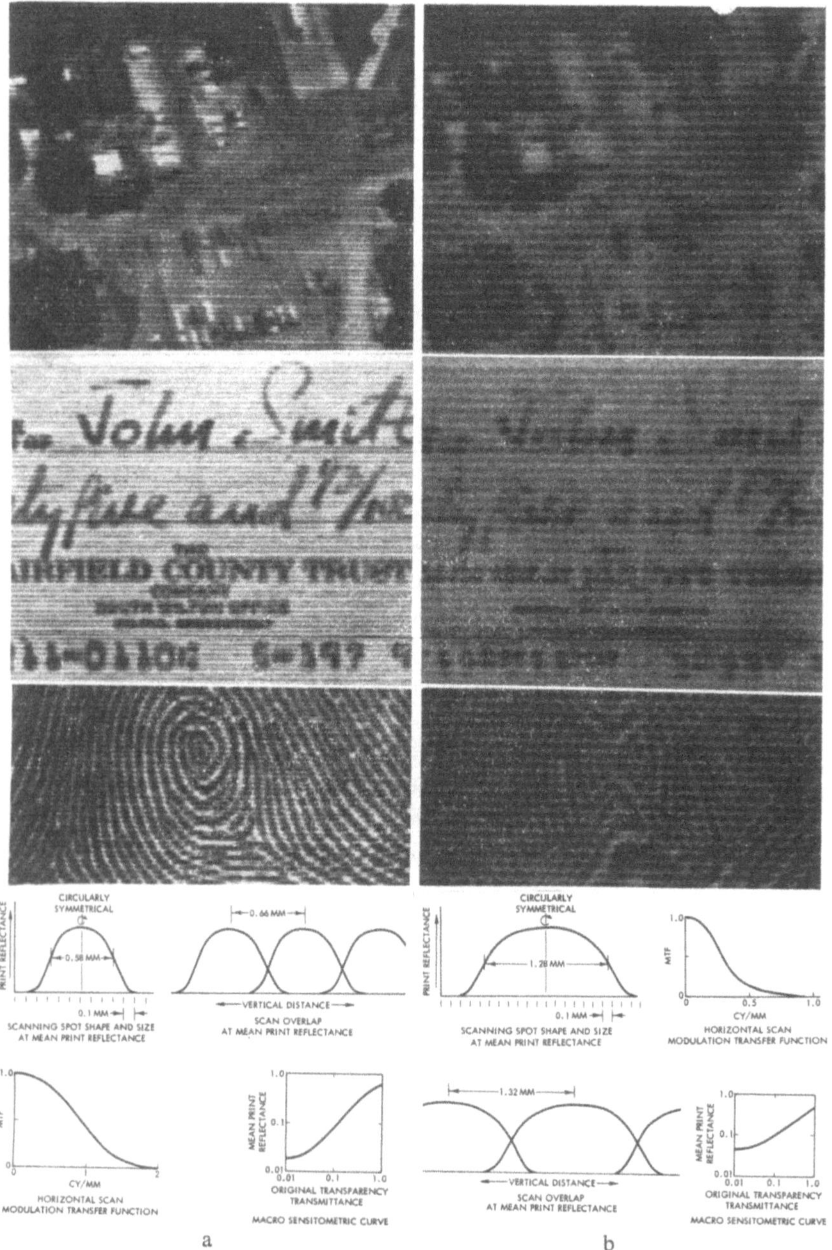

Fig. 2.16. Left: Line-scan image made with LSIG in optical mode; (a) 1.50 scans/mm; (b) 0.76 scan/mm. Right: Line-scan made with LSIG in electronic mode;

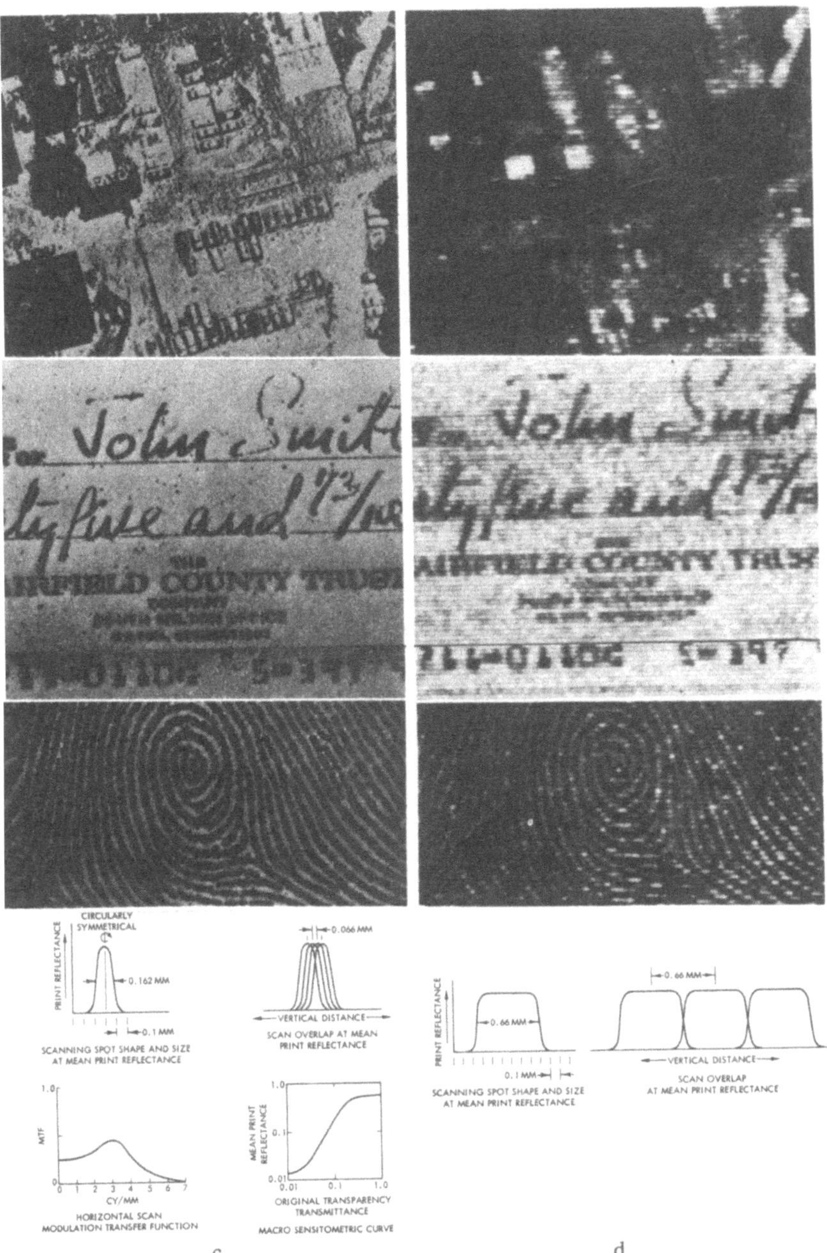

(c) Illustrating low spatial-frequency attenuation; 1.50 scans/mm; (d) Gaussian noise added; 1.5 scans/mm.

display device may well show a *limiting resolution* of well over 1000 lines per picture height in the horizontal direction and be limited by the slightly less than 1000 lines in the vertical. The shift to a 500-line-per-vertical dimension raster *while maintaining the same large signal-to-noise ratio and bandwidth* will merely cause a loss of information in the vertical direction. However, at low signal-to-noise ratios the deliberate choice of a limited-line-number raster together with appropriate bandwidth limitations will produce useful imagery while excluding the high-frequency noise.* A choice of high raster line numbers and wide bandwidth lets in more noise but no more signal, producing a poor, i.e., "snowy," picture.

Although the above was well understood by many, it remained for Scott to produce a line-scan image generator with which he could copy a high-quality photograph and form a line-scanned image with (1) various raster constants, (2) various MTF along a raster line and (3) various signal-to-noise ratios. The variety of image characteristics that result from line scanning are but partially indicated in Fig. 2.16 (F. Scott, 1967b). The raster lines generate a strong visual signal that tends to mask perception of lesser signals at frequencies nearby. For this and other reasons the raster line frequency tends to annoy most viewers.

Early work in the laboratories at Westinghouse by Thompson (1957) showed that when one removed the raster structure, viewers, of their own choice, moved to half their normal viewing distance (when viewing the normal line-structured display) in order to take advantage of the detail they now could see.

Further, one of the major effects of the raster, the formation of aliases, has been overlooked in many of the studies of line-scanned imagery.

Not only is the presence of the raster in itself disconcerting, but further the process of forming an image through line scan sampling

---

* In a poorly designed system, the presence of sharp raster lines may nullify the advantage of low raster frequency. Sophisticated and advanced design criteria will choose high raster line frequencies to achieve flat fields and thus minimize raster line effects, but this requires somewhat sophisticated consideration of band-limiting noise, while providing enough bandwidth to allow the use of the high raster frequencies. In good design, carefully thought out, the use of high raster frequencies reduces the signal-to-noise ratio on each line but allows the eye to sum those low signals back to a useful level. This sort of design requires much careful thought. See Chapters 5, 6, and 7.

processes produces artifacts or false imagery that can help or hinder the interpretation of the imagery.

This topic is discussed fully in Chapters 6 and 7.

2.3.2. Human Factors Experiments with Line Scan Imagery

Sudden interest in low-light-level technology in the early 1960's and particularly in the first of the intensifier-image orthicon television cameras resulted in a flurry of field evaluations which by and large poorly displayed various pictures of various military targets under various circumstances with little or no control of conditions. As new and better tubes and cameras became available the tests were conducted at longer ranges and on darker nights, almost as if to provide a constant level of poor imagery, just enough to show a shapeless blob of appropriate size along with a comment, "That's a Model 47 so and so."

Yet *some* good work was done in which the same levels of "resolution" as were expected from low-light-level television were simulated by projected photographs, but raster effects and noise characteristics were not understood and thus were ignored or were badly simulated. As a result no real relationship was established that would predict, based upon laboratory trials or photographic simulations, how well a man actually using a television set of a given type could perform in various visual tasks.

I am quite careful not to say anything about the relationship between *system quality* and *observer performance*. The reason is that the television systems were specified, ordered, designed, and purchased on the basis of a number of system parameters that had nothing to do with image quality, i.e., the factors that allowed people to see things on a television display.

There were some who attempted to carry out human factors experiments with television earlier, but some of the first quantitative work of interest was carried out under conditions in which television images of relatively high quality were used to test human ability to perform visual tasks, carried out by Erickson and his associates.

In one of the early studies Erickson *et al.* (1968)* found that small targets with a contrast of 18% against a darker background subtended

* Material extracted from the originial papers with permission of the authors.

about four or more TV raster scan lines on the TV monitor before becoming detectable. Erickson uses the following definition for contrast:

$$C = (B_T - B_B)/B_B$$

where B_T is the brightness of the target and B_B is the brightness of the background.

Erickson's work further indicates that targets with -7% contrast against a lighter background subtended about six TV lines to satisfy detection requirements.

He and his associates reported also that they compared observer performance with several TV monitors viewing tasks in daylight and in darkened rooms. The ability to differentiate shades of gray was reduced in daylight. Critical detail in resolution patterns must be 30% larger in order to give the same visual results as patterns seen in a darkened environment.

He conducted a third experiment to compare resolution and symbol legibility on an 875-line TV raster with those on a 525-line raster display. He concluded that 525-line data can be used to predict performance on other spatial frequency rasters if performance is expressed as a function of the number of raster scan lines making up the image and the angular subtense of the image to the observer's eye.

In further experiments Hemingway and Erickson (1969) examined the effects of angular subtense and number of raster lines subtended by symbols.

They reported*:

> "This experiment examined the relative effects of (1) image size and (2) number of TV raster lines making up the image upon an observer's ability to identify 16 different geometric symbols on TV. Four raster-line values per symbol height were each tested at three image angular subtenses. Eight subjects were told to identify 25 symbols for each of the 12 conditions; all had 20/20 near and far visual acuity or better. The forced-choice method was used; no limits were placed on response times. The results showed that (1) at least eight raster lines per symbol height and (2) a symbol subtense of 10 minutes of arc are necessary to obtain good symbol legibility on TV. An equation is

---

* Abridged from *Human Factors* by permission of the copyright owners, Maxwell International Microform Corp., Fairview Park, Elmsford, N. Y.

developed from these and other data which quantifies the tradeoff between line number and angular subtense for different levels of performance."

Erickson and Hemingway (1970b) continued to study the effect of TV parameters through the study of photographs of military vehicles and geometric symbols.

They reported*:

"Four experiments were conducted to measure identification of photographs of geometric symbols and military vehicles both on TV and by direct vision. The number of TV scan lines per image and image angular subtense were systematically varied. The background against which the vehicles were photographed was also varied.

"The mean performance of symbol legibility on TV followed a constant-product rule: The symbol size multiplied by the number of scan lines per symbol was constant for a constant level of performance within a specified range of the variables.

"The vehicle image required at least 10 scan lines per vehicle and an angular subtense of over 14 minutes of arc to ensure a high probability of identification. Vehicle identification performance also varied as a function of the type of background.

"Direct-vision performance was used to estimate the degradation resulting by interposing the TV system between the observer and the photographs. The TV degradation did not vary with image size, but was a function of lines per vehicle and background type."

In still another experiment Erickson and Hemingway (1970d) examined the visibility of raster lines. They reported[†]:

"One of the present unknowns in TV viewing is the effect of this line structure on image legibility. Fink (1957) estimated that adjacent scanning lines cannot ordinarily be distinguished at distances greater than about six times the picture height. This estimate was derived from surveys of observers' viewing preference. The present experiment was conducted to determine the point at which the TV scan-line structure was no longer visible.

"The results indicate that, under most viewing conditions, scan-line structure will be visible on the type of TV system used in this study. For subjects with superior visual acuity, the structure is therefore a relevant parameter and should be considered among the system parameters that affect TV viewing."

---

* Quoted with permission of R. Erickson.
† Abridged quotation by permission of the author and the Optical Society of America.

Johnston (1968) conducted a series of experiments concerned with target recognition as a function of slant range, *horizontal resolution rather than the number of raster lines per object*, and shades of gray. The measured performance parameters were target recognition time and probability.

Ms. Johnston's results showed correlation between the parameters studied. This is a predictable result since *horizontal* resolution is a function of signal-to-noise ratio.

F. Scott *et al.* (1970) carried out a series of simulations of TV imagery through the use of Scott's line scan generator described earlier. This work was like that of Hemingway and Erickson on symbols except that vehicular images were used as inputs and perspective was an important but not consistent parameter.

They reported*:

> "An experiment was conducted to determine the identification and classification accuracy of noiseless, static line-scan images as a function of the number of scans per vehicle (4, 6, 9, 13.5, 20, and 30 scans) and angle of view (nadir and 45-degree obliquity angle). Twenty-five models of military vehicles were photographed on a uniform background at two angles of view and then transformed into line-scan images using Gaussian-shaped scanning spots. Their task was to match line-scan images to the model vehicles. The results showed that identification and classification accuracy, averaged across all vehicles, was slightly better with the oblique view than with the nadir view; but this result was not true for all vehicles. The results also showed that although accuracy increased slightly beyond 13.5 scans, the spread of performance resulting from differences among subjects and among vehicles decreased substantially going from 13.5 to 20 scans per vehicle. It was concluded that the superiority of one angle of view over the other was vehicle specific and that about 20 scans (raster lines) per vehicle is a reasonable number for both a satisfactory level and spread of performance."

In a related work Gaven *et al.* (1970, 1971) studied the effect of shades of gray and the number of scan lines on observer performance. They concluded[†]:

---

* Abstract quoted by permission of the authors and the original publishers, the Society of Photographic Scientists and Engineers.

[†] Abstract quoted by permission of the authors and the original publishers, the Society of Photographic Scientists and Engineers.

> "The informative value of sampled imagery has been studied as a
> function of the number of sampling elements and the number of bits
> used to encode gray levels in the imagery. Informative value was
> equated with human performance in the current identification of objects
> in the test imagery. For noiseless imagery of a test scene consisting of
> an array of vehicle models on a uniform background: A. Informative
> value increase is noted between 3 and 7 bits. B. Above 3 bits informative
> value asymptotically approaches a level determined by the number of
> sampling elements/scene objects. C. Over the range of variables studied
> in this experiment, optimum utility of a fixed total number of bits is
> achieved with fewer sampling elements and more encoding bits/sam-
> pling elements rather than with more sampling elements and fewer than
> 3 encoding bits."

The above series of experiments is not the only work that has been
done but is representative of a class of psychophysical experiment that
relates performance to display parameters that may indeed not be the
independent variables.

The line scan generator of F. Scott (1967) produced a series of line-
scanned imagery illustrated by the few samples in Fig. 2.16. It is obvious
that there are other more basic factors that drive the quality of line
scan imagery, i.e., which produce the dependent variables such as "lines"
and "gray scale."

The raster lines also produce false imagery caused by the interaction
of the signal inherent in the raster line frequency with frequencies
present in the imagery, i.e., sum and difference effects. These are shown
in Figs. 6.10–6.15, Chapter 6. These effects need not occur in TV displayed
imagery.

These factors were predicted before the birth of television as we
now know it. The difficulties due to sampling in a series of discontinuous
lines were clearly set forth. It is of interest to review the thoughtful
paper by Mertz and Gray, published in 1934, thirty years before most
of the papers discussed above.

In the opening remarks of this classic 1934 paper, which must not
only be read but studied by any serious student of information and
imagery, Mertz and Gray (1934) point out that*:

> "In the usual telephotographic or television systems the image
> field is scanned by moving a spot or elementary area along some re-

---

* Quoted by permission of the copyright owner, the American Telephone and Tele-
graph Co. Originally published in *Bell Syst. Tech. J.*

curring geometrical path over this field. In the more common arrangement, this path consists simply of a series of successive parallel strips. Imagining the path developed or straightened out (or in the more common case, the strips joined end to end), this method of scanning is equivalent to transmitting the image in the form of a long narrow strip.

"The theoretical treatment of such transmission has usually been developed by completely ignoring variations in brightness across the image strip, assuming the brightness to have a uniform distribution across this strip. This permits the image to be analyzed as an ordinary one-dimensional or single Fourier series (or integral) along the length of the strip; and the theory is then developed in terms of the one-dimensional steady state Fourier components. Such a method of treatment naturally gives no information in regard to the reproduction or distortion of the detail in the original image across the direction of scanning, nor, as will appear below, does it give any detailed information in regard to the fine-structure distribution of energy over the frequency range occupied by the signal.

"The need of a more detailed theoretical treatment originally arose in connection with studies of the reproduction of detail in television photographic systems, especially in comparisons of distortion occurring along the direction of scanning with that across this direction.

"Furthermore, the existence of signal energy at odd multiples of half the scanning frequency will indicate the existence of a characteristic in the picture which repeats itself in alternate scanning lines. It is to be expected that such detail in a picture cannot be transmitted without accurate registry between it and the scanning lines and that when the detail spacing or direction or both differ somewhat from the scanning line spacing and direction, beat patterns between the two will be produced in the received picture which may be strong enough to alter considerably the reproduction of the original.

"These phenomena are exactly what is observed, and will be treated in more quantitative fashion in the discussion below."

The above quotations are abridged. It is unfortunate that in a book of this size we can only lead the reader to the older literature and encourage him to do serious reading while using most of our space to present that which is as yet unreported. We thus must elect to forego a complete discussion of the work of Mertz and Gray and summarize most briefly by quoting their abstract:

"By the use of a two-dimensional Fourier analysis of the transmitted picture a theory of scanning is developed and the scanning system related to the signal used for the transmission. On the basis of this theory a number of conclusions can be drawn:

1. The result of the complete process of transmission may be divided into two parts, (a) a reproduction of the original picture with a blurring similar to that caused in general by an optical system of only finite perfection, and (b) *the superposition on it of an extraneous pattern not present in the original, but which is a function of both the original and the scanning system*.*

2. Roughly half the frequency range occupied by the transmitted signal is idle. Its frequency spectrum consists of alternating strong bands and regions of weak energy. In the latter the signal energy reproducing the original is at its weakest, and gives rise to the strongest part of the extraneous pattern. In a television system these idle regions are several hundred to several thousand cycles wide and have actually been used experimentally as the transmission path for independent signaling channels, without any visible effect on the received picture.

3. With respect to the blurring of the original all reasonable shapes of aperture give about the same result when of equivalent size. The sizes (along a given dimension) are determined as equivalent when the apertures have the same radius of gyration (about a perpendicular axis in the plane of the aperture).

4. With respect to extraneous patterns certain shapes of aperture are better than others, but an aperture arrangement is presented which almost completely eliminates extraneous pattern while about doubling the blurring across the direction of scanning as compared with the usual square aperture. From this and other examples the degradation caused by the extraneous patterns is estimated.

Mertz and Gray clearly predict the raster effect on image quality. Twenty years later after patenting means to remove some of the injurious effects predicted by Mertz and Gray, Schade (1953) published a complete quantitative treatment supplementing the paper of Mertz and Gray. This is reviewed and updated in Chapter 6.

Schade commented:

"I recall that the B.B.C. constructed or had built special receivers† with 'spot wobble' to eliminate the line structure which is quite pronounced at the then used 405 lines in England. This originated the

---

* My italics.

† British Patent 535,905. See Hollows (1950) or the report in *Electronics* (Anon, 1952).

'spot wobble' technique. The problem at this early time was that the 'vertical' deflection linearity must be very good to give a uniform field with spot wobble. This is no longer a serious problem but requires careful attention because the spot wobble requires a fixed amplitude for a given vertical picture dimension to avoid overlapping (white lines) at any point of the raster."

Four years after Schade's paper, Thompson (1957) reported on personal preferences in viewing television imagery and the effect that removing raster structure had upon these preferences. His report is abridged below*:

"It has been reported that the most popular vertical viewing angle for movies is approximately $17°$ while that for television is $8°$. Previous data which was verified by experiments described in this paper indicate that the average eye is just able to resolve lines which subtend one minute of arc at the eye. This value of one minute of arc is equivalent to a vertical viewing angle of 7.8 degrees for a picture composed of 480 active lines.

"Experiments indicate that television viewers tend to select a viewing angle at which the line structure just begins to disappear. Under these circumstances it is easy to see one reason why the trend toward larger television pictures has declined. The average living room cannot easily provide the minimum $10\frac{1}{2}$ foot viewing distance desired for 24 inch receivers.

"It would be desirable to suppress or eliminate the line structure in television pictures thereby allowing the larger receivers to be comfortably viewed at reduced distances. Experiments indicate that the same viewers that chose 8 degrees vertical viewing angles for a conventional picture prefer 10 to 16 degree angles for a reduced structure television picture.

"*Fundamental Considerations*

"There are several methods of reducing the appearance of line structure in television pictures. Each must accomplish the same basic task; filling in the dark spaces between the scanning lines with information which does not contrast with the scanning lines.

"The desired result may be obtained by increasing the vertical dimension of the scanning spot."

Thompson described a variety of means for accomplishing such a vertical increase in spot size, and of these chose to sinusoidally oscillate the spot at a frequency very high compared to the horizontal scan

---

* Quoted with the author's permission.

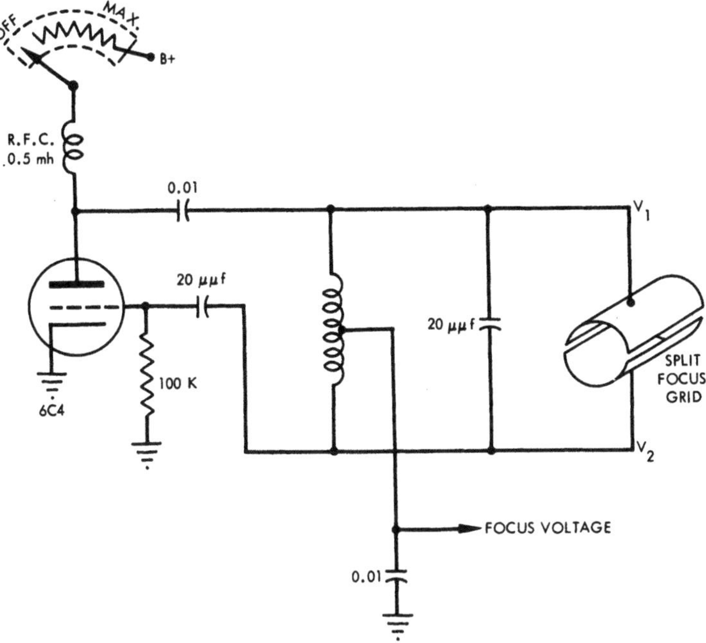

Fig. 2.17. Spot wobble oscillator circuit.

frequency. He accomplished this by placing a pair of split cylindrical electrodes about the CRT as an auxiliary dither-producing deflection circuit. This is illustrated in Fig. 2.17.

Thompson conducted a series of experiments with various raster frequencies. The results strongly confirm each other and thus we quote only his results obtained with a 525-line raster:

> "A group of fifty viewers was used to determine the distance from a 24-in. receiver at which the line structure becomes barely resolvable. The test was conducted using a standard 525-line television raster with no video modulation at a brightness of 20 footlamberts. The viewers backed away from the receiver until the lines just blended together. Distances were recorded for the conventional raster and the wobbled raster using a 13.2 Mc wobble deflection voltage. The number of observers sitting closer than a given distance is plotted in [Fig. 2.18] for the conventional and spot wobble raster. The ratio of the viewing distance for the spot wobble raster to the conventional raster is plotted in [Fig. 2.19].

"The average distance for the conventional raster, 10.6 feet, corresponds to a vertical viewing angle of 7.6°. This result is close to the 7.8° value calculated from the one minute of arc resolution. The average distance for the wobbled raster, 6.1 ft., corresponds to a optical viewing angle of 13.1°.

"A further experiment was conducted using a live telecast on this same 24-in. receiver. A number of the same viewers were seated in a chair on casters and asked to move about and pick the location from which they preferred to view the receiver. When a conventional line picture was shown, they chose approximately the same distance that they had selected in the previous experiment with a conventional raster. When the wobble was added they moved closer to the receiver and chose approximately the same location at which the wobbled line structure had disappeared.

"This experiment indicates that the 8° viewing angle chosen by television viewers was determined by the line structure. Viewers can be expected to choose larger viewing angles with reduced structure television."

Sixteen years after Thompson's experiment and twenty years after Schade's (1953) SMPTE paper, there still is a determined effort to get more information to pilots in aircraft through television or similar

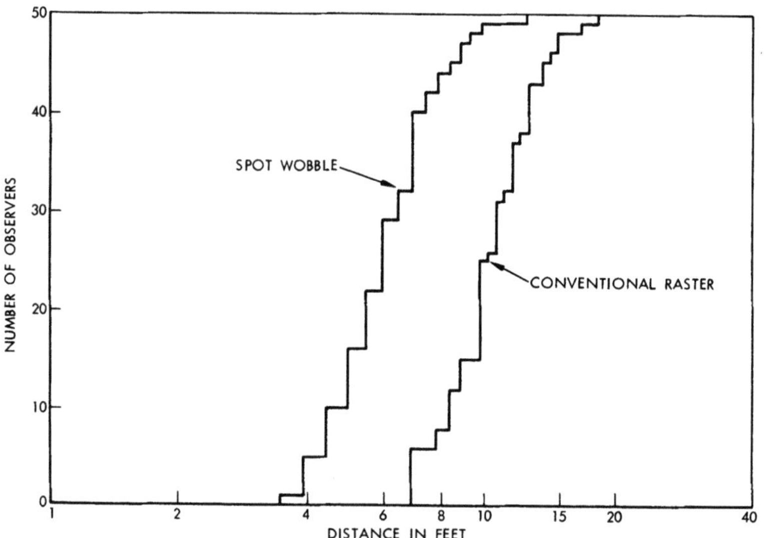

Fig. 2.18. Observers sitting closer than a given distance (525-line raster on 24-in. receiver).

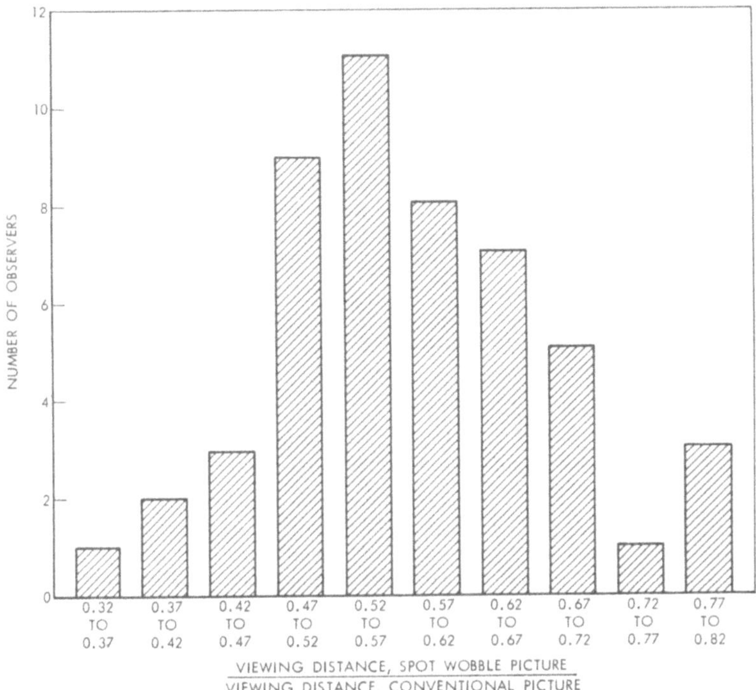

Fig. 2.19. Effect of spot wobble on viewing distance (525-line raster).

image-forming equipment. Though Schade has done the calculations and Thompson has done an irrefutable experiment, system designers still strive for greater amounts of "limiting resolution" in camera tubes and ignore the effect of the raster in the camera tube and *especially* in the display.

Recently I was about to embark on a study of spot wobble and the transfer of information to an observer. As is my custom when facing a problem with which he is familiar, I sought the advice of Schade.

He replied*:

> "The first requirement is excellent interlace, raster stability, and linearity of the vertical scan. I use two single turns lined up exactly *under* the vertical deflection winding of the yoke, as in [Fig. 2.20]. The axial wires of the turns form a 60° angle to get a uniform field.

* Private communication, quoted with permission of the author.

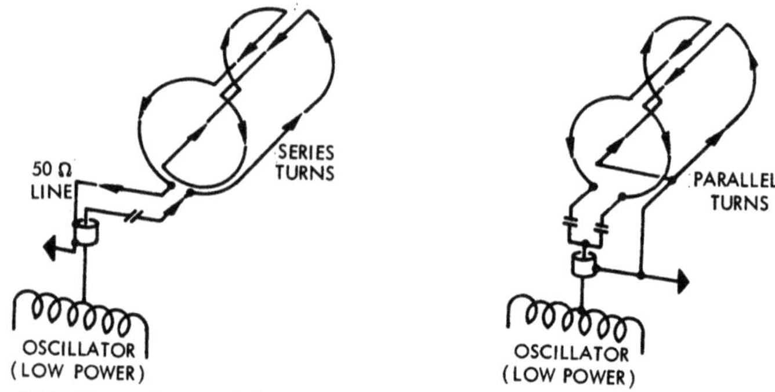

Fig. 2.20. Setup for studying spot wobble (Schade, private communication). Note: Coils subtend 120°.

I have to use parallel turns to get the frequency up to 140 MHz. You can use series turns for 20–40 MHz. The wires may be flat ribbon glued on a thin paper form which slips between the yoke and the tube neck. The coils together with a small series capacitor (30–100 pF) tune the oscillator to the desired frequency and are coupled via a 50-ohm line to one or two turns of the oscillator. The capacitors are essential to prevent loading of the vertical deflection coils by shorted turns.

"The video driver of the CRT should have a sharp cutoff filter with f_c less than the spot wobble frequency to prevent interference modulation. I don't think that normal camera tubes need spot wobble because their MTF is low at 490 cycles or 980 TV lines. I have not used regular vidicons with spot wobble and do not know if the metal wall electrode is an eddy current shield at high frequencies."

In private correspondence A. Oberg of Westinghouse Electronic Tube Division pointed out that*:

"We have made several varieties of electrostatic microdeflection electron guns. The first was the spot wobble tube made by splitting the G4 cylinder of a low-voltage focus gun. Only a few tubes were made.

"Another system used two (2) plates, 1 cm square, in the G5 cylinder of a low-voltage focus gun, with deflection voltages centered around the anode voltage. Neck pins were used since stem leads could not handle Eb2. Production difficulties made this system expensive.

"By far the most successful method is the use of small-deflection plates in the focus cylinder of a high-voltage focus gun. Deflection plates are either 85 or 125 mil apart, depending on beam diameter,

* Quoted with permission of the author.

and 160–200 mil in the Z direction. We have made these with both single- and double-axis microdeflection. Deflection voltages center about focus voltage and are supplied though stem leads. We have made one type with high-frequency deflection (one axis only) supplied by capacitive coupling through the neck.

"Celco makes writing yokes which mount on the neck right behind the regular yoke. Where this is to be used, we would add about 1 in. to the neck length to keep the gun out of the magnetic field. For example, our 16ALP is a variation of the 16AKP in which 1 in. is added to the neck length."

Schade in commenting on spot wobble use in cameras and/or displays says in personal correspondence:

"I do not use spot wobble in the camera although I have tried it on my high-resolution return beam tubes with limited success, because of pickup by the multiplier, which can easily overload the first amplifier. I have not pursued it further because there is no real need with these tubes. It should be easier for tubes without a multiplier.

"The need for spot wobble is real for display CRT's having a small spot compared to the raster pitch, giving weak interline spaces. These show up particularly in photographs taken of the display and with magnification. The sketch [Fig. 2.21] illustrates that a large imbalance of horizontal and vertical resolution never makes a good picture of slanting lines. Four conditions are shown in [Fig. 2.21].

"So one starts with $n_r = N_y = N_x$ and a display spot which is not too small because it will give 'staircases' with sharp corners with a simple sinusoidal spot wobble. If the definition in the image is very noisy, all that is necessary for 'optimum filtering' is simply to reduce the magnification to the eye by increasing the viewing distance or by

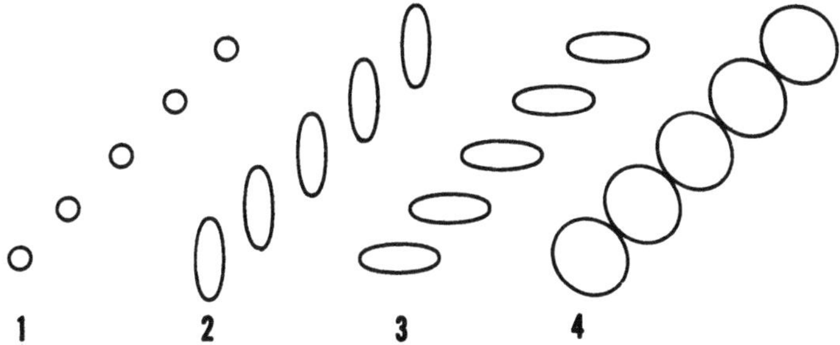

Fig. 2.21. Use of spot wobble. (1) Fine spot, coarse raster high X resolution; (2) vertical wobble leaves gaps; (3) balance between spot and raster dimensions, no spot wobble; (4) line has continuity with spot wobble.

making the display format smaller. The above procedure is satisfactory unless the system saturates (overloads) because of noise. Then the bandwidth and raster line density (n_r) should be reduced to fit the small picture. n_r is the raster line frequency and N_x and N_y are the television line numbers per vertical frame dimension in the x and y directions."

In the experimental material reviewed earlier the ability of the viewer to accomplish certain visual tasks was consistently related to a number of parameters, primarily the number of scan lines and the number of shades of gray in the image.

Thus, though we do not disagree with the conclusions drawn from these particular experiments, we would like to point out that by changing the properties of the spot sizes or "apertures" used with a display raster, but not the raster frequency or basic form, the results would be strongly affected. Actually one must conclude that the angular subtense and number of raster lines do affect the legibility of symbols; however, other basic factors of the raster can have equal or greater effect and thus the raster parameters must be specified.

For the particular conditions of the experiment, the MTF of the camera tube, the effective prefiltering of the lens used, etc. result in a level of aliased signal that causes definite and perceptible deterioration in image quality of the type that Schade (Chapter 6) shows in such examples as Figs. 6.10–6.15. These figures show the effect of the raster on symbols (type) of constant size. The small type is legible when the raster is not pronounced but is not legible when the raster is sharply apparant.

Experiments have been conducted and results reported on the effect of an apparant and common fault of most existing television systems. This fault, the visibility of the raster lines, is not an inherent property of either the television camera or of its display. In well-designed and adjusted TV systems, such as those discussed by Schade in Chapter 6, the raster lines can and should be invisible even at close viewing distances. The degree to which this is not the case is a measure of the departure from ideal of the design, construction, and/or adjustment of the display primarily, though some adverse affects are introduced by the same nonideal conditions in the camera.

It therefore is a difficult task to ascribe visibility to rasters in general when such an effect is highly related to the design and adjustment of an individual specific television system. It is clear that the raster visibility will be an uncertain, variable, and real problem as long as there is little or no concern about its existence or degree of interference—and presently there is almost no such concern, since by and large the buyers of television sets designed for domestic entertainment not only accept as fundamental the presence of the raster lines but often choose a set on the supposition that a clear, sharp, highly visible raster is an indication of a well-designed set in good adjustment.

Much thought needs to be given about the real findings of the experiments discussed earlier. As is shown in Chapters 4 and 5, modulation or signal-to-noise ratio can be used to indicate the limitations in both "resolution" and "shades of gray" in imagery.

It is clear that the recognition process and related visual processes are related to "resolution" and "shades of gray" but it is equally clear that both of these are dependent upon the signal-to-noise ratio in the imagery as a function of spatial frequency.

It is interesting to note that shades of gray represent steps of $\sqrt{2}$ in luminance of the image or resolution element. Obviously with a limited signal-to-noise ratio the number of shades of gray is automatically limited; two shades of gray call for a dynamic range in S/N of $(\sqrt{2})^2$ or 2:1, while eight shades of gray call for $(\sqrt{2})^8$ or a S/N dynamic range of 16:1.

The interaction between object size and shades of gray is highly predictable. Since the response of a television system decreases in a predictable manner as image sizes become smaller, signal-to-noise ratios decrease as images become smaller, until finally the signal-to-noise ratio drops to a just detectable level when the size of the target reaches the "limiting resolution" of the system. This point is a function of the radiance of the object and its contrast against the background and the characteristics of the TV equipment. At limiting resolution there normally is no gray scale. There is less than one shade of gray difference between target and background; for anything smaller the signal-to-noise ratio drops below a detectible value or, as is more usually expressed, below a discernible contrast level at the given level of display brightness.

If the properties of the experimental equipment were known, the results obtained should be predictable from these parameters as well as from the derived parameters of resolution and shades of gray, which are the more subjective manifestations of signal-to-noise ratio.

As stated before, a TV system with the lens capped produces the same display raster, with the same bright and dark horizontal stripes, as the raster associated with a bright, clear image. The first has a good signal-to-noise ratio but only at the raster frequency and then only in the vertical direction and thus has no informative content. The picture, on the other hand, has a good signal-to-noise ratio across the frequency spectrum represented by the imagery presented to the camera when its lens cap is removed.

An objective study, it would appear, would begin with an understanding of the signal level in the test imagery as a function of spatial frequency. In most of the television experiments discussed earlier the image size was sufficiently large so that the frequency dependence of the signal was minimal, and the value of signal-to-noise ratio at the display was essentially the same for all objects above the minimum size used in the experiment.

This situation applied because (1) the signal levels were very high, and (2) spatial frequencies were so low that the aperture functions did not yet seriously affect the signal. If one could reproduce these conditions in a given application the performance should match the quoted experiments. More usually one has real factors that affect the signal-to-noise ratio and begin to remove the results, obtained at high value of signal-to-noise ratio, from a useful base of comparison.

If one can in future experiments of this sort determine both the subjective and objective characteristics of the imagery along with the observer's performance, not only would one understand the signal-to-noise ratio as a function of frequency that was required in an image to allow an observer some probability of achieving a given visual function, but one would know whether some other image under somewhat different circumstances and viewed through different equipment with different transfer functions would produce equal, better, or worse visual performance.

It is of interest to examine Rosell's data (in Chapter 5, in which he predicts through a theoretical analysis and confirms in a series of psychophysical experiments that squares of various sizes subtending 4 to 128 525-nominal-TV raster lines per picture height would give the same probability of detection P_d provided that the video signal-to-noise ratio was made inversely proportional to the linear dimension of the square detected. He further goes on to show that for any given probability of detection of not very small squares the probability is a smooth function of $(S/N)_D$:

$$P_d = F(S/N)_D$$

where

$$(S/N)_D = [2t_e(\Delta f_v)a/A]^{1/2}(S/N)_v$$

where t_e is the integration time of the eye for a given display brightness, Δf_v is the video bandwidth, a is the area of the square to be detected, A is the area of the display, and $(S/N)_v$ is the (broad-area) video signal-to-noise ratio.

Since in a square the linear dimension is the square root of the area, we may write

$$(S/N)_D \simeq (s/S)(S/N)_v$$

where s and S are the dimensions of the square and display, respectively. Thus a small s requires a large value of $(S/N)_v$ for a constant $(S/N)_D$ and thus a constant probability of detection.

This relationship between P_d and the signal-to-noise ratio in the image on the display was theoretically predicted and experimentally verified, as shown in the experimental results of Chapter 5.

The above-described experimental and theoretical agreement of Rosell and the theoretical background derived from de Vries, Rose, Schade, Coltman, and others do not directly yield an estimate of the performance of a man in, for example, distinguishing between the body styles of various automobiles. Yet that same set of theories coupled with a knowledge of the equipment characteristics and the scene characteristics allow one to predict the signal-to-noise ratio versus frequency on the display of the given equipment, the raster parameters, and therefore the number of lines on a given image, and the shades of gray in that image.

If the last two determine the informative content of imagery, the combination of a knowledge of the scene and the equipment should allow one to predict the level of achievement of the given visual task which was predictable from the subjective characteristics. Perhaps more to the point, knowing the characteristics of the scene and the observer's requirements, one should be able to specify the level of system specification necessary to achieve a given visual task.

Actually the computational process for determining probability of detection or identification, or the inverse problem of determining system parameters necessary to achieve the required probability, is considerably simpler than it sounds in the above discussion. The process and methods of calculation are presented in a straightforward manner toward the end of Chapter 5.

2.4. SCALE AND TIME

The experiments to relate image quality to physical factors have gone a long way toward indicating the parameters that affect both the personal subjective judgment of image quality and more objective measurements as well. As will be discussed in the next chapter, there can be a strong relationship between subjective and objective evaluations. However, the ability of people to carry out visual tasks is related to more than just image quality.

This point was made strongly in a paper presented by Self (1969) in which he points out that there are a number of parameters in addition to those contributing to image quality which have perhaps even greater impact on observer performance, especially if the time factor is considered in evaluating observer performance viewing *stationary* imagery.

Resolution and Performance. In his paper, Self speaks of several kinds of visual and instrumental resolution and acuity. He has further remarks, relating to indistinction*:

> "1. There are several types or kinds of human visual acuity for two-dimensional images, each of which has significance for the detection and recognition of details: (a) minimum separable acuity (gap resolution), (b) minimum perceptible acuity (spot detection), and (c) vernier acuity (misalignment detection).
>
> "2. Numbers representing limiting gap resolution of sensors, displays, and observers all depend upon the shape or form of the resolution test pattern. The more that image detail patterns deviate from shape correspondence with the test patterns, the less accurate the latter is in describing the former.
>
> "3. The limiting resolution of the human observer varies with both the image brightness and the image contrast, as well

---

* Extract of Self's paper quoted by permission of the author.

as with the form of the resolution test pattern. Some imaging sensors also vary in effective resolution with both contrast and scene brightness.

"4. No matter how measured, limiting resolution in an image varies with location in the scene or image: Image-forming sensors do not resolve uniformly across the total picture. In addition, resolution in different directions at any point in a two-dimensional image is usually different.

"5. It is often a matter of conjecture as to what constitutes significant detail in target objects, and as to how much resolution is required to adequately record or perceive such detail. How much resolution is needed is further complicated by whether or not the detail of concern appears in an appropriate or expected part of the target.

"6. The significant details of a target may differ in contrast (hence in resolution) from the average contrast of the target with its immediate surrounding. The background of the significant detail may even be the target.

"7. When measured by time to detect or recognize a target, increased resolution increases performance for a while, but is a matter of diminishing returns. A point will be reached beyond which increased resolution does not improve performance.

"8. Attaining, when viewing time is unlimited, some given probability of recognizing a target by form alone, i.e., without briefing or contextual cues, requires some minimum number of resolution elements across the maximum dimension of the target. The higher the desired probability, within limits, the more resolution than detection. The number of resolution elements required depends upon the critical details so is different for different target objects. Resolution required also depends upon the shape of competing nontarget objects.

"9. When viewing time is not unlimited the same factors must be taken into account and, in addition, the dependency of required resolution upon the time limits or desired reaction time must be taken into account."

Self does not indicate that resolution is a function of other parameters. We point out here and in Chapters 4 and 5 specifically that the ability of the observer depends strongly on the modulation present in the image to be resolved and the illumination level of that image, or, from another point of view, the signal-to-noise ratio of the image. These comments do not conflict with those of Self, but their application can make his remarks more quantitative.

TABLE 2.3

Factors Influencing Target Detection and Recognition*

The scene (or total picture)

1. The size of the picture or displayed image.
2. Numbers, sizes, shapes, and scene distribution of areas contextually likely to contain the target object.
3. Scene objects: numbers, shapes and patterns, achromatic and color contrasts, colors (hue, saturation, lightness), acutance, amount of resolved details, all both absolutely and relative to the target object.
4. Scene distribution of objects.
5. Granularity, noise.
6. Total available information content and amount of each type of information. This is one way of summing up 1–5 plus other elements.
7. Average image brightness or lightness.
8. Contextual cues to target object location.

The target object

1. Location in the image format.
2. Location in the scene.
3. Shape and pattern.
4. Size, color, resolution(s), acutance, lightness or brightness.
5. Type and degree of isolation from background and objects.

The test subject (Observer)

Training, experience, native ability, instructions and tasks briefing, search habits, motivation, compromise on speed versus accuracy, assumptions.

* Self (1969), Table IV-1. Not all factors listed in each group are independent of other factors under the same heading and the list is neither systematic nor complete.

Scale and Time. If time is an important factor, then scale becomes one of the most important, if not *the* most important, of the factors.

Table 2.3 gives Self's (1969) listing of the factors found in his work in studying the detection, recognition, and identification of objects in the general context of such Air Force tasks as target finding, in terms of a wider variety of parameters than those we have discussed.

Self pointed out the importance of scale (size) in the tasks of target detection and recognition. His evaluation states:

"Scale Factor

"1.   If no *a priori* (or briefing) information is available, a target object's image must subtend not less than about 12 minutes of visual angle to be identified by form alone, and 20 minutes of arc is not unreasonable under operating or field conditions. The values depend upon the probability of accomplishing the task, and will be larger when the task must be completed within limited time periods.

"2.   With complete *a priori* target intelligence, targets often are found and 'recognized' with few resolution elements, sometimes with less than one element. If a 'blob' is in the right location (contextually) in an image, it has to be the target.

"3.   With a display of a given size, scale factor determines terrain coverage and, for moving image displays, how long a target image will be on the display.

"4.   Some 'empty' magnification is beneficial to quickness of response and viewing ease.

"5.   Clearly, required search time to find a target object depends, among other things, upon both the size and scale of the image or display. Either size or scale can be too large or too small for obtaining the shortest time scores. Thus, when one resolution element subtends over three minutes of arc, recognition can be hindered.

"6.   Ideally, scale factor should match the expected size of the target and the kind and quality of *a priori* information about the target."

Contrast and Detection-Recognition Performance. Self then lists comments about the role of contrast in the detection-recognition tasks. He concludes:

"Contrast

"1.   An object (or image) is not visible unless some minimum contrast is present. This minimum depends upon both image and observer characteristics, as well as how zero visibility is determined.

"2.   Within limits, high contrast of a target image facilitates identification. However, if target contrast is known in advance, low contrast targets can be more quickly found in a complex background than can target objects with an intermediate amount of contrast.

"3.   Only when both target and background are without internal details or contrasts, i.e., when they are each of a different but uniform lightness or brightness, can a single ratio or contrast number be a unique and exhaustive description of the contrast of the target with its background.

"4. When percentage of targets detected, or portion of responses that are correct are used as performance measures, contrast variation appears to have little effect as long as very low contrasts are not involved.

"Acutance: Acutance is a measure of edge gradient and is related to the subjective impression of sharpness, but is independent of resolution.

"1. Pictures with resolution near that of the eye appear sharper when acutance is high and resolution low than when the opposite is true.

"2. Edge sharpness or image enhancement studies show that when pictures appear sharp, objects are more quickly found on them."

Observer's Procedures, Training and Time Requirements. Finally, Self makes some observations based upon the yet unpublished results of some research by him and his associates.

"1. When a target is not quickly found, searchers tend to 'oversearch' (repeatedly search) likely areas and completely avoid areas dismissed as either unsuitable or as suitable but not containing the target. Frequently targets in contextually unlikely places are not found for minutes even though of adequate size, resolution, and contrast for quick recognition when examined.

"2. Despite instructions and training, few observers systematically search a scene until after initial rapid scene-appropriate search fails to find a target. Clearly, search is neither purely systematic nor purely random.

"3. Observers sometimes forget which areas have been searched and assume that they have searched an area when they have not. This leads to large time scores when the target is there.

"4. Other things being equal, target objects closer to the center of the picture tend to be found quicker.

"5. Numerous moving image studies show that subjects under high pressure do hurry to find targets much quicker than those under little or no pressure.

"6. Some observers quickly find targets that others with equal training find only after extended search time or do not find at all. Chance factors, such as looking at the right place early in search, are clearly important. However, some subjects are consistently as much as two to three times faster than others over dozens of targets and scenes, and across studies.

"7. Averaged across many subjects, identically-appearing target images vary drastically in the time required to detect and to recognize them in different backgrounds (scenes). In other words, there is a strong target–background interaction.

"8. When briefing target pictures are rotated relative to the target in the scene, or are of a different size or lightness, target detection and recognition are slower."

Finally Self presented the following conclusions:

"Upon close examination it is seen that many variables or factors influence detection and recognition of objects. The effects become especially apparent when the time to view an image is limited. Even the common image quality measures in use today turn out to be complex in application and in specification of the obtained values. For example, it was pointed out that, at different points in the image and in different directions at any given point, obtained image resolution varies. In making predictions of observer performance, it is clear that the effects of even the simple quality aspects depend upon the state of adaptation, visual capabilities, training, instructions, motivation, etc., of the observers. Even observer search patterns are important. Clues from briefing and/or the image context can make a very large difference in performance. Similarly, time to find targets or the probability of finding them within specified time limits is greatly influenced by 'image complexity' variables, several of which are included in the term 'context.' The influence of target–background interaction effects is clearly established."

Some of the factors that influence target detection and recognition are listed in Table 2.3. Since it includes some factors and approaches not covered here, examination of it will give the reader some food for thought and possibly ideas for research.

The particular image, the particular target, and the particular observer are all important in predicting the elicited performance. Despite the huge amount of research done to date, it may be that only the surface of the prediction problem has been scratched: The end or goal is not yet in sight.

A Simple Time-Related Problem. It is important to consider not only the unhurried ability of the observer to do an ultimate job under laboratory conditions, but we must also consider performance, given limited time of response under stress.

After considering what Self has said, perhaps it would be worth-while to do a simple calculation of scale required for the task of target detection. Now, as yet there is little data on such tasks done through the medium of television imagery, but work like that of Steedman and Baker discussed below permits one to make some estimates even though the conditions of their experiments do not include raster effects. We make a yet unsubstantiated crude allowance by using their *recognition* data without raster for *detection* with raster—a factor of about two, based upon Thompson's (1957) results.

A study conducted by Steedman and Baker (1960) (Fig. 2.22), reveals that the change in search time required to find an object in a field of low clutter and the percentage of errors that occur are rather severe for small objects until they subtend in the vicinity of 12–20 min of arc. Similar findings (Fig. 2.23) were obtained in studies by Miller and Ludvigh (1960), in which they conducted three separate experiments, essentially showing the acquisition time in seconds against the target size in minutes of arc for a field of very limited clutter.

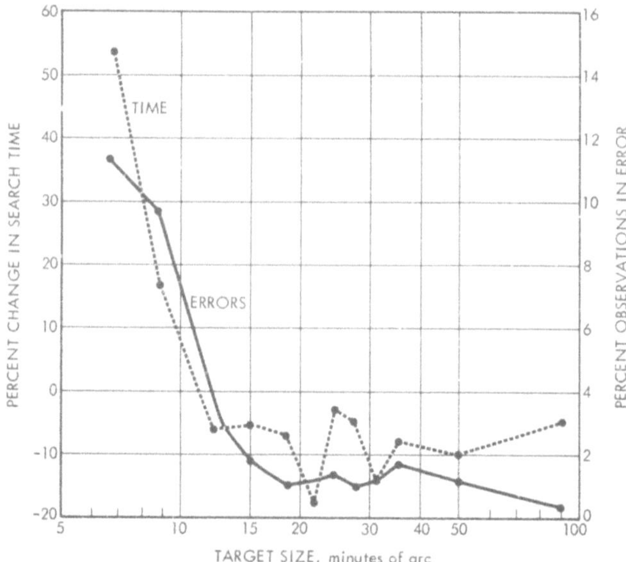

Fig. 2.22. Relative increase or decrease in median-search time and frequency of errors as a function of the visual angle subtense of the targets. Each point was derived from 512 observations. [After Steedman and Baker (1960).]

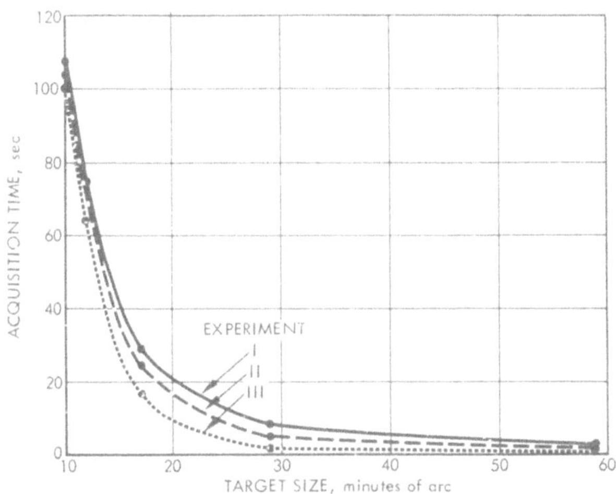

Fig. 2.23. Mean acquisition time of three subjects for three experiments. [After Ludvigh and Miller (1960).]

Let us see what all these data really mean. Suppose one is looking for a vehicle parked on the shoulder of a road at a distance of about two miles. Assume that the vehicle is 20 ft long, 8 ft wide, and 8 ft high. At two miles head-on, the angular subtense of the vehicle is 8 ft/10,000 ft or just less than 1 mrad. A typical field of view of the optical collector might be about 10 deg, which is equivalent to about 180 mrad. Thus, the vehicle subtends about 0.45% of the field of view.

We postulate from the work of Miller and Ludvigh and of Baker and Steedman that the minimum-sized object that can be found in about 10 sec of time is about 12 min of arc. At about a 20-in. viewing distance this requires a displayed image of about 0.08 in. or larger.

From the above we now know that the vehicle occupies 0.45% of the field of view and this must be 0.08 in. or larger. Thus, we can compute the display width $D = 0.08/0.0045$, or about 17.5 in. If the electrooptics is good, an observer can find the vehicle on a smaller screen—it just will take longer, maybe much longer, depending on the display size, brightness, contrast, clutter, etc. One can be sure, however, that any attempt to display the vehicle image on a 4-in. display tube will be rewarded by a combined record of failure or of very slow operator response compared to that obtained with a 17-in. display tube, which the calculations show is the necessary minimum.

Large display *tubes* are not always necessary. Large apparent displays may be used, providing the magnifying optics do not appreciably degrade the overall MTF of the system.

For normal viewing distances of 20 in., in the presence of usual rasters, the observer needs an image size, real or apparent, of at least 0.08 in., say 0.1 in. This means that the observer is not usually going to see anything on the display that is smaller than approximately 1/10 in., and if he is directly viewing a 4-in. display, there are only 40 possible 1/10-in. spots across that 4-in. face, and it is going to be one of those 40 spots that will attract the observer's attention, i.e., we have a 40-line system for the detection process.

The point to be emphasized is that the display size is like a power factor. If the power factor is low, one cannot get the power available in the power line, and if the scope face is small, one cannot get the information that is available on the display face.

Therefore, if space, shape, size, or money limits large display size, then install the size and quality of display allowed by size or money or whatever else the limiting constraint may be but, in any case, scale down the performance of the sensor. It makes no sense to feed an "800-line" sensor to an effective "40-line" display. However, such untenable situations exist in almost every airborne system we have, and the invariable questions arise as to what is really needed in a display, what is needed in a sensor, and what the combination should be. If we consider the curves in Fig. 2.22, it becomes quite clear that, regardless of the contrast—and these data are all for high-contrast situations—the amount of time that is going to be required to find small targets on the face of a conventional 4-in. display is so prohibitive it would be impractical to try and show them.

Determine what the target display size is and how much time will be available, and then design the set accordingly, using either large displays or display magnifications, or else use a lower-performance sensor to match the display.

The glib advice to design the set accordingly is not really glib; the advice is but a few chapters premature. Details about the choice of parameters necessary to achieve a given probability of a given visual task with or without clutter is specifically treated in Chapter 5.

But in any case "design out" the raster effect.

2.5. BIBLIOGRAPHY

Since we do not treat display device technology in terms of storage targets or electron guns, we do append the following bibliography for those who wish to enter the literature of those topics.

Display Devices

Texts and Reference Books

Davis, Samuel (1969). *Computer Data Displays.* Prentice-Hall (EE Series), Englewood Cliffs, N.J.

Kazan, B., and Knoll, M. (1968). *Electronic Image Storage.* Academic, New York.

Luxenberg, H. R., and Kuehn, R. L. (1968). *Display Systems Engineering.* McGraw-Hill, New York.

Moss, Hilary (1968). *Narrow Angle Electron Guns and Cathode Ray Tubes.* Academic, New York.

Sherr, Sol (1970). *Fundamentals of Display System Design.* Wiley— Interscience, New York.

Zworykin, V. K., and Morton, G. A. (1954). "The Kinescope" and "The Electron Gun," Chapters 11 and 12, *Television*, 2nd ed., pp. 383–484. McGraw-Hill, New York.

For general reference see various issues of the *Journal of Information and Display, Journal of the SMPTE, Journal of the Society of Photographic Scientists and Engineers.*

Papers

Johnson, A. D., and Cowden, D. G. (1967). Considerations in Specifying Display System CRT Design Objectives. *Information Display* **1967** (May/June).

Kaestner, Paul T. (1967). Visual Simulation. *Information Display* **1967** (March/April).

Ketchel, James (1969). The Effects of High Intensity Light Adaption on Electronic Display Visibility. *Information Display* (1969) (May/ June).

Ketchel, J. M., and Jenney, L. L. (1968). Electronic and Optically Generated Aircraft Displays: A Study of Standardization Requirements. JANAIR Report No. 68050, 1968.

Langner, Guenther O. (1970). Light Gating Brightens CRT Image for Large Projection Displays. *Electronics*, Dec. 7, 1970.

Slocum, G. K., Hoffman, W. C., and Heard, J. L. (1967). Airborne Sensor Display Requirements and Approaches. *Information Display* **1967** (Nov./Dec.).

Smith, Sidney L. (1963). Color-Coded Displays for Data-Processing Systems. *Electro-Technology* 1963 (April).

Turnage, Rodger Elmo, Jr. (1966). The Perception of Flicker in Cathode Ray Tube Displays. *Information Display* **1966** (May/June).

Walker, Roger S. (1968). Simplified Methods for Determining Display Screen Resolution Characteristics. *Information Display* **1968** (Jan./Feb.).

Weiss, H. (1969). Optimum Spot Size of a Scanned CRT Display. *Information Display* **1969** (Nov./Dec.).

Wurtz, Jim E. (1967). High Resolution Cathode Ray Tubes for the System Designer. *Information Display* **1967** (May/June).

Spot and Resolution Measurement Techniques

Bryden, J. E. (1966). Some Notes on Measuring Performance of Phosphors Used in CRT Displays. In *Proc. 7th Nat. Symp. on Information Display*. Soc. for Information Display.

Constantine, John M. (1966). Two-Slit Spot Analyzer. In *Proc. 7th Nat. Symp. of Information Display*. Soc. for Information Display.

Doyle, R. J., Heiman, F. P., and Kerman, M. (1973). Modulation Transfer Function of Electrical Output Cathode Ray Storage Tubes. *SID Journal,* **1**(4) : 20–22.

Sawtelle, Edward M., and Gonyou, George W. (1968). Dynamic CRT Spot Measurement Techniques. In *Proc. Nat. Symp. Soc. for Information Display, May 1968*.

White, Laurence E. (1959). *Measuring Spot Size in High-Resolution Cathode-Ray Tubes*. Sutton Publishing Co.

Chapter 3

IMAGE QUALITY AND
OBSERVER PERFORMANCE

Harry L. Snyder

3.1. EDITOR'S INTRODUCTION

In earlier material we addressed to the reader our misgivings about the parameters chosen by many systems evaluators. The material in this chapter shows that the area between the MTF and the AIM curves, now called "TQF" or "MTFA" by different people, is a metric that is broadly applicable for general scenes but not necessarily a good metric for specific objects.

The background concepts, equations, and experimental work are described and evaluated for both continuous-tone photography and sampled (TV) imagery. Finally some cautions are stated about the casual use of MTFA without more than casual thought.

The succeeding Chapters 4 and 5 consider the problems of looking for, finding, and recognizing a class of object, and, finally, of identifying a type of object within the class.

It is interesting to note that the concept of MTFA can simply be translated into SNR_D, a concept introduced in Chapter 5, as long as the gamma of the system is unity. For other values of γ the translation becomes complex. This is shown by Biberman *et al.* (1971). (See Chapter 5 by Rosell.)

The MTFA criterion is an evolutionary concept, reliable but not ultimate. The S/N ratio as a function of spatial frequency is introduced in Chapters 4 and 5 and is the criterion we advance for serious consideration and adoption as a more meaningful parameter, over a range at spatial frequencies associated with specific targets at specific distances, upon which to base estimates of human performance under specific conditions with specific objects in mind.

The MTFA is related to the integral, over a range of spatial frequencies, of SNR_D, and either MTFA or SNR_D is a good estimator averaging over many classes of objects.

3.2. NOTATION

A_T	Target area
A_D	Detail area
D	Mean film density
E	Film exposure, usually equal to flux intensity times exposure time
M_D	Detail/background contrast modulation
$M_{D,t}(\nu)$	Target (or object) required contrast modulation at display, normalized
M_0	Target (or object) inherent contrast modulation
$M_t(\nu)$	Adjusted detection threshold at display
MTF	Modulation Transfer Function
MTFA	Modulation Transfer Function Area
$MTFA_D$	The MTFA based upon detail/background modulation
$MTFA_T$	The MTFA based upon target/background modulation
$P_{0.25}$	Physical stimulus eliciting a response with 0.25 probability
$P_{0.50}$	Physical stimulus eliciting a response with 0.50 probability
$P_{0.75}$	Physical stimulus eliciting a response with 0.75 probability
P_c	The probability of correct recognition
$R_0(\nu)$	Sine wave modulation transfer factor
$\bar{R}_s$	The mean slant range to target for correct recognition
$R_{sq}(\nu)$	Square wave modulation transfer factor
γ	System gamma, taken from $D \log E$ characteristic curve
ν_0	Spatial frequency lower limit
ν_1	Spatial frequency at which MTF curve crosses threshold detectability curve
$\sigma_{24\mu}$	RMS film granularity, measured with a 24-μm scanning aperture

3.3. PHOTOMETRIC DISPLAY QUANTIFICATION

The typical input to the eye of an observer searching an electro-optical display consists of a scanned spot of light with a (typically Gaussian) flux density distribution. Many design parameters can affect the width of this spot, including the phosphor selected, the cathode emitting area, the electron lens magnification, the beam current, any focusing aberrations, etc. Detailed discussion of such design tradeoffs is outside the intent of this book; for such, see either Poole (1966) or Luxenberg and Kuehn (1968). What is, however, important from an image quality standpoint is the realization that the luminous output pattern presented at the display surface must be quantified precisely in order to define what is meant by the quality of the image for any given human observer response.

Appropriate quantification of the display surface consists of measurement of the time-varying luminous output pattern in two dimensions. Standard microphotometric instruments can be scanned in either X or Y axes, with the sensing eyepiece either a slit of any specified length, or a finite circular aperture. Figure 3.1 illustrates one measurement arrangement for determining the luminous output pattern of a tribar input, while Figs. 3.2a and 3.2b show an example of the tracing obtained from a single scan of one pattern element in both X and Y directions. (See Chapter 6 for a detailed discussion of raster-sampling effects.) Measurement areas as small as 0.5 μm in width are possible with commercially available equipment, although, as a practical matter, it is unnecessary to measure widths smaller than 50 μm, due to the unimportance of higher spatial frequencies to human vision at normal viewing distances, and also to spot size limitations of high-resolution CRT's.

In many high-quality systems the video voltage input to the CRT can be linearly related to the luminance of the corresponding area on the phosphor surface, although inappropriate settings of the "brightness" and "contrast" controls, present on most displays, by an observer can distort this linearity. Similarly, many of the high-quality complete television systems can be set in such a manner that the luminance of any scene is linearly proportional to the luminance, point by point, on the CRT for any element in the scene. Given such a linear relationship, the X–Y luminance patterns of any given object or area of a scene can then be determined by scanning microdensitometric measurements for

Fig. 3.1. A representative measurement technique for determining photometric display characteristics.

a film transparency input to the television system, or by scanning telephotometric measurements for a given distant scene. As will be shown in a later section of this chapter, such measurements are critical to the prediction of observer performance in recognizing real-world objects in typical cluttered surroundings, a problem totally different from the standard laboratory perceptual problem of recognizing homogeneously luminous forms against a homogeneous unstructured background.

Microphotometric techniques (e.g., Fig. 3.1) can also be used to measure the noise level at the CRT surface. The photomultiplier tube

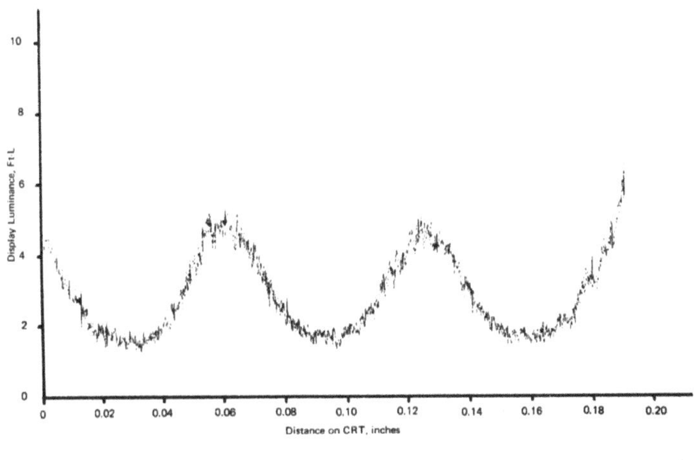

a

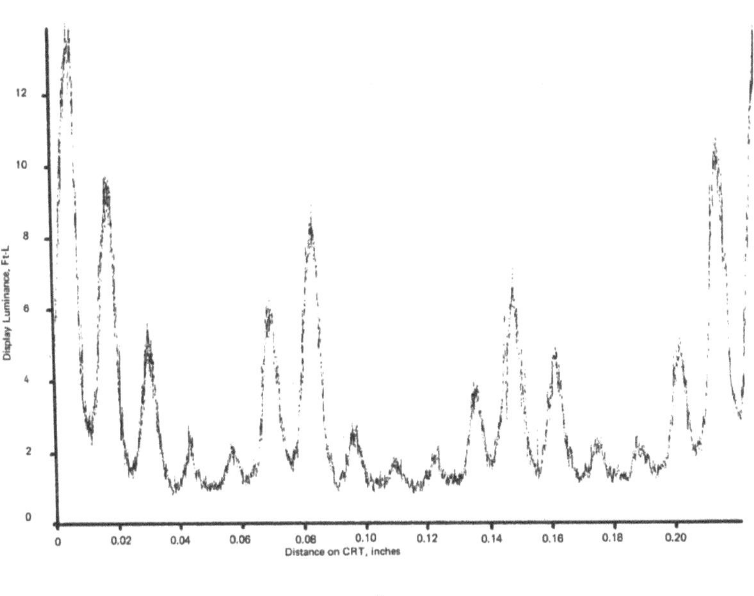

b

Fig. 3.2. (a) Luminous output pattern on CRT for scanning slit perpendicular to raster. Input image is standard tribar pattern. (b) Luminous output pattern on CRT for scanning slit parallel to raster.

output is amplified through a wide-bandwidth amplifier, displayed on an oscilloscope, and photographed (or measured) with, most conveniently, a continuously moving film strip exposed through a narrow slit to produce a long-term trace of the fluctuations in luminance for a given area on the CRT. Optical elements can be selected to measure such fluctuations for either circular or slit areas on the CRT. In this manner both signal levels and noise levels (for any given signal area) can be measured photometrically and related to both observer performance and to analytical derivations of the form developed in later chapters of this book.

3.4. HUMAN PERFORMANCE EVALUATION CONSIDERATIONS

As a critical part of any man–machine system, the human operator receives information, processes it, and takes some specific action upon the system. While we may conveniently think of, and physically measure, such inputted information in appropriate physical and temporal units, it is unfortunately true that the human sensory system does not usually encode and perceive the information in the same physical units (or linear equivalents) or with the same resolution, range, or fidelity. Thus, we find that human perception is typically not homologous with the physical world, and, further, that this lack of correspondence often deteriorates as the human operator becomes overloaded in his task requirements or is presented with information close to his absolute threshold of sensitivity. The relationship between the physical dimensions of an input and the observer's perception of that input (measured on a different, but appropriate, continuum or metric) is the rightful study of psychophysics. Thus we have the psychophysical scales of loudness (sones or phons) as a correlate of sound amplitude, of hue as a psychological correlate of dominant wavelength, and of apparent "brightness" as a correlate of the luminance of a specified visual stimulus.

The estimation of human perceptual thresholds is possible by many of the standard psychophysical methods, each of which can be shown to yield a somewhat different threshold value for a given set of experimental conditions. The reason for these differences lies in the fact that the threshold is, by definition, a statistical concept, and is defined typically as the stimulus value which elicits a 0.50 likelihood of response (corrected for guessing) or, by some investigators, as the value of $(P_{0.75} - P_{0.25})/2$, where $P_{0.75}$ is the physical stimulus value eliciting a 0.75 likelihood of response and $P_{0.25}$ is the physical stimulus value corresponding to a 0.25 likelihood of response. This latter value is often called the interquartile range, and is equivalent to the more standard 0.50 point ($P_{0.50}$) for a symmetric distribution between the $P_{0.25}$ and $P_{0.75}$ points. For a discussion of the methods of average error, limits, and constants, or modifications of these classical psychophysical techniques, the reader is referred to Kling and Riggs (1971). What must be remembered, however, is that such a threshold is a statistical quantity which can vary with the measurement techniques, the form of analysis of the data, and, most importantly, the instructions or criterion level set by the observer. Generalization of one threshold value from a particular experiment conducted in a particular laboratory to another experiment conducted in another laboratory with an entirely different set of equipment and observer instructions may not usually be done with complete confidence.

In evaluating raster-scan displays using some type of observer response as a criterion of display goodness, one is faced with many of the problems inherent in basic psychophysical experimentation. Specifically, the observer's criterion of response must be controlled (or measured by some appropriate means), the experimental or operational environment must be controlled or measured, and all pertinent parameters of the display and its content must be specified. This last requirement will be discussed in a later section of this chapter. To date over 300 laboratory experiments, analytical studies, and field tests have been conducted to relate one or more characteristics of the visual display to the performance of the human observer in obtaining information from a search-type display. In general, these researchers have reported their results in terms of the following classifications of the observer's response to searching for an object in a visual field (Snyder *et al.*, 1967):

1. *Detection* is said to occur when the observer correctly indicates his decision that an object of interest exists in the field of view.

2. *Recognition* is said to occur when the observer correctly indicates to which class of objects the detected object belongs.

3. *Area recognition* is said to occur when the observer correctly indicates that the location of the target is in the field of view.

4. *Discrimination* is said to occur when the observer correctly indicates that the *singular* target of interest is in the field of view; that is, he is correct in separating the single target of interest from the class of recognized targets.

There is also a tendency for the term "target acquisition" to be used in some operational contexts, in which case it typically refers to the set of decisions subsumed above under the terms detection, recognition, and discrimination. It is rarely the case that the observer can consciously separate these classes of response, particularly in a limited-time search situation; further, it is rarely the case that studies purporting to separate these classes of response can demonstrate meaningful differences. Thus, one must be careful that the terms used are not considered to reflect different levels of decision-making if in fact no such differences exist. Consider, for example, the observer searching for an isolated white disk on a homogeneous black background. We normally speak of this as a detection situation. Yet, because the observer *knows* that only one such disk may (or may not) appear on the display, the detection criterion is equivalent to that for recognition or discrimination. For the more conventional situation in which the observer is looking for a particular preselected target among many other objects and only one such example of that target is possible in the display, the tasks of detection, recognition, and discrimination are considered by many researchers to be at least nonindependent, if not equivalent. In the case of the unknown target, as in Johnson's work discussed in Chapter 2, the target uncertainty makes possible the separation of the responses of detection, recognition, and discrimination, provided that there is adequate time for the observer to distinguish among his responses. For an experimental relationship among the observer's level of confidence and these response measures see LaPorte and Calhoun (1966).

Regardless of the nature of the response required by the observer, the operational definition of the response threshold is the point at which a 0.50 probability of response exists. In object recognition research pertaining to air-to-ground operations, however, other, more useful measures of performance have been devised. These include the following.

1. *Response time*: The time between the presentation of the target and the observer's response. This is often converted to *range to target* for the air-to-ground search case.

2. *Per cent or probability correct*: Usually related to detection or recognition criteria, the per cent correct, for a given situation, is an estimate of the asymptotic performance of a number of observers (or a number of trials for a given observer). Often the per cent correct is plotted cumulatively versus range to the target, or versus the time the target is in the field of view. Other measures derived from this score are those of *completeness* (the number of correct responses divided by the number of targets possible) and *accuracy* (the number of correct responses divided by the total number of responses, correct and incorrect). These last two measures are most often used in photointerpretation studies.

Ensuing sections of this and other chapters will use one or more of the above operator performance measures. Unfortunately, it is not always possible to move analytically from one measure to another, particularly if the author of a given article has not adequately defined the constraints of the experimental situation. When, however, the care is taken to assure the resemblance of one experimental situation to another, very high correlations among data sets can be obtained, as in the case of a product–moment correlation of 0.96 between simulator and field-test observer performance for an air-to-ground search situation (Wyman *et al.*, 1968).

3.5 INDIVIDUAL DISPLAY PARAMETERS AND OBSERVER PERFORMANCE

During the past two decades several hundred laboratory and analytical studies have been performed to assess the relationship between variation in line-scan display image parameters and observer performance. Conclusions drawn from critical reviews of these studies (e.g., Snyder *et al.*, 1967; Hairfield, 1970) have indicated that cross-study comparisons are virtually impossible. Variation in specific system design parameters or in the manner by which display image quality is synthetically manipulated is often incompletely controlled, so that concomitant variation in the several contributing sources of image quality results. Table 3.1 lists some of the system variables which have been shown to have a significant (although often inconsistent) effect upon operator information extraction (e.g., target acquisition) performance. It should

TABLE 3.1

A Partial List of Raster-Scan Imaging System Parameters Affecting Observer Performance

Mean luminance	Aspect ratio
Size	Raster direction
Viewing distance	Contrast enhancement
Number of active raster lines	Video bandwidth
Contrast	Detector irradiance level
Scene movement	Target and background characteristics
Gamma	Terrain masking
Signal-to-noise level	

be noted that individual experiments have tended to examine the effects of one, two, and sometimes three such variables. However, due to the inherent interaction (nonindependence) among these variables in their effects upon observer performance, quantitative combination of the results is hazardous even in the presence of good experimental control and measurement. In the absence of such control any *a posteriori* attempt to combine such results has proven useless.

Because of these gross conflicts and inconsistencies in the experimental literature dealing with the effects of individual system parameters, recent efforts have been oriented toward the development of (1) analytical expressions of overall image quality, such as those discussed in later chapters of this book, and (2) experimental evaluations of logically derived summary measures of image quality. The remainder of this chapter will discuss previous research pertaining to one promising summary measure of image quality and its recent application to raster-scan displays.

3.6. THE MODULATION TRANSFER FUNCTION AREA

Originally proposed by Charman and Olin (1965), who termed it the threshold quality factor, and renamed by Borough *et al.* (1967), the MTFA concept has been evaluated in two photointerpretation experiments and demonstrated to relate strongly to the ability of image interpreters to obtain critical information from reconnaissance imagery. In its original form the MTFA was proposed as a unitary measure of photographic image quality which contains "the cumulative effect of the various stages of the atmosphere–camera–emulsion–development–observation process, the 'noise' introduced in the perceived image by photographic grain, and the limitations imposed by the physiological and psychological systems of the observer" [Charman and Olin (1965), p. 385]. While this measure was originally developed for direct photographic systems, its generalization to electrooptical systems is valid and quite useful, as will be shown subsequently.

The MTFA is derived in such a manner as to make use of the modulation transfer function of the imaging system, thereby retaining the analytical convenience of component analysis based upon sine wave response characteristics. In addition, it attempts to take into account other variables critical to the imaging and interpreting problem, such as exposure, the characteristic curve, granularity, the human observer capabilities and limitations, and the nature of the interpretation task. For the electrooptical system the first three of these variables can be considered analogous to detector irradiance level, gamma (typically unity), and noise, respectively.

Figure 3.3 shows that the MTFA is the area bounded by the imaging system MTF and the detection threshold curve of the total system, including the eye. The MTF curve for the imaging system is obtained in the conventional manner, while the detection threshold curve requires

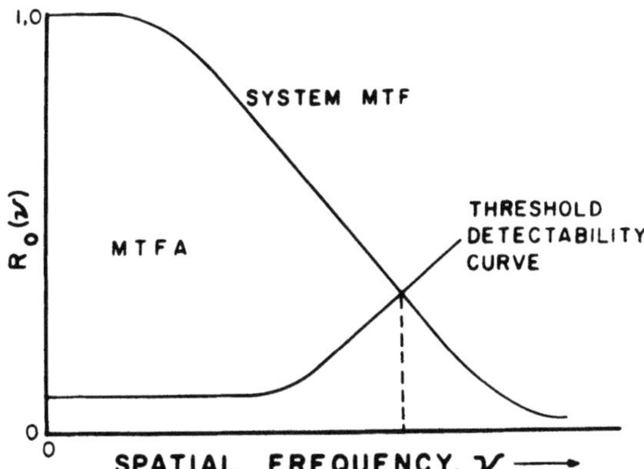

Fig. 3.3. Modulation transfer function area.

several assumptions regarding the human observer. Specifically, it is assumed that the viewing conditions are optimum and that threshold detection of any target in the imaged display is a function of the target image contrast modulation, the noise in the observer's visual system, and the noise in the imaging system exclusive of the observer. It should be noted that the crossover of the two curves in Fig. 3.3 represents the limiting resolution of the system for a sine wave target.

At low spatial frequencies the threshold detection curve is dependent primarily upon the properties of the human visual system, as shown in Fig. 3.4. At higher spatial frequencies the effect of imaging system noise becomes important. For the photographic case this imaging system noise is equivalent to granularity. It is assumed further that under the illumination conditions used to view the imagery the eye's threshold contrast modulation is 0.04, so that a target image contrast modulation of 0.04 must be realized at the display for the target to be detected, regardless of the inherent contrast of the target object.

Figure 3.4 illustrates the normalized detection threshold curve, which must be adjusted both vertically and horizontally for a specific set of conditions. First the curve is positioned vertically by increasing the normalized ordinate scale by $M_{D,t}(\nu)/M_0$, where $M_{D,t}(\nu)$ is the normalized value as shown in Fig. 3.4 and M_0 is the object contrast modulation. Note that the lower portion of the threshold curve (at the lower spatial frequencies) is also adjusted by the system gamma, which, if greater than

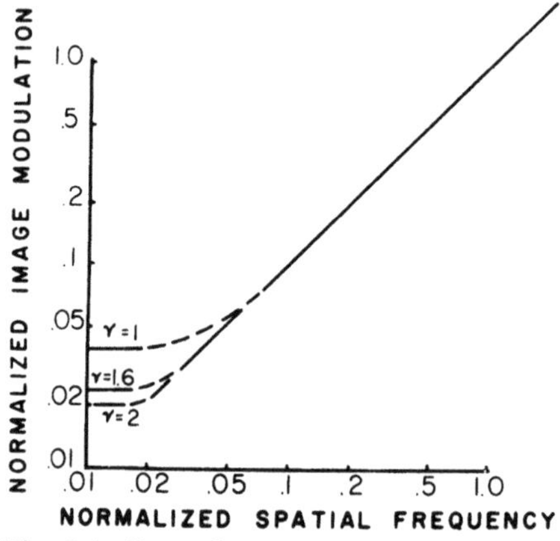

Fig. 3.4. Generalized detectability threshold.

unity, enhances the modulation recorded at the display (e.g., film) so that the minimum detectable threshold modulation decreases by $0.04/\gamma$.

Next the detection threshold curve is positioned horizontally by multiplying the scale of the abscissa in Fig. 3.4 by $2/C\sigma_{24\mu}$, where C is an empirically derived constant [0.03 for fine-grained films and 0.04 for coarser-grained films (Scott, 1963)], and $\sigma_{24\mu}$ is equal to the standard rms granularity measured with a 24-μm scanning aperture.

Algebraically, the detection threshold curve for a photographic system is therefore (Charman and Olin, 1965)

$$M_t(\nu) = 0.034[dD/d(\log_{10}E)]^{-1}(0.033 + \sigma_{24\mu}^2\nu^2S^2)^{1/2}$$

in which ν is any spatial frequency, in lines per millimeter; 0.034 is an empirically derived constant;* D is the mean film density; E is the exposure; 0.033 is an empirically derived constant;* $\sigma_{24\mu}$ is the rms granularity for a 24-μm scanning aperture; S is the signal-to-noise ratio necessary for threshold viewing, assumed to be about 4.5 (Carman and Charman, 1964); and $dD/d(\log_{10} E)$ is the film characteristic slope, including effect of development.

When the MTF curve and the detection threshold curve are plotted on log-log coordinates (Borough et $al.$, 1967) the expression for the MTFA becomes

$$\text{MTFA(log-log)} = \int_{\log\nu_0}^{\log\nu_1} [\log R_0(\nu)]\, d\log\nu$$
$$- \int_{\log\nu_0}^{\log\nu_1} \left[\log \frac{M_{D,t}(\nu)}{M_0}\right] d\log\nu$$
$$= \int_{\log\nu_0}^{\log\nu_1} \left[\log \frac{M_0 R_0(\nu)}{M_{D,t}(\nu)}\right] d\log\nu$$

where ν_0 is the low spatial frequency limit, in lines per millimeter; ν_1 is the spatial frequency at which the MTF curve crosses the detection threshold curve (limiting resolution); $R_0(\nu)$ is the MTF value at spatial frequency ν; M_0 is the object contrast modulation; and $M_{D,t}(\nu)$ is the contrast modulation of the target, as measured at the imaged display, i.e., film.

---

* For derivation see Charman and Olin (1964). Generation of these values is considered unimportant in the present context.

When the MTF curve and the detection threshold curve are plotted on linear coordinates the area of interest is given by (Borough *et al.*, 1967)

$$\text{MTFA(linear)} = \int_0^{\nu_1} \{R_0(\nu) - [M_{D,t}(\nu)/M_0]\}\, d\nu$$

The linear form computation utilizes no lower-frequency cutoff, whereas the log-log formulation employs an arbitrary cutoff at, say, 10 lines/mm. The reason for this difference is simply because the log-log plot integration would place an inappropriately large weight upon integration over the lower spatial frequencies were this cutoff eliminated. The nature of the linear plot avoids the need for such an arbitrary cutoff.

It might also be noted, parenthetically, that the detection threshold curve as described here is akin to such concepts as contrast sensitivity (Campbell and Green, 1965), sine wave response (DePalma and Lowry, 1962; Lowry and DePalma, 1961; Bryngdahl, 1966), and the demand modulation function, as mentioned previously in this book.

3.7. EVALUATION OF THE MTFA FOR PHOTOGRAPHIC IMAGERY

To date two empirical evaluations of the MTFA concept have been conducted using photographic imagery. In the first study (Borough *et al.*, 1967) the MTFA was related to subjective estimates of image quality obtained from a large number of trained image interpreters. In the second of these experiments actual information-extraction performance data were obtained, as well as subjective estimates of image quality, and both measures were compared with the MTFA values of the imagery. While it is desirable from an application viewpoint to have a quick judgment of subjective image quality to serve as an indicant of the quality of any source of imagery for, say, rapid screening purposes, the critical measure of goodness of any imaging system is the ability of the observer to perform the required information-extraction tasks.

In the first study to evaluate MTFA the purpose was to determine whether a strong relationship existed between MTFA and subjective image quality. This limited evaluation was imposed simply to reduce data collection costs in the event that the MTFA measure proved fruitless. Nine photographic reconnaissance negatives were used as the

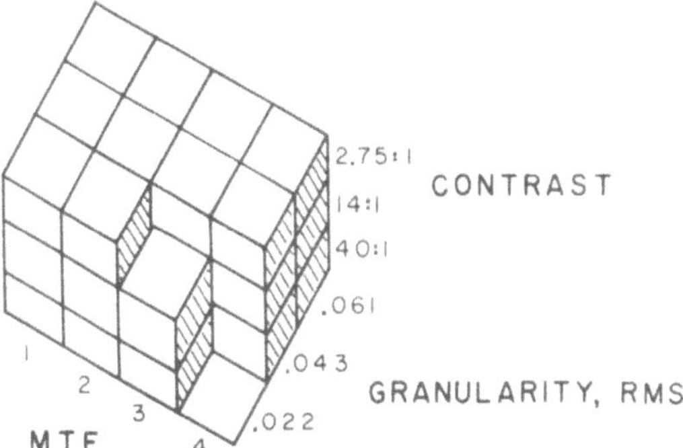

Fig. 3.5. Generation of MTFA values.

basis for laboratory-controlled manipulation of image quality. Each of the scenes was printed in 32 different MTFA variants, determined by four different MTF's, three levels of granularity, and three levels of contrast, as illustrated in Fig. 3.5. Four cells of the matrix were deleted because their MTFA values corresponded to others in the 32-cell matrix. The MTF curves are illustrated in Fig. 3.6.

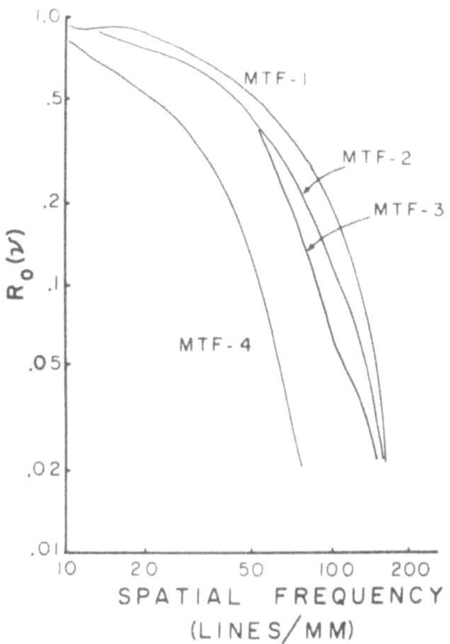

Fig. 3.6. Average modulation transfer functions
measured by edge-response method.

TABLE 3.2

Correlations of Physical Variables with Subjective Image Quality Scale Value

Physical variable	Scene number									Mean* r
	1	2	3	4	5	6	7	8	9	
MTFA (linear)	0.921	0.927	0.900	0.925	0.935	0.919	0.919	0.920	0.913	0.920[†]
Modulation	0.220	0.641	0.511	0.618	0.680	0.699	0.497	0.698	0.632	0.576
MTF	0.698	0.529	0.580	0.660	0.579	0.608	0.697	0.469	0.542	0.601
Granularity	−0.543	−0.632	−0.618	−0.450	−0.516	−0.428	−0.505	−0.589	−0.577	−0.543
MTFA (log-log, 2 cycle)	0.666	0.863	0.866	0.821	0.874	0.890	0.749	0.902	0.876	0.846
MTFA (log-log, 2 cycle)	0.768	0.923	0.923	0.867	0.920	0.921	0.824	0.941	0.920	0.900
Acutance	0.599	0.448	0.526	0.568	0.564	0.599	0.625	0.440	0.602	0.555

* These mean values were determined by transforming the correlations to z values. Such a transformation is necessary when correlations are being combined to obtain a mean correlation.

† This mean value was significantly greater ($p < 0.01$) than all of the other mean correlation values except the value for MTFA (log-log, 2 cycle). This latter value was still significantly less than the MTFA linear value at the 0.05 level of significance.

The resulting 288 transparencies (nine scenes by 32 variants/scene) were used in a partial paired-comparison evaluation by 36 experienced photointerpreters. The subjects were asked to select the photo of each pair that had the best quality for extraction of intelligence information. All pairs were composed of two variants of the same scene; each subject made a total of 256 comparisons, for a grand total for all subjects of $36 \times 256 = 9216$ judgments.

Correlations were obtained between the subjective image quality rating (derived from the paired comparisons) for each of the 32 variants and several physical measures of image quality. Table 3.2 shows the results. Most important to this discussion is the mean correlation of 0.92 between MTFA (linear) and subjective image quality, which indicates that MTFA is strongly related to subjective estimates of image quality.

The next experiment, by Klingberg *et al.* (1970), examined the relationship between objectively measured information-extraction performance and the MTFA values. As a check on the results of the previous experiment, Klingberg *et al.* also obtained subjective estimates of image quality, so that all three intercorrelations among MTFA, subjective rating, and information extraction performance could be obtained.

The imagery used for this experiment was the same as that used by Borough *et al.* (1967). A group of 384 trained military photointerpreters served as observers. Each observer was given one variant of each of the nine scenes and asked to (1) rank the image on a nine-point interpretability scale, using utility of image quality for information extraction as the criterion, and (2) answer each of eight multiple-choice questions dealing with the content of the scene. The interpretability scale values were used to develop a subjective image quality measure for the 288 images, while scores on the multiple-choice interpretation questions were used to measure information-extraction performance.

Figure 3.7 shows the scattergram between information-extraction performance and MTFA for the 32 MTFA values. The resulting correlation, averaged across the nine scenes, is -0.93. (The minus value is due to the use of number of errors as a measure, which is inversely related to MTFA.)

Individual correlations among performance, MTFA, and subjective quality (rank) are shown in Table 3.3. It is apparent that the relationship between MTFA and performance is not as high for some scenes (e.g., 6 and 9) as for others, but that the mean correlation r across scenes

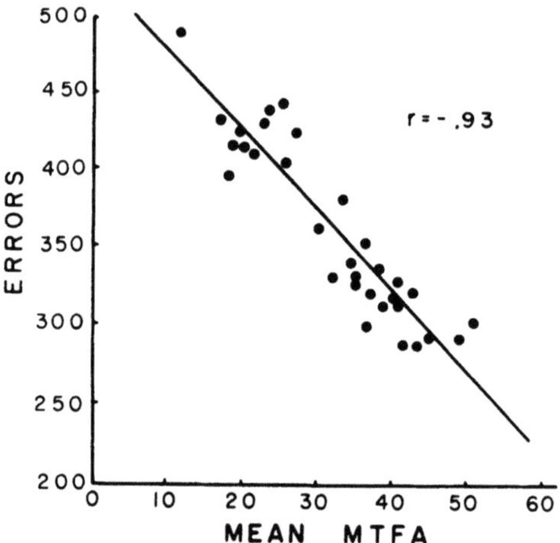

Fig. 3.7. Scattergram of information extraction
performance versus MTFA.

(0.72) is quite high. Further, disregarding scene content and placing
all scenes on a common performance continuum, the correlation of
−0.93 accounts for over 86% of the variance in information-extraction
performance. Further, the 0.96 correlation of MTFA with subjective
quality (rank) agrees quite well with the correlation of 0.92 obtained by
Borough *et al.* Finally, as might be expected, there was a very high correlation ($r = 0.97$) between information-extraction performance and
subjective scale values of image quality.

The results of these two studies have shown that a physical measure
of image quality based upon the excess of MTF over the threshold
detection level, integrated over all functional spatial frequencies, correlates highly with both the ability of the photointerpreter to obtain
critical operational information from the imagery and with his overall
subjective estimate of the quality of the imagery. It must be remembered,
however, that the assumptions of the MTFA concept are not totally
consonant with the typical use of video systems. Specifically, the MTFA
threshold detectability curve assumes both optimum viewing distance
and also adequate (perhaps unlimited) viewing time, neither of which
is usually the case for raster-scan applications, particularly for air-to-
ground search. That is, the observer can rarely change magnification
of the sensor/display system to provide an image of the target which he

TABLE 3.3

Correlations (Pearson r's) among Image Quality, Interpreter Performance, and Subjective Judgments*

	Scene									$\bar{r}$	r_m
	1	2	3	4	5	6	7	8	9		
Performance/MTFA	0.69	0.66	0.80	0.65	0.78	0.55	0.84	0.86	0.46	0.72	0.93
Performance/rank	0.71	0.67	0.89	0.60	0.80	0.42	0.78	0.76	0.42	0.70	0.96
MTFA/rank	0.90	0.87	0.90	0.93	0.94	0.87	0.92	0.86	0.83	0.90	0.97

* $N = 32$ image quality levels (MTFA). $\bar{r}$ is the average of r's using z scores. r_m is the values averaged across scenes before computing scene-by-scene correlations.

believes is optimum in size for inspection of that particular target. Because the same system is used for viewing many targets of many sizes and contrasts, a fixed magnification is typically used. Also, many search situations are dynamically changing, perhaps at a rate much faster than the observer would prefer, and he therefore cannot inspect any given object as long as he might wish or at an image size which is invariant.

For these and other less obvious reasons it is not a totally safe generalization from the MTFA as used in photographic imagery evaluation to the evaluation of raster-scan displays. For example, while the photointerpreter has the availability of variable-magnification devices, adjustable-intensity transilluminators, etc., the CRT observer is often faced with such annoying circumstances as variation in surround luminance, glare, vibration, time stress, glint, etc. Thus for a given set of circumstances the threshold detectability curve might be significantly different than that for another situation. In addition, the noncorrelated nature of the noise in the video system makes the generalization from the static photographic noise a very unsafe one. The constants in the derivation of the threshold detectability curve, as defined by Charman and Olin (1965), are not directly translatable into video parameters. Nevertheless, the general MTFA concept appears intuitively meaningful in the context of video systems, and to date our research has supported its validity, as will be shown in the next section.

3.8. THE MTFA AND RASTER-SCAN DISPLAYS

Very recently several experiments have been conducted in our laboratories to evaluate the utility of the MTFA concept as it might be applied to real-time, raster-scan displays. Emphasis has been placed upon careful quantification of the photometric characteristics of the CRT display and upon measurement of meaningful but objective responses of the human observers. In general, the results indicate that the MTFA concept, with some needed modification and redefinition for application to the raster-scan display, is a valid predictor of overall system quality but that prediction of a specific level of observer performance for locating and identifying a specific target under a specific set of operational circumstances is still a poorly understood problem.

The difficulty of generalizing the MTFA concept as originally proposed by Charman and Olin (1965) and later modified by Borough *et al.* (1967) to raster-scan displays lies in the specification of the threshold detectability curve for the given viewing conditions. It was previously shown that the threshold detectability curve was defined by a generalized curve (Fig. 3.4) which was shifted vertically and horizontally as a function of such parameters as system gamma, target inherent contrast, some constant C dependent upon the typical graininess of the film, and the rms granularity. Of course, target contrast can be defined (however, see a later section of this chapter) and system gamma can be set, typically at unity, for any electrooptical system. More critical, however, is the selection of the constant C and the translation into the electrooptical system analog of the rms granularity term. Preliminary analysis did not permit us to translate these two terms confidently into measurable or useful parameters. Further, we felt that the addition of noise to a video signal might easily serve to change the overall shape of the threshold detectability curve, not merely to shift the curve of Fig. 3.4 to the left and upward. Thus a preliminary experiment, the results of which are illustrated in Fig. 3.8, was conducted.

As Fig. 3.8 indicates, increases in noise level lead to increases in the contrast modulation required to detect, using a 0.50 threshold

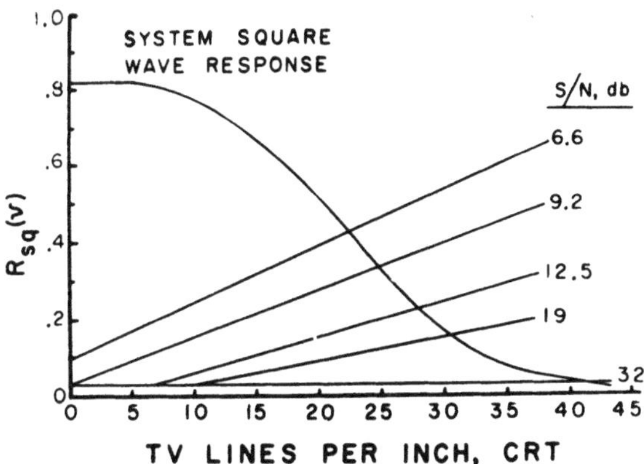

Fig. 3.8. Threshold detectability curves and system MTFA. System line rate, 945; video bandwith, 16 MHz; vertical raster on 17-in. (diagonal) CRT; viewing distance, 40 in. MTF curve and threshold detectability curves are for square-wave bar patterns.

criterion, a standard tribar pattern of any spatial frequency. Further, the effect of noise level upon the required threshold modulation is greater for higher spatial frequency patterns. While these data are very preliminary, their use in defining the MTFA of any particular raster-scan system may be evaluated by the results to be presented below. As this book goes to press, we are conducting a more complete experiment to define these threshold detectability curves for a variety of line rates, video bandwidths, noise passbands, etc. As noted on Fig. 3.8, the obtained data pertain only to a 945-line, 16-MHz system having the given MTF.

Using the multiparameter video system illustrated in Fig. 3.9, an experiment was conducted to determine the likelihood of recognition and air-to-ground slant range of recognition of 25 different real-world targets on a typical midwestern United States terrain. The content of the imagery and the means of obtaining the imagery are described in detail by Snyder *et al.* (1967). In general, a high-fidelity 3000:1 scale terrain model was "flown over" by a 35-mm camera at controlled speeds and altitudes and with a fine-grain film exposure time of 1/30 sec. Various camera lens fields of view and depression angles were combined factorially with pertinent mission parameters. For the data to be reported here the films represented an altitude of 10,000 ft, a groundspeed of 300 ft/sec, a camera depression angle of 45°, and a vertical field of view of 35°. A 4:3 (vertical:horizontal) film camera format and display format were used; thus in the video playback the raster lines were vertical.

A total of 55 observers was used, each of which had at least 20/22 near and far vision, and no measurable visual defects as measured with a Bausch and Lomb Orthorater. Each of five groups of 11 observers searched for the 25 different targets in a preselected order. Each observer was shown an isolated photograph of the target and studied the target book sufficiently to recognize each target and to distinguish it from its background under time-limited search conditions typical of the air-to-ground search situation.

Each of the five groups of observers was tested under a different video noise level. All other parameters of the test situation were constant, including the overall video level, which automatically compensated for the changes in video voltage induced by the noise levels. Thus the mean video signal level into the CRT was independent of the particular noise level. Microphotometric monitoring of the CRT verified overall intensity

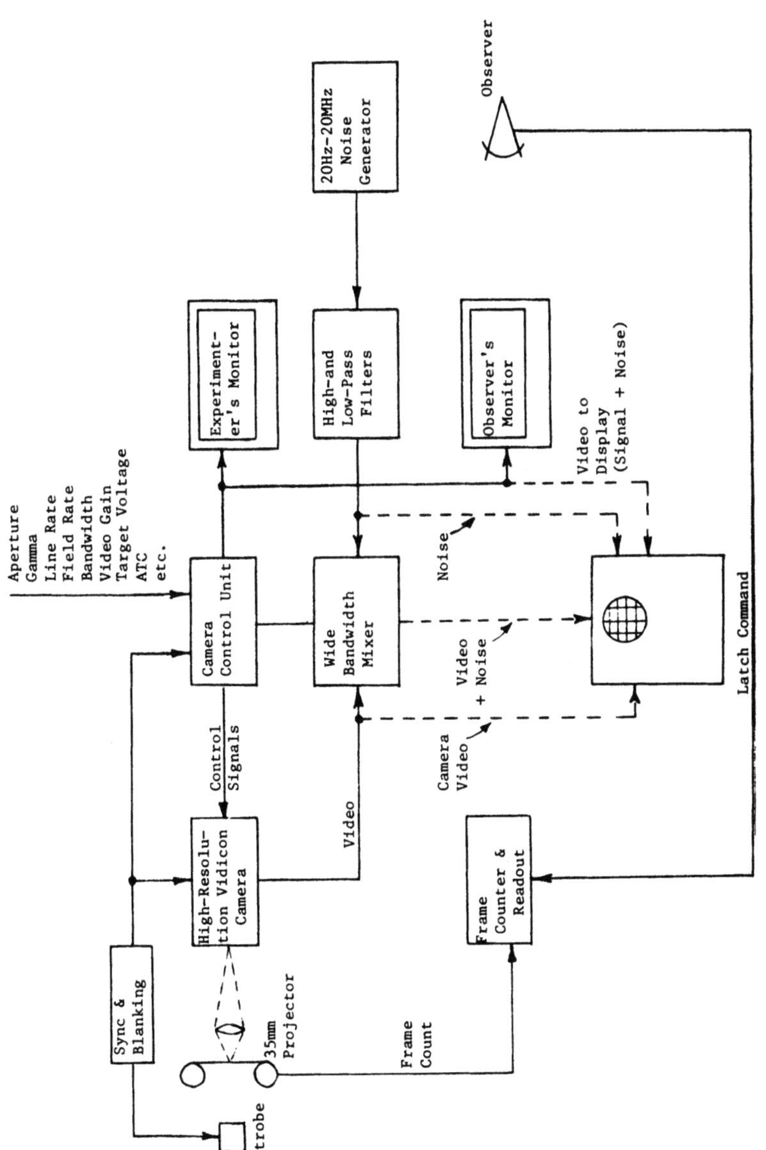

Fig. 3.9. Multiparameter raster-scan equipment block diagram.

levels, display contrast, and total-system MTF, the latter measured under zero-noise levels.

For each of the five noise levels the total system MTF was used to compute the MTFA using the threshold detectability curves given in Fig. 3.8. Observer performance was measured in two ways. First the probability of correct recognition was defined as the total number of correct recognition responses divided by the total number of targets presented (25 per observer). A correct recognition response was credited the observer when he responded by depressing a discrete response button when the target was in the field of view and he was able to indicate in which fourth of the display the target was located. Second, for each correct response a slant range to the target was computed. Mean slant ranges were computed by observer, by condition, and by target.

Figure 3.10 illustrates the relationship between the system MTFA, based upon a unity modulation input as shown in Fig. 3.8, and the probability of correct recognition summed across all observers and all targets. This very high correlation of 0.97, based upon the linear MTFA, indicates that the overall goodness of a system, as defined by the linear MTFA using these preliminary threshold detectability curves, can predict 93% of the variability among systems, averaged over a variety of targets. Using a log-log MTFA measure, the same correlation is 0.96, which is not surprising when it is found that the correlation between linear MTFA and log-log MTFA for these five conditions is 0.998.

When the MTFA is used to predict differences among the five systems in slant range of recognition, the prediction is not nearly as good. Using the linear MTFA the product–moment correlation is 0.76, while

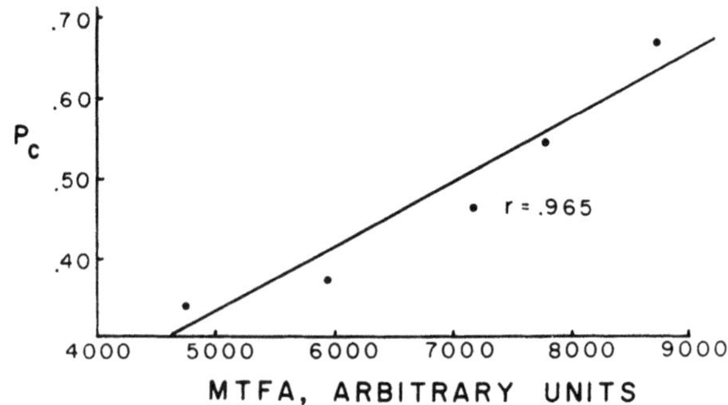

Fig. 3.10. Prediction of probability of recognition from MTFA.

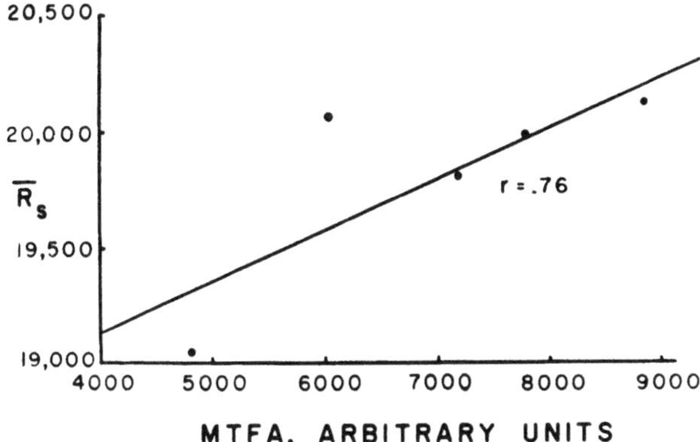

Fig. 3.11. Prediction of recognition range from MTFA.

it is 0.75 using the log-log MTFA computation. As shown in Fig. 3.11, the mean slant range for the 9.2-dB S/N level was illogically high; an inspection of the data suggested that several "wild" guesses were made correctly but had to be included because of our *a priori* scoring procedures. Using a larger data sample, it is likely that a higher correlation would be obtained.

Both the MTFA concept and the SNR_D concept (to be presented in Chapter 5) depend upon the contrast modulation of the target for prediction of observer performance in recognizing a particular target under a particular set of conditions with a given system. In order to obtain more precise prediction among targets, we made carefully calibrated microdensitometric traces of each of our targets. Because seven of these 25 targets were also used in a previous experiment, quantification of their luminance characteristics (or transmission characteristics on the film input) was made in greater detail, and is presented here as shown in Table 3.4. As this table indicates, mean target luminance, mean background luminance, inherent target/background contrast modulation, target area, mean target detail luminance, mean background luminance of the target detail, and detail/background contrast modulation were measured from the microdensitometric traces. Following Zaitzeff (1971), the background for both the target and the target detail was defined as the area to either side of the target (or the detail) to a distance of 25% of the width of the target (or the detail). A target detail was defined as the part of the target that was considered most conspicuous and re-

TABLE 3.4

Characteristics of Seven Targets Used in Target Prediction Analyses

	#4 Nine-car train	#21 Six oil tanks	#23 Three boats	#26 Airfield	#36 Missile site	#40 Ammunition bunkers	#46 Small buildings	#49 Harbor complex
Mean target luminance, ft-L	23.32	30.77	14.02	26.37	13.51	9.11	19.09	22.31
Mean background luminance, ft-L	21.80	22.82	19.77	25.69	15.03	3.02	15.20	18.76
Mean detail luminance, ft-L	7.93	41.77	16.89	45.49	28.06	14.02	30.77	36.86
Mean background luminance, ft-L	25.02	12.50	15.20	8.94	7.93	5.05	12.16	16.56
Target length, ft	675	375	342	4212	340 (diam.)	246	412	2170
Target width, ft	21	225	25	792	340 (diam.)	180	330	1396
Detail length, ft	85	75 (diam.)	90	4212	120 (diam.)	66	48	100
Detail width, ft	21	75 (diam.)	25	262	120 (diam.)	32	30	33
Target/background modulation	0.033	0.146	0.143	0.013	0.052	0.474	0.111	0.085
Detail/background modulation	0.507	0.532	0.051	0.663	0.548	0.453	0.426	0.375

cognition of which was necessary to define the target. Then MTFA's were calculated by multiplying the system MTF (Fig. 3.8) by the target/background modulation to obtain a target-specific MTFA for each of the five noise conditions. Similarly, MTFA's were calculated using the target detail/background contrast modulation as the multiplier of the system MTF curve.

The resulting MTFA's based upon overall target luminance and upon target detail luminance were correlated with both probability of recognition and recognition slant range. The results are indicated in Table 3.5. The detail/background MTFA correlates much more highly with probability of recognition than does the target/background MTFA (0.56 versus 0.09). Similarly, the target detail MTFA correlates only slightly with slant range (0.34), while the target MTFA is *negatively* correlated (-0.28) with slant range.

While these data may suggest that the MTFA is not valid for discrimination among targets, a moment's reflection will indicate that no measure based primarily upon target/background overall contrast will

TABLE 3.5

Product-Moment Intercorrelations among MTFA, P_c, $\bar{R}_s$, and Other Target Metrics*

		9	8	7	6	5	4	3	2
1.	MTFA$_T$	-0.33	-0.16	-0.26	-0.14	0.95	-0.28	0.09	0.01
2.	MTFA$_D$	0.40	0.03	0.48	-0.05	0.08	0.34	0.56	—
3.	P_c	0.71	0.54	0.56	0.45	-0.07	0.61	—	—
4.	$\bar{R}_s$	0.85	0.32	0.84	0.21	-0.48	—	—	—
5.	$M_0 M_D$	-0.33	-0.13	-0.27	-0.11	—	—	—	—
6.	$A_T M_T$	0.46	0.99	-0.03	—	—	—	—	—
7.	$A_D M_D$	0.88	0.10	—	—	—	—	—	—
8.	$A_T M_D M_0$	0.57	—	—	—	—	—	—	—
9.	$A_T M_D$	—	—	—	—	—	—	—	—

* MTFA$_T$ is the MTFA based upon target/background modulation, MTFA$_D$ is the MTFA based upon detail/background modulation, P_c is the probability of correct recognition, $\bar{R}_s$ is the mean slant range to target for correct recognition, M_0 is target/background contrast modulation, M_D is detail/background contrast modulation, A_T is target area, and A_D is detail area.

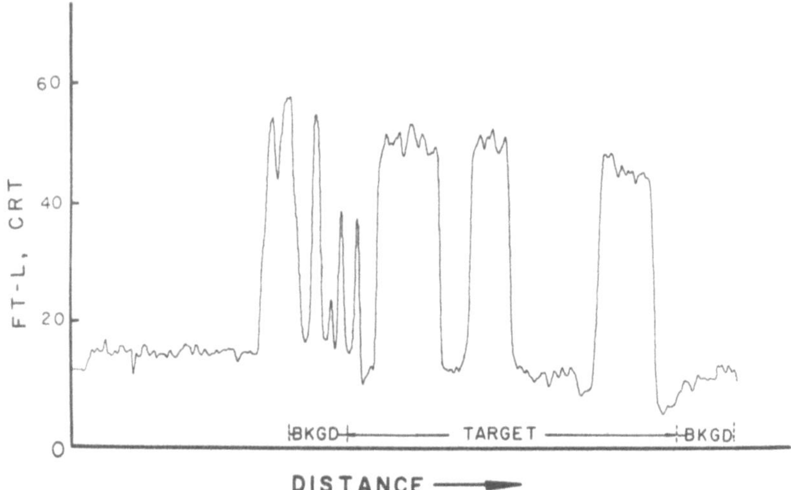

Fig. 3.12. Luminance tracing across target 26, airfield.

be very predictive for *real-world* textured targets in a structured heterogeneous background. Most targets of real-world interest have textural patterns that vary considerably in luminance, against a background that similarly varies in luminance. To take the mean of one such object (the target) and compare it with another (the background) is akin to asking what is the contrast between a black-and-white checkerboard against a medium-gray background. We can clearly see the checkerboard, even though its mean luminance is equal to that of the background and the target/background contrast is therefore *zero*. It is to be expected that the primary use of the target/background contrast metric would yield high predictability only for such homogeneous targets as geometric shapes against a homogeneous background and for artificially painted or real targets whose luminance is relatively constant over the entire target surface (e.g., olive drab trucks). When, however, the targets have heterogeneous luminance patterns, such as an airfield (our target #26, Table 3.4), then to speak of mean luminance contrast is futile. Figure 3.12 illustrates the luminance pattern of one tracing across the airfield, a very large and easily recognized target, even though its mean target/background contrast modulation is 0.013, which is below the accepted threshold modulation of 0.04. Figure 3.13 illustrates the CRT image of this easily recognized airfield at five different noise levels.

Fig. 3.13. Photographs of CRT illustrating effect of noise level upon image quality. (a) Zero noise added, $S/N = 32$ dB; (b) $S/N = 19$ dB; (c) $S/N = 12.5$ dB; (d) $S/N = 9.2$ dB; (e) $S/N = 6.6$ dB.

While further research on target-specific metrics has been urged for years by several researchers (e.g., Snyder *et al.*, 1967; Wyman *et al.*, 1968), only recently has it been shown by Zaitzeff (1971) that a stepwise linear regression equation which includes such parameters as target size, target contrast, detail contrast, background complexity, etc. has high predictive importance for a specific target in a specific background. Using this approach, some very simple combinations of these types of variables were made and correlated with the observer performance measures of our latest experiment. These correlations, also indicated in Table 3.5, show, for example, that the product of the area of the target detail A_D and the detail/background modulation M_D correlates highly with slant range ($r = 0.84$). If the target area is substituted for the detail area in this product, the resultant correlation is 0.85. Similarly, a product of target area A_T and M_D correlates highly ($r = 0.71$) with the probability of correct recognition, much more highly than does the MTFA based upon detail modulation alone.

These results clearly indicate that the contrast modulation of an object of interest against its immediate background, calculated from the mean luminance of each, is not a very usable predictor of object recognition performance. Most real-world objects have texture, luminance irregularities, unique patterns, and other particular qualities that are more important to the prediction of observer performance than is the simple measurement of overall luminance contrast modulation. To adequately predict the observer's object recognition performance, we must know a great deal more about the object *and* its background than merely its size and contrast.

3.9. CONCLUSIONS AND CAUTIONS

The above data show quite conclusively, I believe, that the MTFA concept applies as well to raster-scan systems as it has been shown to apply validly to photographic systems. Of course, the inherent convenience of using an MTF-based metric is indisputable, although the requirement that we define adequately for all systems the detectability threshold curve must still be met.

The consideration and choice of a unitary measure of image quality must be considered in view of its application, for two distinctly different types of application exist, and different metrics appear most suitable for each. First, one is often concerned with selecting the best imaging system for a broad range of applications, with the objects to be recognized undefined, the viewing conditions only loosely specified, the environmental considerations unpredictable, and the future possible uses of the system perhaps totally unknown. In such an application one is interested in determining the performance to be expected from each of several candidate systems, any of which would be used for a myriad of purposes. For this application the MTFA is a highly suitable metric of *overall* image quality which does predict, very highly, operator performance in recognizing various objects of interest.

The second application is one in which we still have considerable difficulty in prediction. In this application we are trying to predict the typical performance of an observer to recognize a *particular* object of interest, having real-world contrast patterns and texture, in a real-world background. For this situation any metric, including the MTFA, which is based largely upon the overall luminance of the target object versus that of its immediate background will yield poor prediction. This is so largely because people do not look for and make decisions with regard to the overall luminance pattern of an object, but rather with regard to the patterns and variations in luminance *within* the object. Thus, to predict the recognizability of a single object, one must take into account such variables as target detail size and luminance, target detail contrast, target size, detail size, background complexity and ambiguity, etc. A start has been made in this direction but much more research must follow before we shall have great confidence in our predictions.

3.10. EDITOR'S POSTSCRIPT

As the manuscript of this book was being readied for publication, a letter arrived from Prof. Snyder regarding a paper he was to give at the SPIE meeting on "Security, Surveillance, and Law Enforcement," 21 Sept. 1972. In that letter he reported:

> "Incidentally, I think you will be pleased with the results that I will be discussing at this meeting. In essence, we have achieved very high correlations between the probability of recognizing a face, among many, and the MTFA of a video system, when compared across many

such system configurations. Similarly high correlations exist between the time it takes a subject to recognize the face and the MTFA. Perhaps more importantly, we have found that our (very good)* monitor is far from a linear device, even for very large targets when the video bandwidth and line rate are increased to specified large values. Thus, I feel the problem of photometric measurement, rather than merely electrical measurement at the CRT input is extremely important and cannot be ignored, as others have done in the past. Again, we shall discuss this, I am sure, at the New York meeting."

* Editor's, not author's, description.

ANALYSIS OF NOISE-REQUIRED CONTRAST AND MODULATION IN IMAGE-DETECTING AND DISPLAY SYSTEMS

Alvin D. Schnitzler

4.1. INTRODUCTION

The purpose of an electrooptical display system is to convert electrical messages into luminous messages for transmittal to the retina of the eye. In the retina photodetection occurs and the signal processing is initiated which, continued in the brain, permits the observer to make decisions based on the acquired visual sensory information. The overall performance of this display–observer system is measured by the probability of the observer making a correct decision in a given visual task. Clearly the overall performance depends on the interaction between the parameters of the display and eye and on the parameters of the human visual system as well as on the parameters of the display itself. Therefore the choice of display design parameters to ensure optimum overall performance depends on a clear understanding of the operation of the visual sensory system and the interaction between it and the display.

An excellent method for analyzing the overall performance of the display–observer system is provided by statistical communication theory. Since messages and noise are the principal commodities in communication systems, it is important that they are adequately described. The messages transmitted to the display consist of spatially modulated electric current densities which correspond to the image of a scene. The messages transmitted to the observer consist of spatially modulated luminous flux densities emitted by the display and focused by the eye lens to reconstruct an image of the scene on the retina. In the retina the messages are converted into spatially modulated electrochemical potentials for transmission to the brain. At each stage of the display–observer system, in general, the messages are corrupted by both spatial and temporal dispersion as well as by the addition of noise, which consists of random spatial and temporal fluctuations in the electric current and luminous flux densities.

The general objective of statistical communication theory is to develop a statistical description of messages and noise in order to predict the detection and false alarm probabilities that occur when their sum exceeds some predetermined threshold. A more limited objective is pursued in many cases where it is assumed that a known message is transmitted and a statistical description of only the noise is needed in order to predict the probability of detecting the message. It is well known in communication theory to be more convenient in practice to relate the statistical description of the noise to its power density spectra rather than deal directly with the probability density functions of statistics. Representation of messages and noise by their spectra permits the use of temporal and two-dimensional spatial frequency response functions to describe the successive alterations of messages and noise by the cascaded components of the display–observer system.

The process of conversion of electrical messages into luminous messages by a display may be either spatially and temporally continuous or discrete in one or both spatial coordinates as well as time. In the former case, represented by x-ray and/or night vision image intensifier–observer systems, spectral analysis is readily applied to predict the alterations of the messages and noise by the components of the systems. The latter case is typically represented by line scanners such as television systems where the conversion process is continuous in one and discrete in the other spatial coordinate as well as in time. Since the use of spectral analysis is limited to a continuous interval of the coordinates, in order to predict the alteration of messages and noise by the components of a discrete sampling system, it is convenient to obtain the result for each sampling interval and then sum over the total measurement interval in space and time coordinates.

The peculiar effects (spurious response to periodic patterns in the scene, etc.) of discrete sampling on the reproduction of images on displays and the precautions required to avoid them were mentioned earlier and are more thoroughly discussed in Chapters 6 and 7.

The present chapter, emphasizing the combined display–observer system, will parallel the historical evolution of the theory of image-detecting systems from the ideal photon counter model and the concept of detective quantum efficiency to the modern use of power spectral density analysis. The detailed analysis included here is limited in application to the background- or shot-noise-limited regime of the visual system, direct-view image intensifier, low-light-level television and infrared scanner systems, although the analysis could be extended to the system-noise-limited regime where internally generated thermal noise is dominant.

4.2. HISTORICAL REVIEW OF THE SIGNAL-TO-NOISE RATIO THEORY OF VISUAL PERFORMANCE

Before presenting a complete analytical development of the signal-to-noise ratio theory of the visual performance of electrooptical display–observer systems a brief historical review of the development of key concepts is presented here. This review contains by no means a complete bibliography of the many papers which have considered visual performance as a signal-to-noise ratio problem. Rather a selection has been made of papers each of which it seems to this author represents a benchmark in the advancement of the theory and analytical methods.

The earliest extensive discussions of the concept that the performance of the human visual system, like the performance of an electrical communication system, must be limited by statistical fluctuations accompanying the signal power were published in the early 1940's by Hecht *et al.* (1942), Rose (1942), and de Vries (1943). They recognized that the absorption of luminous flux by the photoreceptors of the retina is quantized, obeying the Poisson probability law for random discrete events. If an average of N quanta are counted in each of a large number of experiments involving identical exposure times, then the standard deviation in the number will equal the square root of N. Hecht *et al.* applied Poisson statistics to explain the probability of detecting small-size, short-duration light flashes at the absolute threshold (dark-adapted eye, no background luminance). Both Rose and de Vries hypothesized that statistical fluctuations in the absorption of quanta even at high background luminances would limit the contrast sensitivity and acuity of the visual system.

Rose (1948a) proceeded to introduce the concept of an ideal image-detecting device in order to compare the performances of real systems such as the visual system, television systems, and photographic film with ideal performance. The ideal image-detecting device is a simple photon counter which counts all the absorbed quanta, introduces no additional statistical fluctuations in excess of those inherent in the absorbed quanta, functions as a perfect integrator of the signal over both the area of the target image on the sensor and the duration of the exposure, and possesses a spatial point spread and a temporal impulse response both described by a delta function. Obviously the phenomenon of diffraction prohibits the physical realization of such an ideal device.

Rose noted that the only parameter peculiar to the device alone and which differentiates the performance of one ideal device from another is the quantum yield of the photoabsorption process. He showed that the quantum yield required by an ideal photon counter to achieve a certain degree of performance in the detection of targets of circular area is given by*

$$Q = 5k^2 \times 10^{-3} / EC^2\alpha^2 D^2\tau \tag{1}$$

where k is the signal-to-noise ratio and C is the contrast ($\Delta E/E$) required to achieve a specified detection probability, E is the background luminance in foot-Lamberts, α is the angular size of the target in minutes, D is the aperture diameter in inches, and τ is the effective exposure time.

In an experiment to determine the value of Q for a real image-detecting device such as the human visual system the values of E, C, α, and D are readily measured. However, some uncertainty persists even today in the proper value of τ if the exposure time exceeds the equivalent integration time of the eye. Both Rose and de Vries were convinced that the proper value of τ is approximately 0.2 sec for an observer viewing a stationary scene. It is now generally believed that this is a maximum value which exists only at very low values of background luminance.

If the performance of a real device were equal to that of an ideal photon counter, the value of Q determined with Eq. (1) would be the actual quantum yield of the real device. However, if the performance of

---

* The derivation of Eq. (1) is given in Section 4.3.3.

a real device is degraded, for example, by its introduction of fluctuations in excess of those inherent in the absorbed quanta or by diffraction and aberrations in the optical components [terms which would account for these effects are not included in Eq. (1)], then the value of Q calculated with Eq. (1) would be less than the true quantum yield. Thus Rose pointed out that the parameter Q calculated from experimental data with Eq. (1) can be viewed as an index to performance of real image-detecting devices. It is a measure of the efficiency with which a real image-detecting device utilizes the radiant power to detect a target. The range of Q is from unity to zero. Unfortunately, because the effect of diffraction was excluded from consideration in the definition of an ideal image-detecting device, the value of Q for any physically realizable image-detecting device is not unique. It will depend on the angular size of the test target, varying from a maximum value determined by detecting large targets to zero as the size of the target decreased until its image size is determined by diffraction. A credible definition of an ideal image-detecting device and index to performance should take into account the effects of diffraction by the optical aperture.

The value of the parameter k in Eq. (1) was found by Rose (1948a) to equal five in some experiments in which he used an observer as a decision-maker. He treated this value of k as a universal constant in his calculation of the value of Q for a number of television systems, photographic films, and the human visual system. However, it was the original intent for the parameter Q to be an index to performance of a real image-detecting device relative to an ideal one. Hence the value of k used to calculate the Q ought to be a value appropriate to an ideal image-detecting device—not a value dependent on the decision-making performance of a human observer.

Following Rose, considerably later, R. C. Jones (1959) elaborated on the concept of the index to performance of image-detecting devices, calling it the detective quantum efficiency (DQE), and extended the concept of an ideal image-detecting device to include an ideal decision-making device as well. This eliminated the uncertainty in the proper value of the required signal-to-noise ratio k to be used in calculating the DQE. Two kinds of image-detecting tasks were analyzed by Jones. In the first instance he assumed that the ideal device was faced with the task of deciding whether the amplitude of the input power is such that it implies that a signal is present. He then assumed that the ideal device has a threshold such that if the amplitude of the power is above the threshold, it is concluded that a signal is present, and otherwise that just noise is present. Thus, in order to be precise in the calculation of the DQE in Eq. (1) the value of k must correspond to the false alarm and detection probabilities observed in the determination of the threshold contrast.

Jones proceeded to calculate the DQE of the human visual system using some flash perception data which he obtained from Blackwell and McCready (1958). He used the Blackwell and McCready data in order to avoid the uncertainty in the effective integration time of the eye. The observer was given the task of deciding in which of four successive 2.5-sec intervals a signal of duration τ was presented. Jones

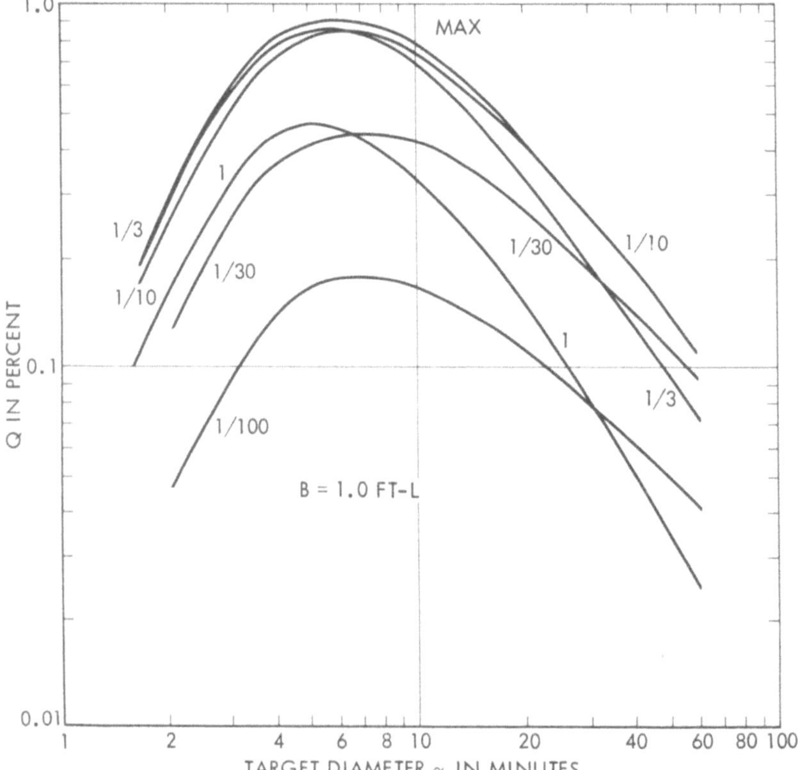

Fig. 4.1. Detective quantum efficiency versus angular size with flash duration as a parameter [after R. C. Jones (1959)].

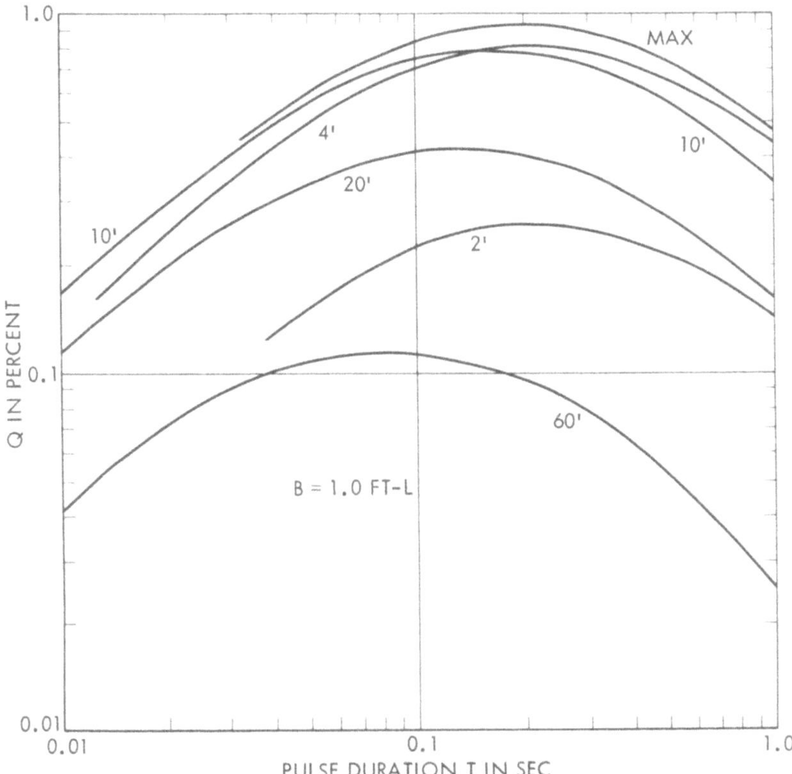

Fig. 4.2. Detective quantum efficiency versus flash duration with angular size as a parameter [after R. C. Jones (1959)].

observed that this multiple-choice type of visual task is considerably more complicated to analyze than the threshold type described above, and attributed a solution to Birdsall and Peterson.

A value of k equal to 1.22 was found to be appropriate for the Blackwell and McCready data which was reported at 50% detection reliability and involved a four-channel multiple-choice task.

The detective quantum efficiency of the human visual system calculated by Jones using the Blackwell and McCready data obtained at a background luminance equal to 1 ft-L is shown in Figs. 4.1 and 4.2. The first figure is a plot of the detective quantum efficiency Q versus the target diameter α in minutes of arc at several flash durations ranging from 1/100 to 1.0 sec; the second figure is an analogous plot of Q versus the flash duration τ at several target diameters ranging from 2 to 60 min. These figures plainly illustrate several deficiencies in the performance of the visual system relative to an ideal image-detecting device. At small values of both α and τ dispersion of the signal occurs spatially and temporally, respectively. The result is that the excitation of the retina by the signal is spread out in area and lengthened in duration to encompass excitations from a larger area and for a longer duration of the background than would occur with an ideal device with its delta-function type of point spread and impulse response functions. Thus the required contrast C to be substituted into Eq. (1) to determine the DQE of the visual system is higher than that required by an ideal device to achieve the same signal-to-noise ratio and the DQE is lower. As α and τ increase, the relative effects of the dispersions upon the required contrast decrease and the DQE increases. However, both figures show that the DQE is a maximum at an intermediate value of α and τ, respectively. Thus, as the effects of the dispersions decrease with increasing α and τ, the effects of another deficiency increase. This latter deficiency is due to the failure of the visual system to function as a perfect integrator of the excitations of the retina as the area of the image of the target on the retina and the flash duration become too large.

In a series of papers Schade (1951, 1952, 1953, 1955, 1956), by employing the newly developed modulation transfer function theory of image formation, extended the signal-to-noise ratio analysis of the performance of image-detecting devices to include the effects of spatial and temporal dispersion. He extended the concept of the noise equivalent passband from the temporal to the spatial frequency domain and introduced the effective sampling area defined by $\bar{a} = 1/N_e^2$, where $N_e = \int_{-\infty}^{\infty} T^2(v_x)\,dv_x$ is the spatial noise equivalent passband and $T(v_x)$ is the modulation transfer function at the spatial frequency v_x of a sine wave-modulated test pattern periodic in the x coordinate. The effective image area $\bar{a}_i$ of the response of the visual system to a target is approximately given by $\bar{a}_i = \bar{a}_0 + \bar{a}_1 + \cdots + \bar{a}_n$, where $\bar{a}_0$ is the area of the response to the image of the target by an ideal image-detecting device and $\bar{a}_1, \ldots, \bar{a}_n$ are the effective sampling areas of the components of a complete image-detecting system including the display and the human visual system.

Schade calculated the signal-to-noise ratio by applying Poisson statistics to the number of absorbed quanta or, in the case of an electrooptical image-detecting device, statistical units* within the effective image area and integration time of the visual system. Such a procedure will largely take into account the effects of spatial dispersion upon the performance of an image-detecting system. However, it ignores the reduction in the statistical fluctuations caused by spatial dispersion of the statistical units themselves (especially within electrooptical image-detecting devices) and thus actually overestimates the degradation of performance. An equivalent explanation of this reduction in statistical fluctuations is that it is due to attenuation of high-spatial-frequency components of the power spectral density. This phenomenon is discussed in Section 4.5.

The most accurate and exact technique for evaluating and calculating the performance of real image-detecting systems is that provided by statistical communication theory with its employment of autocorrelation techniques and representation of signal and noise by their power density spectra. R. C. Jones (1955) strongly advocated the employment of statistical communication theory to analyze the performance of photographic systems in the statement that, "The only adequate way to describe the granularity of photographic materials is by means of a film noise spectrum." The first realizations that statistical communication theory could be usefully employed to analyze the performance of photographic systems were made by Fellgett (1953) and Elias (1953). They were followed by R. C. Jones (1955) and Zweig (1956), who publish extensive discussions of both the theoretical and experimental applications of the theory.

The application of statistical communication theory to analyze the performance of image-detecting systems other than photographic systems has begun more recently. The basic principle of the representation of statistical fluctuations by their power density spectra and the general application of statistical communication theory to analyze the performance of electrooptical image-detecting systems were enunciated by Lawson (1968, 1971). In a subsequent paper Kornfeld and Lawson (1971) applied the theory to analyze the performance of the human visual system. The theory was applied by Schnitzler (1971b) to analyze the performance of image intensifier systems.

The practice of ignoring the limitations imposed on the performance of image-detecting systems, especially the human visual system, by

---

* A statistical unit is an aggregate of quanta or electrons which act as a unit.

statistical fluctuations accompanying the signal power is still widespread despite the fact that it has been three decades since Hecht *et al.* (1942), Rose (1942, 1948a, 1948b), and de Vries (1943) published their accounts and two decades since Schade (1951, 1952, 1955, 1956) published his monumental works. The wide divergence in approaches to the performance of the human visual system is well illustrated in a recent issue of *Progress in Optics* in which, on the one hand, an article by Fry (1970) reviews and interprets a number of threshold measurements of the performance of the human visual system while ignoring the physical limitations imposed by statistical fluctuations, and, on the other hand, an article by Levi (1970) points out the need for one to know the complete noise power density spectrum in order to be able to interpret threshold data.

4.3. THE IDEAL PHOTON COUNTER MODEL OF AN IMAGE-DETECTING SYSTEM

As a first approximation to an image-detecting system we will consider the ideal photon counter model introduced by Rose (1948a, 1948b). In a discussion of the performance of an image-detecting system incorporating the human visual system and brain one must note that the only accessible output is a response to a visual stimulus. The output signal-to-noise ratio is not accessible by direct measurement. Hence the discussion must begin with decision theory. It is followed by an analysis of the output signal-to-noise ratio, the detective quantum efficiency, and the noise-required contrast. The latter topic will be the major subject of this section.

4.3.1. Elementary Decision Theory

The final output of any image-detecting system is an observer's response (the ringing of a bell, the verbalizing of a yes, the flashing of a light, etc.) signifying that he sees a target in the field of view of the system. In the ideal image-detecting system of Rose (1948a) and R. C. Jones (1959) introduced in Section 4.2 a response results if, and only if, a physical decision is made that the output amplitude of the target, signal power, and/or the noise power exceed a threshold value called the decision criterion. If the target is present, the probability P_R of a response will equal the probability P_D (called the detection probability) that the sum of the signal and noise power amplitudes exceeds the threshold value. If the target is not present, P_R will equal the probability P_F (called the false alarm probability) that the noise power amplitude exceeds the threshold value. In any real image-detecting system employing the human brain as the decision-making device, if the target is not present, P_R will equal the sum of P_F and a spurious response probability P_S due to guesses.

Our development of elementary decision theory will include the derivation of (1) the relationship connecting the detection probability and the threshold and the output signal-to-noise ratios, (2) the relationship between the false alarm probability and the threshold signal-to-noise ratio, and (3) a discussion of multiple-choice testing to quantify the spurious response probability. For this analysis it will be assumed that the scene within the field of view of the system consists of a simple test target projected onto a screen of uniform background radiance caused by diffuse reflection of thermally produced radiant power.

The derivation of the detection probability begins with the observation that the number of quanta N counted within the area of the target image in a series of exposures of equal duration is a random variable with a Poisson probability density.* If the mean value is not too small, the Poisson probability density of a random variable is indistinguishable from a Gaussian probability density, the variance remaining equal to the mean value. Thus the probability density of the number of quanta collected and counted from the background may be represented approximately by the Gaussian function

$$p(\Delta N)\, dN = (2\pi)^{-1/2}\sigma_B^{-1} \exp[-(\Delta N)^2/2\sigma_B^2]\, dN \tag{2}$$

where $\Delta N = N - \bar{N}_B$ and $\sigma_B{}^2 = \bar{N}_B$. The normalized Gaussian probability density is obtained by letting $X = \Delta N/\sigma_B$. Then we have

$$p(X)\, dX = (2\pi)^{-1/2} \exp(-X^2/2)\, dX \tag{3}$$

---

* The Poisson probability density is strictly correct only if the incident radiant power is coherent. If it is chaotic and the detection of successive quanta by a photoreceptor occurs within the response time, then a bunching effect known as the Hanbury-Brown and Twiss (1956) effect will alter the probability density, increasing its variance. See Lachs (1968) for a discussion of statistical fluctuations in photoelectric detection.

Likewise the probability density of the number of quanta collected and counted from the target is given approximately by

$$p(N - \bar{N}_A)\, dN = (2\pi)^{-1/2}\sigma_A^{-1} \exp\{[-(N - \bar{N}_A)^2]/2\sigma_A^2\}\, dN \qquad (4)$$

where $\sigma_A^2 = \bar{N}_A$. By adding and subtracting $\bar{N}_B$ within the quantity $(N - \bar{N}_A)$, noting that $\sigma_A^2 = (1 + C)\sigma_B^2$, where

$$C = (\bar{N}_A - \bar{N}_B)/\bar{N}_B \qquad (5)$$

is the target-to-background contrast at the output of the device,* and again letting $X = \Delta N/\sigma_B$, we obtain

$$p(X - k_o)\, dX = [2\pi(1 + C)]^{-1/2} \exp[-(X - k_0)^2/2(1 + C)]\, dX \qquad (6)$$

where $k_o = S_o/\sigma_B$ and S_o is the output signal, equal to $C\bar{N}_B$. If the contrast is small compared to unity, then

$$p(X - k_o)\, dX = (2\pi)^{-1/2} \exp[-(X - k_0)^2/2]\, dX \qquad (7)$$

and to this approximation $k_0 = S_0/\sigma_B$ is the output signal-to-noise ratio.

In the ideal image-detecting system a response occurs if and only if the signal and/or noise power exceed a threshold value, i.e., if and only if $\Delta N \geq \Delta N_T$. If we divide by σ_B, the condition becomes $X \geq k_T$, where $k_T = \Delta N_T/\sigma_B$ is the threshold signal power measured in units equal to the standard deviation of the counted background quanta, i.e., the threshold signal-to-noise ratio.

The relationships of the probability densities of the counted background and signal quanta to each other and to the threshold are shown in Fig. 4.3. The detection probability is given by the area under the

---

* It should be noted that by the definition of an ideal image-detecting system the contrast is invariant throughout the system. Contrast, like signal-to-noise ratio is treated throughout this chapter as a positive number. Thus, the precise definition of contrast is expressed by $C = |\bar{N}_A - \bar{N}_B|\, \bar{N}_B$. This definition is in conformity with all those authors whose works were reviewed in Section 4.2. Sometimes contrast is defined by $C = |\bar{N}_A - \bar{N}_B|/(\text{the greater of } \bar{N}_A \text{ or } \bar{N}_B)$. Such a definition is clumsy to handle analytically, resulting in two sets of equations, one for $\bar{N}_A > \bar{N}_B$ and another for $\bar{N}_B > \bar{N}_A$.

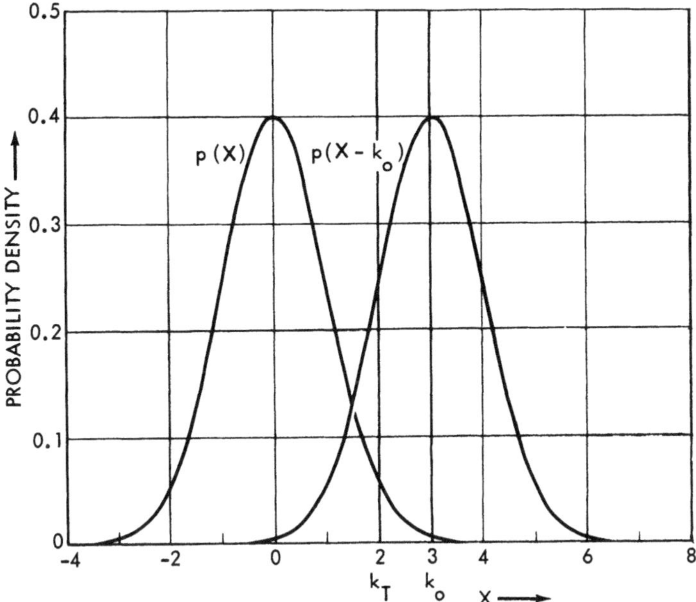

Fig. 4.3. Normalized probability densities. Note that k_T and k_0 are threshold and output signal-to-noise ratios, respectively.

$p(X - k_o)$ curve to the right of $X = k_T$. Thus P_D is given by the integral

$$P_D = \int_{k_T}^{\infty} p(\chi - k_0)\, d\chi \tag{8}$$

and noting that $p(X - k_o)$ is symmetric about $X = k_o$, we have

$$P_D = \int_{-\infty}^{2k_0 - k_T} p(\chi - k_0)\, d\chi \tag{9}$$

$$= \int_{-\infty}^{k_0 - k_T} p(u)\, du \tag{10}$$

$$= \mathrm{erf}(k_0 - k_T) \tag{11}$$

where

$$\mathrm{erf}(k_0 - k_T) = \int_{-\infty}^{k_0 - k_T} (2\pi)^{-1/2} \exp(-u^2/2)\, du$$

is the error function. It should be noted that since $\mathrm{erf}(0) = 0.5$, the threshold signal-to-noise ratio equals the output signal-to-noise ratio that is required to achieve 50% detection probability.

The false alarm probability is given by the area in Fig. 4.3 under the $p(X)$ curve to the right of $X = k_T$. Thus

$$1 - P_F = \int_{-\infty}^{k_T} p(\chi)\, d\chi \tag{12}$$

$$= \mathrm{erf}(k_T) \tag{13}$$

As a numerical example let us assume that $P_F = 0.023$. Then from Eq. (13) and a table of error functions we have $k_T = 2$. If we also assume that $P_D = 0.841$, then from Eq. (11) we have $k_o = 3$. This example is illustrated in Fig. 4.3. In general if P_D and P_F are specified, then k_o is given by

$$k_o = \mathrm{erf}^{-1}(P_D) + \mathrm{erf}^{-1}(1 - P_F) \tag{14}$$

In any real image-detecting system employing a human brain as the decision-making device, in addition to false alarms, spurious responses result from guessing that a target is in the field of view when in fact it is absent. Thus in this case the probability of a response when the target is present is given by

$$P_R = P_D + (1 - P_D)P_S \tag{15}$$

where P_D is the probability of detection due to a physical decision and P_S is the probability of a spurious response. From Eq. (15) the detection probability is given by

$$P_D = (P_R - P_S)/(1 - P_S) \tag{16}$$

If the target is not present the probability of a response is given by

$$P_R = P_S + P_F \tag{17}$$

where P_F is the probability of a false alarm due to an erroneous physical decision.

Two observations to be made about P_S are that (1) it is much greater than P_F and (2) in single-choice (yes or no response) testing it is extremely variable from one observer to another. Therefore to measure the performance of a real image-detecting system it is preferable to employ forced, multiple-choice testing, for then P_S is simply the reciprocal of the number of choices.

Two alternative models of the decision process in multiple-choice testing with human observers may be considered. The two models differ according to whether the assumption is or is not made that a threshold is involved. R. C. Jones (1959) assumed in interpreting the Blackwell and McCready (1958) flash perception data discussed in Section 4.2 that the observer examines the amplitude of the signal in each temporal interval and selects in each trial the interval in which the amplitude is largest. In this situation no fixed threshold exists. However, this seems unlikely since the background fluctuations which occur continuously (only the target is flashed on and off) are not perceptible. This suggests that the signal must exceed some threshold equal to several times the standard deviation in the background fluctuations. Thus we will assume that a threshold is involved in the decision process in both single-choice and multiple-choice testing and hence the detection probability is always related to the threshold and output signal-to-noise ratios by Eq. (11).

Since it has been observed that the probability of a response by an observer when a target is absent is predominantly due to guesses rather than physical decisions due to fluctuations which exceed the threshold, it is not possible to employ Eq. (13) to determine k_T from the false alarm probability. Instead, we must determine the probability of a response P_R when a target is present as a function of target and background characteristics. Then with the above values of P_R and the spurious response probability P_S we may use Eq. (16) to determine the detection probability P_D as a function of target and background characteristics. Finally, if either by direct measurement or by calculation we knew the values of k_0 as well as the above values of P_D as a function of the target and background characteristics, we could use Eq. (11) to deduce the value of k_T. Unfortunately, k_0 is not accessible to direct measurement in the human visual system. Therefore the only way to fully describe any real image-detecting system employing a human brain as the decision-making device is to derive an expression for k_0 as a function of target and background characteristics. However, calculation of the values of k_0 from such an expression will be necessarily approximate since the values of the visual system parameters and even the principle of operation of the visual system have not been fully and accurately determined.

4.3.2. Output Signal-to-Noise Ratio

As a first approximation which will be improved upon throughout the remainder of this chapter, the derivation of the output signal-to-noise ratio of a real image-detecting system will be based upon the ideal photon counter model. It should be recalled that an image-detecting system conforms to the ideal photon counter model if it introduces no additional statistical fluctuations in excess of those inherent in the absorbed quanta, functions as a perfect integrator of the absorbed quanta over the area of the target image and the duration of the exposure, and possesses spatial point and temporal impulse responses described by delta functions. Therefore, according to the ideal photon counter model the output signal-to-noise ratio of an image-detecting system is equal to the signal-to-noise ratio at the primary sensor.

At the primary sensor the signal is equal to the number of quanta absorbed per exposure due to the differential radiant power of the image relative to the background within the area of the image. Analytically, this input signal S_I is given by

$$S_I = \eta(\bar{n}_{AI} - \bar{n}_{BI})a_I\tau \tag{18}$$

where η is the responsive quantum efficiency of the primary sensor (i.e., the fractional number of an electron per incident quantum), $\bar{n}_{AI}$ and $\bar{n}_{BI}$ are the mean values of the irradiance of the primary sensor in quanta per second per unit area due to the target and background radiances, respectively, a_I is the area of the perfect image of the target formed on the primary sensor, and τ is the exposure in seconds. In terms of the input contrast, which is defined by $C_I = (\bar{n}_{AI} - \bar{n}_{BI})/\bar{n}_{BI}$, the input signal is given by

$$S_I = \eta C_I \bar{n}_{BI} a_I \tau \tag{19}$$

At the primary sensor the noise, expressed as the standard deviation σ_I in the number of quanta absorbed per exposure due to the total radiant power incident on the area of the perfect image of the target, is given by

$$\sigma_I = (\eta \bar{n}_{AI} a_I \tau)^{1/2} \tag{20}$$

since the probability density is Poissonian. If $\bar{n}_{AI}$ is expressed by $\bar{n}_{AI} = (C_I + 1)\bar{n}_{BI}$ and it is assumed that C_I is small compared to unity, then Eq. (20) reduces to

$$\sigma_I = (\eta \bar{n}_{BI} a_I \tau)^{1/2} \tag{21}$$

Since the output signal-to-noise ratio k_0 is equal to the signal-to-noise ratio k_I at the primary sensor, k_0 is obtained by dividing Eq. (19) by Eq. (21). The result is given by

$$k_o = C_I(\eta \bar{n}_{BI} a_I \tau)^{1/2} \tag{22}$$

If the target and background are projected onto a screen as in the Blackwell and McCready (1958) experiments, the value of k_I and hence k_o can be related to the radiant characteristics of the screen by noting that $\bar{n}_{BI} a_I = \zeta \bar{n}_B a$, where ζ is the collection efficiency of the aperture of the image-detecting system, $\bar{n}_B$ is the mean radiant exitance in quanta per second per unit area due to the background, and a is the area of the projected target. If the distance to the screen is large compared to the diameter of the aperture, ζ is approximately equal to $D^2/4d^2$, where D is the diameter of the aperture and d is the distance to the screen. Thus, with reference to the screen, the output signal-to-noise ratio is given by

$$k_o = (C_I D/2d)(\eta \bar{n}_B a \tau)^{1/2} \tag{23}$$

It should be noted that Eq. (23) applies as well to an object plane containing physical targets as to a screen.

Equations (11), (13), and (23) completely describe the performance of an ideal photon counter image-detecting system.

In the design of an ideal image-detecting system (assuming for the moment that it were physically possible to construct one) one must first decide the false alarm probability that can be tolerated. This determines the threshold signal-to-noise ratio k_T by Eq. (13). Then one must decide the desired detection probability, which along with the above value of k_T, determines by Eq. (11) the required output signal-to-noise ratio k_o.

Finally, for a given set of target and background characteristics (C_I, a, and $\bar{n}_B$) and the above value of k_0 the required diameter D of the objective is determined by Eq. (23).

Presumably η for our ideal image-detecting system would be unity and τ would be determined by consideration of the required speed of response.

4.3.3. Detective Quantum Efficiency

The detective quantum efficiency Q of a real image-detecting system, called by Rose (1948a) an index to performance, was defined by him to be the responsive quantum efficiency η of an equivalent (same size aperture and integration time) ideal image-detecting system. Thus the expression for Q is readily obtained by solving Eq. (23) for η and representing the expression for η by Q. The result is given by

$$Q = 4k_T{}^2 d^2 / C_I{}^2 D^2 \bar{n}_B a \tau \tag{24}$$

where we have substituted the threshold signal-to-noise ratio k_T for the value of k_0 at $P_D = 0.5$. If circular test targets are used to determine the value of Q, then a is approximately equal to $\pi d^2 \alpha^2 / 4$ and Q is given by

$$Q = 16k_T{}^2 / \pi C_I{}^2 \alpha^2 D^2 \bar{n}_B \tau \tag{25}$$

where α is the angular size of the target. This is the form of the equation for Q used by both Rose (1948a) and R. C. Jones (1959).

The detective quantum efficiency of an image-detecting system would be unity if it had unity responsive quantum efficiency, introduced no additional statistical fluctuations in excess of those inherent in the absorbed quanta, functioned as a perfect integrator of the absorbed quanta over both the area of the target image and the duration of the exposure, and possessed a spatial point spread and a temporal impulse response both described by a delta function (no spatial and temporal dispersion). The relative magnitude of the detective quantum efficiency of any real image-detecting system incorporating the human visual system and brain is reduced at large values of α and τ by the failure of the visual system to function as a perfect integrator and at small values of α and τ by spatial and temporal dispersion, respectively. The two last effects may be introduced by optical and electrooptical components as well as the visual system. All four of these effects were shown by Figs. 4.1 and 4.2 to occur in the visual system. In addition, the absolute magnitude of the detective quantum efficiency of any real image-detecting system may be reduced by the imperfect responsive quantum efficiency of the primary photoelectric sensor, by the introduction of noise—especially thermal noise generated in the circuit resistors in television camera systems, and by the imperfect transfer of statistical units between an electrooptical system and the visual system which will occur if amplification in the electrooptical components is not sufficient.

As a rational index to the performance of any image-detecting system involving an observer, detective quantum efficiency has several deficiencies. In the first place, since all of these *image-forming systems** utilize the visual system and brain for decision-making, not all of the quantities upon which Q depends are directly measurable. In Eq. (24), for example, the values of d, C_I, a, D, τ, and $\bar{n}_B$ corresponding to the detection of a target with 50% probability are readily measurable. However, since the output of the visual system is inaccessible, the proper value of k_T must be deduced from some model of an ideal decision-making device.

In one instance, for example, R. C. Jones (1959) calculated the maximum value of Q for the visual system to be only approximately 1%. It is likely that this small value is due not only to the imperfect responsive quantum efficiency of the visual system but also to the small value of k_o chosen by Jones to correspond to $P_D = 0.5$. His deduction that the proper value of k_T to use with the Blackwell and McCready data should be 1.22 was discussed in Section 4.2. If the value of k_T were this small, fluctuations in the background luminance would have been readily perceptible to the human visual system. But the fluctuations were not perceptible. Therefore the threshold signal must be several times the standard deviation, k_T must be greater than 1.22, and the responsive quantum efficiency must be greater than 1%.

A second deficiency in the detective quantum efficiency as an index to performance of real image-detecting systems is based on the observation that since all of these systems exhibit spatial and temporal dispersion and depend on the visual system for spatial and temporal integration, the value of Q is different for each value of target angular size and

---

* Note that we do not include simple nonimage-forming radiant power detectors.

exposure as shown in Figs. 4.1 and 4.2. Therefore, since a scene normally contains a variety of objects of different sizes, an image-detecting system would have several values of Q for a single scene.

A third deficiency in the use of detective quantum efficiency is based on the observation that all image-detecting systems require some kind of objective optics to collect radiant power from a target and to form an image on the primary sensor. One of the most critical parameters determining the performance of an image-detecting system is the diameter of the aperture stop, but Q is independent of it.

The collection efficiency ζ was noted following Eq. (22) to be proportional to the aperture diameter D. The appearance of the factor D^2 in Eq. (25) for Q is deceptive. In an ideal image-detecting system, for example, Q is equal to unity and thus the product $C_I^2 D^2$ is independent of D and ζ. In the human visual system as D increases, the radiant power is utilized less efficiently due to the Stiles–Crawford effect, and thus the product $C_I^2 D^2$ increases slightly, providing a tendency for Q to actually decrease slightly with increasing D. Thus the increase in the performance of an image-detecting system provided by an increased aperture diameter is not reflected in the value of the detective quantum efficiency.

A fourth deficiency in the use of detective quantum efficiency is that it is an inadequate measure of the rate at which information can be transmitted via an image-detecting system. The rate at which information can be transmitted via an image-detecting system is proportional to the product of the number of independently detectable image elements per unit area and the area of the image of the field stop (usually the primary sensor), but Q is independent of the area of the field stop.

Finally, a philosophical objection to detective quantum efficiency as an index to performance of real image-detecting systems is based on the observation that, since it is a comparison of the measured performance of real image-detecting systems with the predicted performance of an ideal image-detecting system only limited in performance by shot noise, its continued use may serve as a deterrent to an understanding of other phenomena such as spatial dispersion which, in addition to shot noise, limits the performance of real image-detecting systems. A more rational approach is to employ the iterative process of developing both a detailed model of a real image-detecting system including the visual system and the system performance function, guided by comparisons of predicted performance and experimental performance data. The ideal image-detecting system is merely the first step in the iterative process.

4.3.4. Noise-Required Input Contrast

The noise-required input contrast C_{IN} of a given system is defined as the contrast of the target (or displayed target on a screen) required for 50% detection probability.

The response probabilities given by Eqs. (11) and (13) could have been derived in terms of the output contrast rather than the number of absorbed quanta, since a fluctuation in the number of absorbed quanta is equivalent to a fluctuation in output contrast. The standard deviations σ_B and σ_C of the fluctuations in the number of absorbed quanta and the contrast, respectively, are related by $\sigma_B = \bar{N}_B \sigma_C$, where $\bar{N}_B$ is the mean number of absorbed background quanta. It was noted following Eq. (6) that the output signal S_o and output contrast C_o are related by $S_o = \bar{N}_B C_o$. Thus the variable X in the probability density equations may be expressed as $X = C_o(t)/\sigma_c$, where $C_o(t)$ is the fluctuating value of the output contrast in successive exposures. Similarly, the output and the threshold signal-to-noise ratios in the normal probability integrals of Eqs. (11) and (13) are given by $k_o = C_o/\sigma_c$ and $k_T = C_T/\sigma_c$, respectively, where C_T is the threshold contrast. Thus the detection and false alarm probabilities are normal probability distribution functions of the contrasts as well as the signal-to-noise ratios.

It was noted following Eq. (11) that at 50% detection probability $k_o = k_T$. It follows from the above expressions for k_o and k_T that at 50% detection probability $C_o = C_T$ and, in addition, from the definition of the noise-required input contrast the target contrast C_I equals C_{IN}. Hence, if $C_o = C_I$, then $C_T = C_{IN}$. In general C_o is less than C_I and thus C_T is less than C_{IN}. However, in an ideal image-detecting system $C_o = C_I$ and $C_T = C_{IN}$.

To the approximation of the ideal photon counter model, the expression for the noise-required input contrast can be derived from Eq. (23) by setting k_o equal to k_T, C_o equal to C_{IN}, and solving for C_{IN}. The result is given by

$$C_{IN} = 2k_T d/D(\eta \bar{n}_B a\tau)^{1/2} \tag{26}$$

While the detective quantum efficiency completely excludes a major system parameter which strongly affects detection probability, Eq. (26) reveals that the noise-required input contrast is an explicit function of the aperture diameter and collection efficiency. Moreover, the chief physical attribute of a target for detection, at least besides its size and the background radiance, is its contrast. If an image-detecting system is designed and constructed to provide a specified detection probability P_D at a given background radiance, target size, and contrast C_I, then another target of the same size but contrast greater than C_I will be detected with a probability greater than P_D, and conversely. Thus, the noise-required input contrast provides a natural and totally adequate measure of the performance of an image-detecting system.

4.3.5. Noise-Required Input Contrast of the Visual System

The largest collection of visual detection data available for comparison is that obtained by Blackwell (1946). The data were obtained by projecting disk-shaped targets onto a uniformly illuminated white screen and individually testing the responses of a group of observers who sat approximately 60 ft from the screen. More than two million responses of 19 subjects having normal vision were recorded and about 25% were statistically analyzed. The exposure was effectively determined by the integration time of the visual system. For each target size and background luminance the single-choice responses of each observer were recorded at five values of contrast, which yielded a set of detection probabilities ranging from 10 to 95%. The statistical analysis of the data revealed that the relationship between the probability of detection and the contrast is the normal probability distribution function, in agreement with the assumption made in the derivation of the response probability equations above that the probability density of the number of absorbed quanta per exposure is essentially Gaussian.

For disk-shaped targets a is approximately equal to $\pi d^2 \alpha^2/4$ and hence the noise-required input contrast, to the approximation of the ideal photon counter model, is given by

$$C_{IN} = 4k_T/\alpha D(\pi\eta\bar{n}_B\tau)^{1/2} \tag{27}$$

A comparison of experimental and theoretical curves of required contrast for 50% detection probability versus the reciprocal of the angular size of a disk is shown in Figs. 4.4a and 4.4b. Each solid curve is the locus of data points after Blackwell (1946), while the broken lines drawn tangent to the experimental curves represent the theoretically linear relationship predicted by Eq. (27) for the ideal photon counter.

It is clear from Figs. 4.4a and 4.4b that only general agreement exists between the performance of an ideal photon counter and that of the human visual system. Both the theoretical and the experimental curves portray C_{IN} at a given background luminance increasing with increasing values of $1/\alpha$. However, disagreement in detail exists at both low and high values of $1/\alpha$. As $1/\alpha$ approaches zero, the experimental curves do not approach zero and beyond the point of tangency they increase more and more steeply with increasing $1/\alpha$ instead of at a constant rate.

The disagreement between the theoretical and experimental curves at both low and high values of $1/\alpha$ was explained in Section 4.2 as due to omissions in the ideal photon counter model of the visual system. At low values of $1/\alpha$ the visual system fails to function as a perfect spatial integrator of the responses of all the photoreceptors excited by a target. At high values of $1/\alpha$ spatial dispersion in the response of the visual system to targets of small angular size becomes important. Spatial dispersion causes the output contrast to decrease while, in addition, the fluctuations in the contrast are increased due to the increased area of the background covered by the target image. Both effects reduce the output signal-to-noise ratio and increase the contrast required on the screen to equal the threshold contrast. The analysis of spatial dispersion including inhibition in the photoresponse of the retina is discussed in Section 4.4.

The threshold signal-to-noise ratio can be estimated from the slopes of the theoretical curves in Figs. 4.4a and 4.4b to the approximation that the experimental values of C_{IN} at the point of tangency are little affected by the phenomena neglected by the ideal photon counter model. To make the estimate we must obtain an independent estimate of the visual parameters D_E, η_E, and τ_E contained in Eq. (27), where the subscript E indicates that we are now discussing the eye rather than a general image-detecting system. The diameter D_E of the aperture of the eye as a function of background luminance is available from the work of Reeves (1920). The integration time τ_E of the visual system is still not known accurately. Both Rose (1948a) and de Vries (1943) estimated

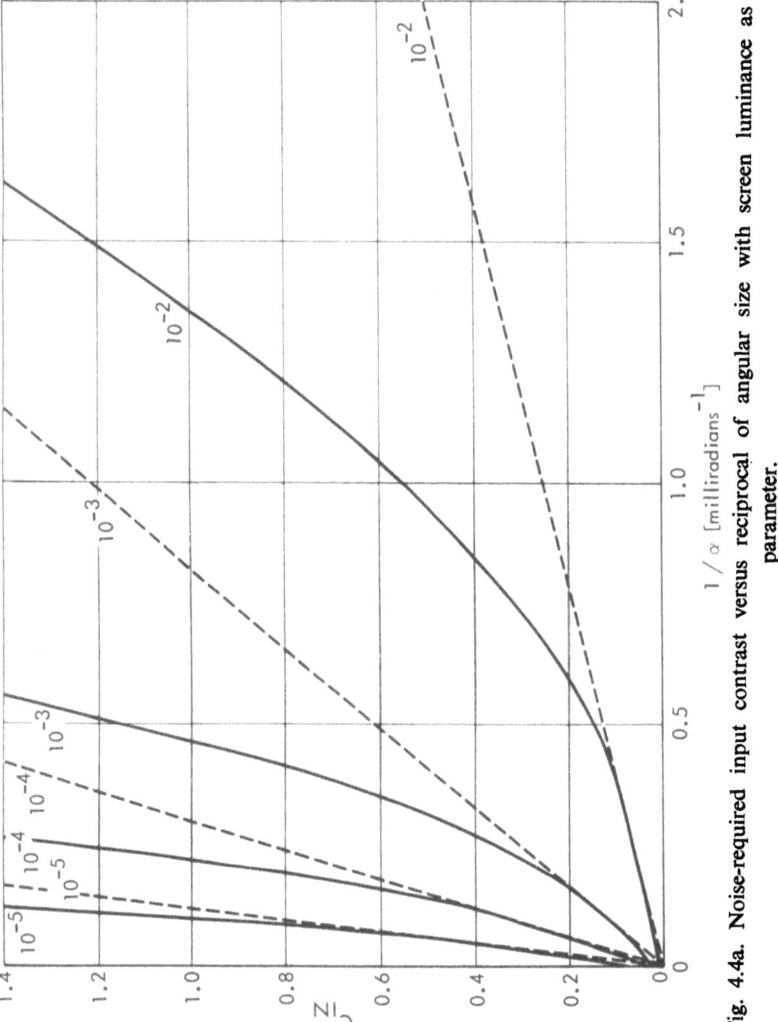

Fig. 4.4a. Noise-required input contrast versus reciprocal of angular size with screen luminance as a parameter.

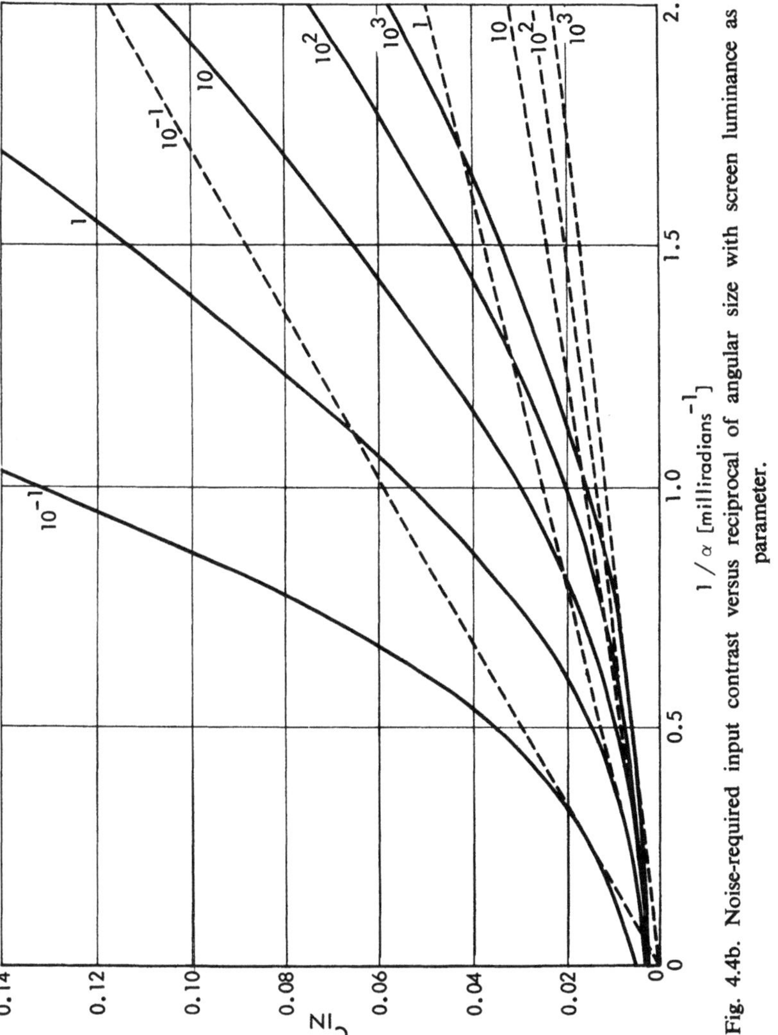

Fig. 4.4b. Noise-required input contrast versus reciprocal of angular size with screen luminance as a parameter.

τ_E as equal to 0.2 sec. This is now generally considered too high except at low background luminance. Schade (1956) estimated that τ_E varies from 0.2 sec at 10^{-6} ft-L to 0.05 sec at 10^3 ft-L. Due to the lack of a precise knowledge of τ_E we will use a value of $\tau_E = 0.1$ sec independent of luminance.

An upper limit to the responsive quantum efficiency η_E of the eye can be estimated from the product of the transmittance of the ocular media, the transmittance of the luminosity filters of the receptors, and the Stiles–Crawford factor. The transmittance of the ocular media in the 0.4–0.73-μm passband, measured by Ludvigh and McCarthy (1948), was found to be approximately 50% for white light. The transmittance of the luminosity filters of the receptors (rods, red and green cones) in the 0.4–0.73-μm passband is approximately 30% for white light. The magnitude of the Stiles–Crawford factor ξ depends on the aperture diameter D_E, which in turn depends on the background luminance. Values of ξ are available from the review by Moon and Spencer (1944). The product of the above three factors yields the estimated upper limit to the responsive quantum efficiency of the visual system presented in Table 4.1.

It is convenient to calculate the threshold signal-to-noise ratio with the following expression derived from Eq. (27):

$$k_T = 1.7 m D_E (\eta_E \tau_E \bar{E})^{1/2} \times 10^3 \tag{28}$$

where $m = C_{IN}/(1/\alpha)$ is the slope of a theoretical curve, α is the angular size of a disk in milliradians, D_E is the diameter of the eye aperture in centimeters, η_E is the responsive quantum efficiency, τ_E, the effective integration time of the eye, is approximately 0.1 sec, $\bar{E}$ is the background luminance in foot-Lamberts, and one lumen of white light is equivalent to 1.4×10^{16} quanta/sec emitted by a blackbody at $5400°$K in the passband 0.4–0.73 μm. The results of the calculation are presented in Table 4.1, where both k_T and the false alarm probability P_F determined with Eq. (13) are tabulated.

Several observations are worth making about these results. First, the values of k_T and P_F are quite plausible, for they are in accord with the observation that little or no conscious perception of fluctuations in the target luminance on the screen is experienced by observers. Second, the relatively small values of k_T for $\bar{E}$ less than 10 ft-L indicate, over a

TABLE 4.1

Threshold Signal-to-Noise and False Alarm Probability As a Function of Luminance

$\bar{E}$, ft-L	m, mrad	D_E, cm	ξ_E	η_E	k_T	P_F
10^{-5}	8.4	0.73	0.50	0.075	2.8	2.6×10^{-8}
10^{-4}	3.3	0.71	0.52	0.078	3.5	2×10^{-4}
10^{-3}	1.2	0.74	0.55	0.082	4.4	$< 10^{-4}$
10^{-2}	2.5×10^{-1}	0.70	0.58	0.087	2.8	2.6×10^{-8}
10^{-1}	5.9×10^{-2}	0.61	0.66	0.099	1.9	2.9×10^{-2}
1	2.5×10^{-2}	0.50	0.76	0.11	2.1	1.8×10^{-2}
10	1.6×10^{-2}	0.40	0.84	0.13	3.9	$< 10^{-4}$
10^2	1.3×10^{-2}	0.28	0.93	0.14	7.4	$< 10^{-4}$
10^3	1.2×10^{-2}	0.25	0.96	0.14	19	$< 10^{-4}$

limited range of target sizes corresponding to the points of tangency in Fig. 4.4a and 4.4b, that indeed the visual system does function very nearly like an ideal noise-limited photon counter with a variable signal threshold set by the background luminance at two to four times the standard deviation σ_B. Third, an explanation for the increase in the calculated values of k_T at $\bar{E}$ greater than 1 ft-L is contained in a theory due to Bouman *et al.* (1963). According to their theory, the visual system exhibits a refractory time, i.e., a dead period of approximately 10^{-3} sec following the absorption of a quantum during which any subsequent quanta do not produce electrical excitations. Consequently, when $\bar{E}$ becomes great enough for the absorption of more than one quantum/msec by a photoreceptor the value of η_E begins to decrease. Calculation of the mean number of quanta absorbed per rod in a millisecond yields 0.9 at $\bar{E} = 1$ ft-L and 6.9 at $\bar{E} = 10$ ft-L, in agreement with the refractory theory at $\bar{E}$ greater than 1 ft-L. Thus, since the increase in calculated values of k_T for $\bar{E} \geq 10$ ft-L is probably spurious, a more plausible procedure to follow is to assume that the value of k_T, equal to approximately 2, at $\bar{E}$ equal to 10^{-1} and 1 ft-L is valid at $\bar{E} \geq 10$ ft-L as well. If k_T is set equal to 2, then η_E equals 0.034, 0.010, and 0.0016 at $\bar{E}$ equal to 10, 10^2, and 10^3 ft-L, respectively.

Finally, it should be observed that the calculated values of k_T at $\bar{E}$ less than 10 ft-L do not permit much decrease in the estimate of the product of τ_E and η_E, although our estimate of η_E was an upper limit. The estimate did not include the fractional absorption of quanta by the visual purple of the photoreceptors. If the fractional absorption were significantly less than unity, the value of τ_E would have to be increased by a compensating factor to maintain a plausible value of k_T. For example, if the fractional absorption were 1/3 as considered by Wald *et al.* (1963), then τ_E would have to be approximately 0.3 sec, corresponding to an eye fixation time. Such a possibility cannot be excluded, for the values of the visual parameters have yet to be accurately determined.

4.4. MODIFICATIONS OF THE IDEAL PHOTON COUNTER MODEL

The discrepancies between the theoretical and experimental curves of noise-required input contrast versus reciprocal of angular size shown in Figs. 4.4a and 4.4b at low and high values of the reciprocal angle were ascribed to failure of the visual system to function as a perfect spatial integrator and to spatial dispersion in the response of the visual system, respectively. In any real image-detecting system employing the visual system the spatial integration function is performed by the visual system. Thus if we develop an analytical description of the spatial integration function of the unaided visual system, then we can apply it in our analysis of the coupling between the display and the observer in an electrooptical–observer image-detecting system.

On the other hand, spatial dispersion occurs in all optical and electrooptical components as well as the visual system. Thus it is necessary to develop an analytical description of spatial dispersion in each component and in a convenient form such that the contributions of the components can be readily combined to account for the total spatial dispersion in the system. A convenient analytical description is provided by the well-known modulation transfer function (i.e., spatial frequency response function in analogy to the temporal frequency response function of electronics) developed by the employment of Fourier analysis.

4.4.1. Fourier Analysis of Spatial Dispersion

Spatial dispersion in optical systems is most familiar to users of telescopes and microscopes, where it is noticed that the image of a bright, point source is not a point but is spread out over a small area with more or less circular symmetry. The spatial distribution of the irradiance in the area of the image of a point source is represented by the point spread function. If the effects of both diffraction and aberrations were absent in optical systems, no spatial dispersion would occur and the point spread function would be the delta function. If the effect of aberrations alone were absent, the point spread function would have to describe the familiar Airy diffraction pattern formed when radiant power is transmitted through a circular aperture. However, in most image-detecting systems both aberration and diffraction degrade the image of a point source.

In the components of photoelectronic imaging systems aberrations are not absent but, on the contrary, they are the major sources of spatial dispersion. Aberrations are dominant in the low-f-number objectives required by most low-light-level image-detecting systems to provide both high collection efficiency and large field of view as well as in the electron lens used in electrooptical components. Other sources of spatial dispersion in image-detecting systems are lateral spread of electrical charge in various electrooptical structures used for photoelectric detection, charge amplification, charge storage, and electron-to-photon conversion. In most cases simple analytical expressions for the point spread functions originating from lens aberrations and electrical charge spread in electrooptical structures are not available.

When an extended object is imaged by a linear image-forming system the image is represented by the superposition of all the point spread functions corresponding to the points in the object. If the optical system is composed of several components, the process is repeated for each component analogous to the process of ray tracing in geometrical optics. Quite clearly, to follow such a procedure in a complex electro-optical image-detecting system would be a formidable task. Hence it is highly desirable to develop a simpler procedure, analogous to that provided in electronic communications by the transfer function theory, such that the overall effect of spatial dispersion in such a system can be obtained by merely multiplying the separate effects of individual components. Such a procedure may be obtained by application of two-dimensional (three-dimensional if the radiance of the scene is time dependent) Fourier analysis to the superposition process.

Mathematically the superposition process may be represented by the convolution integral given by

$$\bar{n}_0(x, y) = \int\limits_{-\infty}^{\infty}\!\!\int g(x, y; u, v)\bar{n}_I(u, v)\, du\, dv \qquad (29)$$

where $\bar{n}_o(x, y)$ is the mean output irradiance, $\bar{n}_I(x, y)$ is the mean irradiance in the perfect (in the sense of geometrical optics) image of the input, and $g(x, y; u, v)$ is the point spread function. That is,

$$g(x, y; u_1, v_1) = \int\limits_{-\infty}^{\infty}\!\!\int g(x, y; u, v)\, \delta(u - u_1)\, \delta(v - v_1)\, du\, dv \qquad (30)$$

is the output irradiance of an image-forming component responding to an impulse whose perfect image is a point (u_1, v_1). If the image-forming component is isoplanatic, the point spread function is spatially invariant. It follows that if a local coordinate system is chosen with origin at any point (u_1, v_1), then

$$g(x, y; u_1, v_1) = g(x - u_1, y - v_1; 0, 0) \qquad (31)$$

and

$$\bar{n}_0(x, y) = \int\limits_{-\infty}^{\infty}\!\!\int g(x - u, y - v)\bar{n}_I(u, v)\, du\, dv \qquad (32)$$

Real image-forming components are not isoplanatic. However, if the departure from isoplanatism is spatially slowly varying compared to the extent of the point spread function, the image plane can be divided into nearly homogeneous regions where Eqs. (31) and (32) are reasonable approximations as indicated by Linfoot (1964).

It is shown in monographs concerned with applications of Fourier transforms, such as that by Papoulis (1968), that the Fourier transformation of a convolution integral such as Eq. (32) yields

$$H_o(v_x, v_y) = G(v_x, v_y)H_I(v_x, v_y) \qquad (33)$$

where $H_o(v_x, v_y)$, $G(v_x, v_y)$ and $H_I(v_x, v_y)$ are the two-dimensional Fourier transforms of $\bar{n}_o(x, y)$, $g(x - u, y - v)$, and $\bar{n}_I(u, v)$, respectively; v_x and v_y are the spatial frequencies. The function $G(v_x, v_y)$ is known as the optical transfer or frequency response function. In general, Fourier transforms such as $H_o(v_x, v_y)$ and $H_I(v_x, v_y)$ do not represent physical quantities but are only mathematical conveniences introduced to reduce a series of convolutions to a simple multiplication process.

For example, let us assume that we have two linear image-forming components with point spread functions g_1 and g_2 and optical transfer functions G_1 and G_2, respectively. Then, since the output $\bar{n}_{o1}$ of the first component is the input $\bar{n}_{I2}$ to the second component, by applying Eq. (29), we have

$$\bar{n}_0(x, y) = \int\limits_{-\infty}^{\infty}\!\!\int g_2\left[\int\limits_{-\infty}^{\infty}\!\!\int g_1\bar{n}_I(u, v)\, du\, dv\right] du'\, dv' \tag{34}$$

where $\bar{n}_I(u, v)$ is the mean irradiance in the perfect image in the image plane of the second component. Similarly, since $H_{o1} = H_{I2}$, by applying Eq. (33), we have

$$H_o(v_x, v_y) = G_2 G_1 H_I(v_x, v_y) \tag{35}$$

The spatial distribution of the output irradiance $\bar{n}_o(x, y)$ can be obtained by performing the inverse Fourier transformation of Eq. (35). In both cases it is necessary to suitably scale the coordinates in the object and image planes to account for magnification.

It is of interest to note that for incoherent radiant power a linear, memoryless (in the spatial as well as temporal sense) image-forming system always functions as a low-pass filter with maximum frequency response at zero frequency. Otherwise, as indicated by Levi (1970), the point spread function would have to be negative in some regions. But this is impossible since both the input and output radiant power are necessarily described by positive real functions.

Of special interest for measuring the performance of image-forming systems is an input irradiance for which the perfect image may be represented in the complex image plane by

$$\bar{n}_I(x, y) = \bar{n}_{BI} + \hat{n}_I(v_x)e^{2\pi j v_x x} \tag{36}$$

where $\hat{n}_I(v_x)$ is the real amplitude of a cosine distribution of irradiance.

The mean output irradiance according to the convolution integral Eq. (29) is given by

$$\bar{n}_0(x, y) = \bar{n}_{BI} \iint\limits_{-\infty}^{\infty} g(x - u, y - v)\, du\, dv$$

$$+ \hat{n}_I(\nu_x) \iint\limits_{-\infty}^{\infty} g(x - u, y - v) e^{2\pi j \nu_x u}\, du\, dv \tag{37}$$

If we substitute $w = x - u$ and $z = y - v$ in the second integral of Eq. (37) and interchange the order of integration, we obtain

$$\bar{n}_0(x, y) = \bar{n}_{BI} \iint\limits_{-\infty}^{\infty} g(w, z)\, dw\, dz$$

$$+ \hat{n}_I(\nu_x) \left[\int_{-\infty}^{\infty} l(w) e^{-2\pi j \nu_y w}\, dw \right] e^{2\pi j \nu_x x} \tag{38}$$

where

$$l(w) = \int_{-\infty}^{\infty} g(w, z)\, dz \tag{39}$$

The function $l(w)$, known as the line spread function, is the response of an image-forming component to a line impulse $\delta(w)$ along the $w = 0$ axis. The bracketed integral in Eq. (38) is recognized as $L(\nu_x)$, the Fourier transform of the line spread function.

By defining the point spread function in Eq. (30) as the response to the irradiance of the perfect image of an impulse rather than to the radiance of an impulse in the object plane, we have avoided the questions concerning collection efficiency and magnification. Consequently, the mean background irradiance $\bar{n}_{BI}$ is identical in both $\bar{n}_I(x, y)$ and $\bar{n}_o(x, y)$ and the double integral of the point spread function is unity. It follows from Eq. (31) that the integral of the line spread function and hence $L(0)$ are also both equal to unity.

By making use of the above observations we have

$$\bar{n}_0(x, y) = \bar{n}_{BI} + \hat{n}_o(\nu_x) e^{2\pi j \nu_x x} \tag{40}$$

where

$$\hat{n}_o(\nu_x) = L(\nu_x) \hat{n}_I(\nu_x) \tag{41}$$

Thus we see that if the radiant exitance of the object plane is a cosine distribution of a single spatial frequency v_x superimposed on a uniform background, the output image irradiance is a cosine distribution of the same frequency superimposed on a uniform background. At $v_x = 0$ the amplitudes of the perfect and the real cosine images are equal, but as v_x increases, the amplitude of the real cosine image decreases while the amplitude of the perfect cosine image remains unaffected by the image-forming component.

If we define modulation as the ratio of the amplitude of a cosine distribution to the mean background, then by Eq. (41) we have

$$M_o(v_x) = L(v_x)M_I(v_x) \qquad (42)$$

where $M(v_x) = \hat{n}(v_x)/\bar{n}_B$. The Fourier transform $L(v_x)$ of the line spread function now may be identified as the well-known one-dimensional modulation transfer function.

In general $L(v_x)$ and hence $M_o(v_x)$ are complex functions. We can represent $L(v_x)$ by $T(v_x)e^{j\varphi(x)}$, where $T(v_x) = |L(v_x)|$ is the magnitude of the complex modulation transfer function and $\varphi(v_x)$ is the phase which in the spatial domain represents a displacement along the x axis. From the definition of $L(v_x)$ in reference to Eq. (38) we note that if $l(X)$ is symmetric about the X axis, $L(v_x)$ is real. More generally, if $g(x, y)$ is circularly symmetric, then $L(v_x)$ and $L(v_y)$ are identical and real. It is sufficient for most purposes to assume that $L(v_x)$ and $L(v_y)$ are real and represented by $T(v_x)$ and $T(v_y)$, respectively.

The relationship between the two-dimensional optical transfer function $G(v_x, v_y)$ and $L(v_x)$ is readily shown, by setting $v_y = 0$ in the Fourier transformation of $g(x, y)$, to be

$$G(v_x, 0) = L(v_x) \qquad (43)$$

and hence

$$T(v_x, 0) = T(v_x) \qquad (44)$$

where $T(v_x, v_y)$, the modulus of the two-dimensional optical transfer function, is the two-dimensional modulation transfer function.

Often in practice it is assumed that

$$G(v_x, v_y) = L(v_x)L(v_y) \qquad (45)$$

However, Eq. (45) is not valid unless

$$g(x, y) = l(x)l(y) \tag{46}$$

In particular, if Eqs. (45) and (46) are valid and the point spread function has circular symmetry, then it is indicated by Papoulis (1968) that the point spread and optical transfer functions are Gaussians, i.e.,

$$g(x, y) = (1/\pi r_0^2) \exp[-(x^2 + y^2)/r_0^2] \tag{47}$$

and

$$G(\nu_x, \nu_y) = \exp[-\pi^2(\nu_x^2 + \nu_y^2)r_0^2] \tag{48}$$

where r_0 is an arbitrary constant.

Often the point spread functions of electrooptical image-forming devices and lens systems (except when the central portion of the aperture is obscured) can be reasonably approximated by Gaussian functions so that Eq. (45) is a reasonable approximation. However, it should be kept in mind that it is an approximation and may not be valid in every particular case.

The practical importance of the cosine-modulated background exitance as an input test signal is now apparent. By measuring the output modulation at constant input modulation as a function of spatial frequency we can employ Eq. (42) to deduce the one-dimensional modulation transfer functions $T(\nu_x)$ and $T(\nu_y)$. Then we can use Eq. (45) to obtain approximately the two-dimensional modulation transfer function $T(\nu_x, \nu_y)$, which we showed via Eq. (33) related the Fourier transforms of the perfect and actual images of the input.

4.4.2. Noise-Required Input Modulation

In practice, except for simple targets which can be represented by areal rectangular pulses bounded by circles, squares, or rectangles, the detailed signal-to-noise ratio and noise-required input contrast analysis of targets of complex shape and structure is too complex to be performed rigorously. Fortunately, Johnson (1958) discovered some empirical criteria relating the probabilities of detection, recognition, and identification of complex targets to the probability of detection of simple periodic test patterns. Specifically, he discovered that detection occurs (presumably with 50% probability) if the angular width α of a complex target equals the angular width of the period of the highest detectable spatial frequency v_R in cycles per unit angle of a periodic test pattern, i.e., $\alpha \simeq 1/v_R$. Similarly, recognition and identification were discovered to occur if $\alpha \simeq 4/v_R$ and $\alpha \simeq 6.5/v_R$, respectively. The criteria are clearly limited in applicability to low-contrast targets, since presumably the modulation of the periodic test pattern must be chosen to correspond to the contrast of the complex target and values of modulation are limited to less than unity. The criteria provide a means for predicting the probability of detection, recognition, and identification of complex targets either by test bench measurements of the probability of detection of periodic test patterns or by applying the signal-to-noise ratio analysis to periodic test patterns.

Since it was shown in Section 4.4.1 that the output modulation is linearly related to the input modulation of a cosine-modulated background exitance, periodic test patterns of this form are most convenient to analyze. In the analysis of cosine-modulated periodic test patterns, it is convenient to introduce the concept of noise-required input modulation in analogy to the noise-required input contrast of aperiodic targets. Noise-required input modulation is defined as the modulation M_{IN} of a cosine-modulated test pattern at a given frequency v_o such that at the decision level in the visual system the output signal-to-noise ratio k_o equals the threshold value k_T, and hence the detection probability P_D equals 50%.

In order to analyze the detection of cosine-modulated test patterns it is necessary to hypothesize the strategy followed by the visual system. In the case of the detection of disk-shaped targets the problem is simpler since the targets are either brighter or darker than the background and the boundary is sharply defined. However, cosine-modulated test patterns may be considered to be either parallel strips of half-period width, alternately brighter and darker than the mean background luminance, with no clearly defined boundary between them, or as parallel strips of width equal to a period brighter than the background. However, clues do exist upon which to formulate a detection strategy for cosine-modulated test patterns.

It seems clear from the general agreement between the ideal photon counter model of the visual system and the Blackwell disk detection data that the visual system functions like an adaptive areal matched filter (of course, limited in range) with an adaptive decision-making device at the output. The visual system automatically adjusts the summation area of the photoreceptor field to match the area of the image of the target and automatically adjusts the decision threshold (or, equivalently, a variable-gain mechanism between the receptors and a fixed threshold) as a function of the mean background illumination to be two to three times the standard deviation of the background fluctuations within the area of the target image.

While the data analyzed in Section 4.3.5 were for targets of positive contrast, Blackwell reported essentially the same data for negative-contrast targets. Thus the above remarks regarding adaptive areal matched filtering and adaptive decision threshold apply to negative-contrast targets as well as the positive ones explicitly considered here. That is, if the illumination within the area of the image of a disk falls below the mean background illumination, the visual system responds when the (negative) differential response of the photoreceptors exceeds two to three times the standard deviation of the background fluctuations within the area.

In discussing the detection of targets of large angular size it was hypothesized that the spatial integration capability of the visual system is limited and that responses of photoreceptors roughly subtending an angle α_1 at the pupil plane in the vicinity of the edge of the target image are summed and compared to the decision threshold. The entire disk is perceived in this case by virtue of signal processing at a higher level in the nervous system in response to the separate inputs from the several lower level nerve cells, each associated with a photoreceptor field defined by α_1 in the vicinity of the periphery, in which the signal-to-noise ratio threshold is set. The important things to note about edge detection with disks are that the angle subtended by the width of the transition from the illuminance of the disk to the illuminance of the background is nil compared to α_1 and for each detection experiment the differential illuminance is either positive or negative relative to the background.

However, if periodic test patterns are used, areas of positive and negative differential illuminance occur simultaneously in adjacent half-periods of the test pattern. In this case it is hypothesized that the responses of the elementary photoreceptors in the positive half-period and the responses of the elementary photoreceptors in the negative half-period add constructively at the decision level where the threshold signal-to-noise ratio is set, i.e., the visual system acts like a difference detector when subjected to excitation by adjacent areas of opposite contrast. Moreover, as the period of the test pattern increases so that $1/v_o$ exceeds α_1, the maximum width of the photoreceptor field over which summation occurs becomes less than a period of the test pattern. Then the maximum difference signal due to two adjacent halves of the photoreceptor field occurs when the field is centered on the line of maximum gradient. Hence it is hypothesized that as v_o becomes less than $1/\alpha_1$, the visual system follows the optimum detection strategy and responds to the difference signal due to the two adjacent halves of the photoreceptor field centered on the line of maximum gradient.

At the transition frequency, in the vicinity of $v_o \simeq 1/\alpha_1$, the visual system changes from detecting the radiant power difference between adjacent half-periods at high frequencies (an area detector) to detecting the radiant power difference between two adjacent strips centered on the line of maximum contrast gradient at low frequencies (a contrast gradient detector). Since the maximum gradient of a cosine test pattern is proportional to v_o while the half-period area is proportional to $1/v_o$, the signal-to-noise ratio goes through a maximum and the noise-required input modulation through a minimum in the vicinity of $v_o \simeq 1/\alpha_1$. Furthermore, since α_1 is a function of the mean background luminance, the transition frequency is a function of the mean background luminance. A comparison of a curve drawn through the values of $1/\alpha_1$ deduced by application [Schnitzler (1973)] of the modified photon counter model to the Blackwell disk detection data and the experimental values of v_o at the minimum in M_{IN} obtained by Schade (1956) is shown in Fig. 4.5. The general agreement lends credence to the above hypotheses of the strategy followed by the visual system in the detection of periodic test patterns.

To derive the expression for the noise-required input modulation, let the irradiance of the primary sensor be represented by

$$\bar{n}_I(x, y) = \bar{n}_{BI} + \hat{n}_I(v_o) \cos(2\pi v_o x) \tag{49}$$

In accordance with the above hypotheses of the strategy followed by the visual system in the detection of the pattern represented by Eq. (49), the difference input signal is given by

$$S_I = 2M_I \bar{n}_{BI} \eta_s \tau_s h_I [\sin^2(\pi v_o w_1/2)]/\pi v_o, \qquad v_o \leq 1/w_1 \tag{50a}$$

$$S_I = 2M_I \bar{n}_{BI} \eta_s \tau_s h_I/\pi v_o, \qquad v_o \geq 1/w_1 \tag{50b}$$

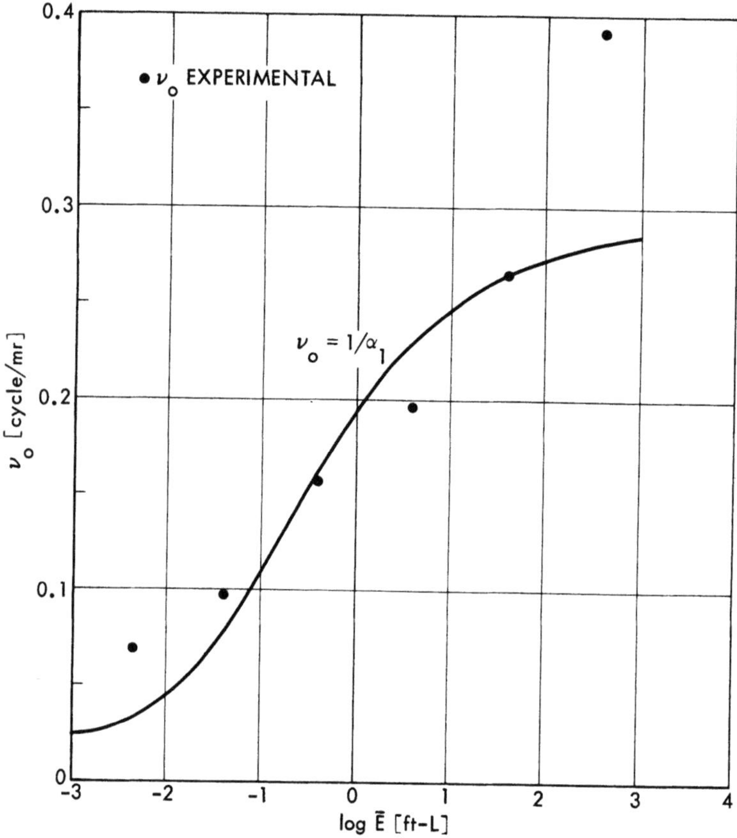

Fig. 4.5. Spatial frequency at minimum noise-required input modulation versus logarithm of screen luminance in ft-L. Note experimental points after Schade (1956).

where h_I is the effective extent of the test pattern along the y axis and w_1 is the width of the two parallel strips, bordered by the line of maximum contrast gradient at $x = \pm 1/4\nu_o$, over which spatial integration effectively occurs. There is evidence that the visual system is more effective at summing the responses of photoreceptors along the length of a line than across the width. Hence a second parameter h_1 must be introduced to take into account the maximum effective length that the visual system can integrate. If the length of the test pattern is greater than h_1, then h_I must be replaced by h_1 in Eqs. (50a) and (50b).

From Eqs. (50a) and (50b) the output signal, equal to the product of the modulation transfer function and the magnitude and the number of statistical units at the retina, is found to be given by

$$S_o = (G_D \zeta \eta_E) T_s(\nu_o) T_E(\nu_o) S_I(\nu_o)(M_T/M_I) \tag{51}$$

where M_T is the modulation of the test pattern.

For a background-limited electrooptical image-detecting system the standard deviation in the fluctuations at the retina is given by either

$$\sigma_o = (G_D \zeta \eta_E)(\bar{n}_{BI} \eta_s \tau_s h_I w_I)^{1/2} \qquad \nu_o \leq 1/w_1 \tag{52a}$$

or

$$\sigma_o = (G_D \zeta \eta_E)(\bar{n}_{BI} \eta_s \tau_s h_I/\nu_o)^{1/2} \qquad \nu_o \geq 1/w_1 \tag{52b}$$

If we divide Eq. (51) by Eqs. (52a) and (52b), respectively, set $k_o = k_T$ so that M_T equals the noise-required input modulation M_{IN}, and solve for M_{IN}, we obtain

$$M_{IN} = \pi \nu_o \alpha_w k_T / (\bar{n}_B \eta_s \tau_s \alpha_h \alpha_w)^{1/2} D_s T_s(\nu_o) T_E(\nu_o) \sin^2(\pi \nu_o \alpha_w/2)$$
$$\nu_o \leq 1/\alpha_w \tag{53a}$$

$$M_{IN} = \pi k_T(\nu_o/\bar{n}_B \eta_s \tau_s \alpha_h)^{1/2}/D_s T_s(\nu_o) T_E(\nu_o) \qquad \nu_o \geq 1/\alpha_w \tag{53b}$$

where α_w and α_h are the angles subtended by w_1 and h_I at the pupil plane of the objective and ν_o is in cycles per unit angle subtended at the pupil plane of the objective.

If the image-detecting system is the visual system, the parameters η_s, τ_s, and D_s are replaced by η_E, τ_E, and D_E, respectively, and $T_s(\nu_o)$ is set equal to unity in Eqs. (53a) and (53b). Then it is worth noting that if $\nu_o \ll 2/\pi \alpha_w$, by Eq. (53a) and since $T_E(\nu_o)$ is approximately unity, M_{IN} is proportional to $1/\nu_o$. This prediction for the low-frequency dependence of M_{IN} is in agreement with carefully obtained experimental data by Davidson (1968).

In the vicinity of $\nu_o = 1/\alpha_w$, $T_E(\nu_o)$ is still essentially unity and by Eq. (53a) or Eq. (53b) M_{IN} is proportional to $\sqrt{\nu_o}$. Thus a minimum in $M_{IN}(\nu_o)$ is predicted to occur, in general agreement with experimental observations. As the value of ν_o increases beyond $1/\alpha_w$, the slope of $M_{IN}(\nu_o)$ would decrease with increasing ν_o except that spatial dispersion in the visual system increases and hence the term $T_E(\nu_o)$ in the denominator decreases, causing the slope to increase.

It is important to note that in Eqs. (53a) and (53b) ν_o appears explicitly as well as implicitly in $T_E(\nu_o)$. Therefore it is quite incorrect to assume that the frequency dependence of the modulation transfer function of the visual system is identical to the frequency dependence of the modulation sensitivity $1/M_{IN}(\nu_o)$ as many researchers have assumed. To make such an assumption is to neglect the capability of the visual system to function as a gradient detector with limited spatial frequency filtering at low test pattern frequencies and as an adaptive spatial frequency (or areal) filter at intermediate and high test pattern frequencies. It is not legitimate to include the frequency dependence due to the detection strategy and spatial filtering in the modulation transfer function since it is highly unlikely that the output at the decision level is an electrochemical spatial replica of the test pattern illuminance of the retina. At a detection probability of 50% only the gradients are sensed at low frequencies and the radiant power difference between positive and negative half-periods at higher frequencies. To assume that detail is sensed at the decision level would imply that an infinitesimal signal-to-noise ratio would be sufficient to make a decision, contrary to the analysis of the Blackwell disk detection data and the absence of the perception of fluctuations and a physical false alarm (other than due to guessing) probability. Of course detail becomes perceptible if the signal-to-noise ratio, and hence probability, for detection becomes high enough.

If Eqs. (53a) and (53b) are multiplied by $T_s(\nu_o)$, then they represent the noise-required modulation M_{DN} on the display of an electrooptical image-detecting system. If, in addition, the length of the test pattern is scaled in proportion to $1/\nu_o$ as sometimes is done in practice, then over a limited range of ν_o greater than $1/\alpha_w$ but less than the value where $T_E(\nu_o)$ decreases significantly, M_{DN} is proportional to ν_o. This case, which is of considerable practical importance, has been discussed in detail elsewhere [Schnitzler (1971b)].

By application of a little algebra, from Eqs. (53a) and (53b) it can be shown that either the noise-required input modulation to a photo-electronic image-detecting system or the noise-required modulation on the display of the system can be expressed in terms of the noise-required input modulation $M_{IN}(\bar{E}_D, \nu_{OD})$ to the unaided visual system corresponding to the display luminance, where ν_{OD} is the angular frequency at the pupil plane of the eye. Such a procedure offers the advantage that the dependence of M_{DN} on the visual system parameters k_T, τ_E,

α_h, α_w, and $T_E(\nu_o)$, providing they are independent of the magnitude of a statistical unit at the retina at a given value of display luminance, is implicitly contained in $M_{IN}(\bar{E}_D, \nu_{OD})$. Thus, if sufficient detection probability data on cosine-modulated test patterns were available to establish a comprehensive set of "standard observer" curves of $M_{IN}(\bar{E}_D, \nu_{OD})$ over the entire range of useful values of display luminance, then the curves of M_{DN} for any combination of photoelectronic image-detecting system parameters, test pattern modulation, and background luminance values could be readily deduced. Unfortunately, reliable detection probability data on cosine-modulated test patterns are virtually nonexistent. The scattering in the literature of "threshold" modulation data of unspecified detection probability and usually for one or two observers serves only as a rough indicator of the standard observer noise-required input modulation curves for the visual system.

However, from either Eq. (53a) or Eq. (53b) it can be shown that the noise-required modulation on the display of a background-limited electrooptical image-detecting system is related to the noise-required input modulation to the visual system at the display luminance by

$$M_{DN}(\bar{E}, \nu_o) = M(\bar{n}_{BD}/\bar{n}_B)^{1/2}(\eta_E/\eta_s)^{1/2}(D_E/D_s)M_{IN}(\bar{E}_D, \nu_{OD}) \qquad (54)$$

where M is the value of the subjective magnification of the photoelectronic image-detecting system. In order to illustrate the significance of Eq. (54) in spite of the unavailability of a standard observer curve for $M_{IN}(\bar{E}_D, \nu_{OD})$, a hypothetical curve postulated from a perusal of the literature is presented in Fig. 4.6 labeled "display limit" for $\bar{E}_D = 10^2$ ft-L and for a subjective magnification equal to $4.5\times$, corresponding to a typical LLLTV system. Assuming that the values of the parameters η_E, η_s, D_E, and D_s are 38 μA/L, 5 mA/W with a 2854°K lamp, 2.8 mm, and 203 mm, respectively, the coefficients of $M_{IN}(10^2, \nu_o/4.5)$ in Eq. (54) at screen luminances of 10^{-5}, 10^{-4}, 10^{-3}, 10^{-2}, and 10^{-1} ft-L are 77, 24.4, 717, 2.44, and 0.77, respectively. The corresponding curves of $M_{DN}(\bar{E}, \nu_o)$, except for $\bar{E} = 10^{-1}$ ft-L, are shown in Fig. 4.6. The curve for $\bar{E} = 10^{-1}$ ft-L falls on the "display limit" curve since the coefficient in Eq. (54) is less than unity.

The curves representing the noise-required modulation on the display combined with the modulation transfer function curve $T_s(\nu_o)$ can be employed to define the useful spatial frequency response bandwidth of

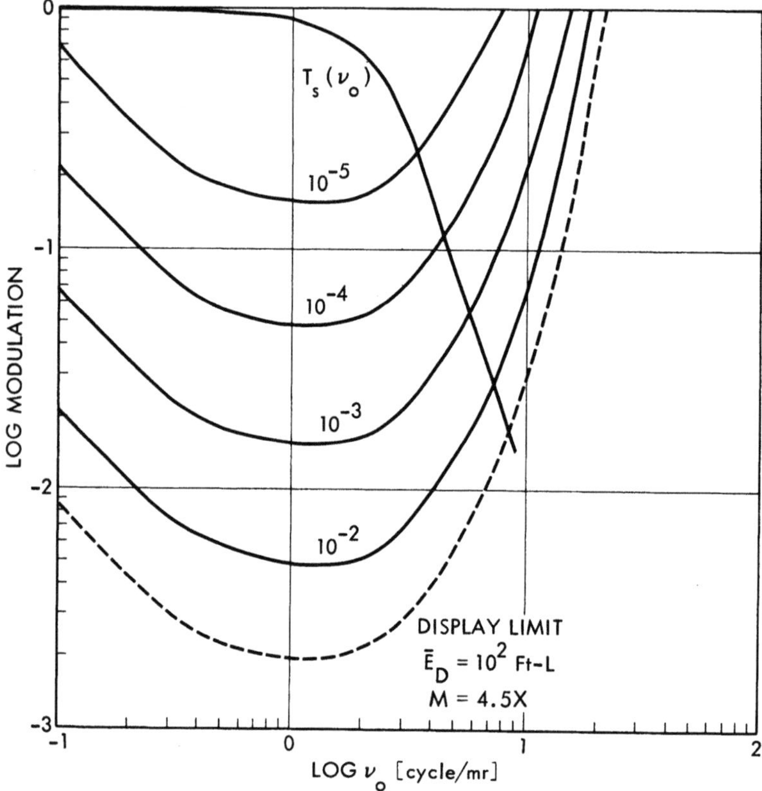

Fig. 4.6. Logarithm of noise-required modulation on the display, with screen luminance as a parameter, and logarithm of system modulation transfer function $T_s(\nu_0)$ versus logarithm of test pattern frequency in cycles/mrad in object space. Note resolution frequencies at intersections of curves.

an electrooptical image-detecting system. For example, in addition to the curves representing $M_{DN}(\bar{E}, \nu_o)$, a curve representing a hypothetical modulation transfer function $T_s(\nu_o)$ is shown in Fig. 4.6. This curve represents the maximum obtainable value of the modulation on the display which only results if the test pattern modulation on the projector screen is 100%. Therefore the value of the test pattern frequency ν_o at the intersection of the curve of $T_s(\nu_o)$ and a curve of $M_{DN}(\bar{E}, \nu_o)$ is the upper limit of the spatial frequency response of the complete electro-optical–visual system.

For the hypothetical example shown in Fig. 4.6 it was assumed the modulation transfer function of the electrooptical system is a Gaussian given by $T_s(\nu_o) = \exp(-\nu_o{}^2/\nu_L{}^2)$ with ν_L equal to 3 cycles/mrad. The resulting useful frequency response bandwidths (zero to ν_o) at $\bar{E}$ equal to 10^{-5}, 10^{-4}, 10^{-3}, and 10^{-2} ft-L and for the "display limit" are 3.4, 4.5, 5.8, 7.3, and 8.5 cycles/mrad, respectively.

Since for a detection probability of 50% the differential radiant power from an entire aperiodic target is required for sufficient signal-to-noise ratio, no detail is perceptible and an aperiodic target can be approximated roughly by a rectangle of the same dimensions. In this case the one-dimensional spatial frequency spectral density corresponding to the angular width α_T of the target is proportional to $\operatorname{sinc}(\pi\nu\alpha_T)$. But $\operatorname{sinc}(\pi\nu\alpha_T)$ is unity at $\nu = 0$, decreases monotonically to zero at $\nu = 1/\alpha_T$, and then undergoes rapidly damped oscillations about the ν axis. Moreover, it is reasonable to expect that detection of an aperiodic target would occur if the useful bandwidth of the image-detecting system approximately equals the extent of the spatial frequency content corresponding to the width of the aperiodic target, i.e., if $\nu_R \simeq 1/\alpha_T$. This is recognized as Johnson's (1958) empirical detection criterion. Therefore, from a knowledge of the curve of $M_{IN}(\bar{E}_D, \nu_{OD})$, the subjective magnification M, the ratios of the parameters η_E/η_s, D_E/D_s, and $\bar{E}_D/\bar{E}$, and the modulation transfer function of the electrooptical image-detecting system, the minimum angular width (maximum range if the linear width is known) of an aperiodic target for a detection probability of 50% can be predicted without explicit knowledge of the modulation transfer function of the visual system.

However, it should be noted that an analysis of electrooptical image-detecting systems based only on the number of statistical units counted at the retina is still, in general, only approximate. The standard deviation in the output response may be increased by additional electrical fluctuations introduced in electrooptical systems so that they are no longer background limited. On the other hand, while the standard deviation in the number of quanta absorbed or the number of output statistical units is equal to the mean number, the standard deviation in the output response may be reduced by the smoothing action on each statistical unit by spatial dispersion in electrooptical image-detecting systems. In any ideal image-detecting system no spatial dispersion would occur and the output statistical units would be points. However, in real image-detecting systems the output statistical units are spread over finite areas which overlap, resulting in a reduction in the fluctuations in the output response. The analysis of this phenomenon requires the use of the generalized Fourier analysis of random processes which will be discussed in Section 4.5.

4.5. NOISE POWER DENSITY SPECTRAL ANALYSIS

The use of noise power density spectral analysis is limited here to the derivation of the noise-required modulation on the display or the input modulation of cosine-modulated periodic test patterns, although to be rigorous it should be used in the derivation of the noise-required input contrast of aperiodic targets as well. The image-detecting system is assumed to be functioning in the background- or shot-noise-limited regime. Thus the expressions derived in Section 4.4.2 for the input signal given by Eqs. (50a) and (50b) and the output signal given by Eq. (51) are applicable in this section. The concern here is for a rigorous derivation of the standard deviation in the output fluctuations of the visual system.

The derivation proceeds by noting that according to Lee (1960) the input noise power density spectrum $\Phi_I(\nu)$ of a Poisson distribution of unit impulses, describing the random excitation of unit electrical charges at the primary sensor, is given simply by the mean number of unit impulses per unit area per second, i.e.,

$$\Phi_I(\nu) = \eta_s \bar{n}_{BI} \tag{55}$$

where ν represents the temporal and two spatial frequencies.

The output noise power density spectrum $\Phi_o(\nu)$ is given by

$$\Phi_o(\nu) = (G_D \zeta \eta_E)^2 T_D^2(\nu) T_E^2(\nu) R_E^2(\nu) \Phi_I(\nu) \tag{56}$$

where $(G_D \zeta \eta_E)$ is the magnitude of a statistical unit at the retina as it was in Section 4.4.2, $T_D(\nu)$ is the modulation transfer function from the primary sensor to the display, and $R_E(\nu)$ is an expression in frequency space corresponding to spatial integration by the visual system. At low frequencies $R_E(\nu)$ is approximated by

$$R_E(\nu) = [\text{sinc}(\pi \nu_x \alpha_1)] \, \text{sinc}(\pi \nu_y \alpha_h) \qquad \nu_o \leq 1/\alpha_1 \tag{57a}$$

and at high frequencies by

$$R_E(v) = [\text{sinc}(\pi v_x/v_0)]\,\text{sinc}(\pi v_y \alpha_h) \qquad v_0 \geq 1/\alpha_1 \qquad (57\text{b})$$

In order to proceed with the derivation of the standard deviation in the output fluctuations from the expression for the output power density spectrum given by Eq. (56), it is necessary to make use of some results from statistical communication theory. If $f(t)$ is a random function of time, then the autocorrelation function of $f(t)$ is defined by

$$\varphi(\tau) = \lim_{T \to \infty} \int_{-T}^{T} f(t)f(t + \tau)\,dt \qquad (58)$$

In addition, according to the Wiener–Khintchine theorem, the autocorrelation function and the noise power density spectrum are Fourier transform pairs. Therefore

$$\varphi(\tau) = \int_{-\infty}^{\infty} \Phi(v)e^{2\pi j v \tau}\,dv \qquad (59)$$

and

$$\varphi(0) = \int_{-\infty}^{\infty} \Phi(v)\,dv \qquad (60)$$

But from Eq. (58) it is observed that the variance $\dot{\sigma}^2$ of $f(t)$ is given by

$$\varphi(0) = \lim_{T \to \infty} \int_{-T}^{T} f^2(t)\,dt \qquad (61)$$

Thus, by combining Eqs. (60) and (61) we have

$$\dot{\sigma}^2 = \int_{-\infty}^{\infty} \Phi(v)\,dv \qquad (62)$$

Extension of these results to three dimensions is self-evident if the variables are separable as assumed here.

Application of Eq. (62) to the output noise power density spectrum given by Eq. (56) yields

$$\dot{\sigma}_0^2 = (G_D \zeta \eta_E)^2 \Phi_I \int_{-\infty}^{\infty} T_D^2(v_x)T_E^2(v_x)R_E^2(v_x)\,dv_x$$

$$\times \int_{-\infty}^{\infty} T_D^2(v_y)T_E^2(v_y)R_E^2(v_y)\,dv_y \int_{-\infty}^{\infty} T_D^2(v_t)R_E^2(v_t)\,dv_t \qquad (63)$$

where the integral over the temporal frequency ν_t is recognized as the overall noise equivalent temporal frequency bandwidth of the electro-optical–visual system or the reciprocal of the effective integration time τ_s of the system.

The variance $\dot{\sigma}_0{}^2$ given by Eq. (63) is actually the variance in the rate or number of statistical units detected per unit area per second. However, to calculate the output signal-to-noise ratio in image detection it is necessary to derive the variance in the number of statistical units detected within the effective area of the image and integration time of the system. That is, one should calculate the variance in the mean value of $n_{BI}(x, y, t)$ within the effective area and integration time multiplied by the square of the product of the effective area and integration time. To simplify the analysis this procedure can be approximated by multiplying the variance of $n_{BI}(x, y, t)$ in the infinite interval given by Eq. (63) by the square of the product of the effective area and integration time, providing the spatial and temporal extent of a statistical unit at the retina is small compared to the dimensions of the effective area and integration time. To this approximation, we have

$$\sigma_0{}^2 = h_I{}^2 w_1{}^2 \tau_s{}^2 \dot{\sigma}_0{}^2 \tag{64}$$

If Eqs. (55), (56), (57a), (57b), (63), and (64) for σ_0 are combined with Eqs. (50a), (50b), and (51) for S_0 to obtain k_0, then, by following the definition of the required modulation on the display, we obtain

$$M_{DN} = \pi \nu_0 \alpha_w k_T / (\bar{n}_B \eta_s \tau_s \alpha_h \alpha_w)^{1/2} D_S T_E(\nu_0 \alpha_w / 2) \qquad \nu_0 \leq 1/\alpha_w \tag{65}$$

where use has been made of the observation that at small values of ν_0, $T_D(\nu)$ and $T_E(\nu)$ are approximately unity in the domain of ν where $R_E(\nu)$ is finite and therefore the integrals over ν_x and ν_y are equal to $1/\alpha_w$ and $1/\alpha_h$, respectively. Equation (65) is identical to the product of Eq. (53a) and $T_s(\nu_0)$. Thus, as expected, the use of power density spectral analysis at low test pattern frequencies is unnecessary since spatial dispersion is negligible.

However, at high spatial frequencies the above combination of equations yields

$$M_{DN} = \pi k_T \left[\int_{-\infty}^{\infty} T_D{}^2(\nu_x) T_E{}^2(\nu_x) R_E{}^2(\nu_x)\, d\nu_x \int_{-\infty}^{\infty} T_D{}^2(\nu_y) T_E{}^2(\nu_y) R_E{}^2(\nu_y)\, d\nu_y \right]^{1/2}$$
$$\times [D_S T_E(\nu_0)(\bar{n}_B \eta_s \tau_s)^{1/2}]^{-1} \qquad \nu_0 \geq 1/\alpha_w \tag{66}$$

Equation (66) does not reduce to Eq. (53b) unless ν_0 is small enough for $T_D(\nu)$ and $T_E(\nu)$ to be approximately unity where $R_E(\nu)$ is finite.

If the image-detecting system is the visual system, then at low test pattern spatial frequencies the parameters η_s, τ_s, and D_s are replaced by η_E, τ_E, and D_E, respectively, in Eq. (65). However, at high test pattern frequencies the modification of Eq. (66) required for its application to the visual system is somewhat more complex. In addition to setting $T_D(\nu)$ equal to unity and replacing η_s, τ_s, and D_s by η_E, τ_E, and D_E, respectively, it should be noted that since the photoreceptors of the retina form the primary sensor, the input power density spectrum at the retina is "white" as given by Eq. (55) if η_s is replaced by η_E. Thus in the unaided visual system no attenuation of the noise power density spectrum occurs in the optical components of the eye. However, attenuation may occur between the photoreceptor and the decision levels in the retina. Consequently for the unaided visual system it is necessary to replace $T_E(\nu)$ in Eq. (66) by a modulation transfer function $T_R(\nu)$ representing the frequency response (or equivalently, spatial and temporal dispersion) between the photoreceptor and decision levels of the retina.

If the noise-required modulation on the display of a background-limited electrooptical image-detecting system is expressed in terms of the noise-required input modulation to the unaided visual system at the display luminance, we have

$$M_{DN}(\bar{E}, \nu_0) = M(\bar{n}_{BD}\eta_E\tau_E/\bar{n}_B\eta_s\tau_s)^{1/2}(D_E/D_s)$$

$$\times \left[\int_{-\infty}^{\infty} T_D^2(\nu)T_E^2(\nu)R_E^2(\nu)\,d\nu \bigg/ \int_{-\infty}^{\infty} T_R^2(\nu)R_E^2(\nu)\,d\nu\right]^{1/2}$$

$$\times M_{IN}(\bar{E}_D, \nu_{OD}) \tag{67}$$

where the reader should note that M is the subjective magnification and the integrals are over ν_x and ν_y.

At low test pattern frequencies, since $T_D(\nu)$, $T_E(\nu)$, and $T_R(\nu)$ are approximately unity, where $R_E(\nu)$ is finite, the bracketed term reduces to unity and Eq. (67) reduces to Eq. (54) if, as assumed in Eq. (54), $\tau_s \simeq \tau_E$. As the test pattern frequency increases, the bracketed term decreases from unity and approaches a limit given by

$$\left[\int_{-\infty}^{\infty} T_D^2(\nu)T_E^2(\nu)\,d\nu \bigg/ \int_{-\infty}^{\infty} T_R^2(\nu)\,d\nu\right]^{1/2}$$

The bracketed term, describing the high-frequency attenuation of the noise power density spectrum between the primary sensor and the decision level in the visual system, results in a lesser rate of increase in $M_{DN}(\bar{E}, \nu_o)$ than predicted by Eq. (54) and therefore the resolution frequencies determined by the intersections of the graphs of $M_{DN}(\bar{E}, \nu_o)$ and $T_s(\nu_o)$ are higher than predicted in Fig. 4.6.

4.6. EDITOR'S POSTSCRIPT

The above analysis by Schnitzler leading to Fig. 4.6 is a beautifully self-consistent and rigorous analysis of the physics and physiology of vision that leads to an understanding of such things as the detection process. It clearly indicates that many existing data are questionable and that many concepts such as DQE do little to aid in solution of real problems.

This chapter can well serve as a guide for future psychophysical and psychological research in vision leading to data which will enable a user to compute rigorously with Schnitzler's equations.

Presently a lack of good data makes such calculations difficult. Thus we continue with Chapter 5, a parallel approach leading to a table and a graph which together allow one to determine the signal-to-noise ratio of a single bar of a target-equivalent bar pattern necessary to detect, recognize, or identify an object at any level of probability.

This method by Rosell is simple, direct, elegant, and easy to apply. Because it does employ an equivalent bar pattern, it answers many, if not all, of Snyder's reservations, due to "internal contrast structure," about equivalent bar pattern models.

RECENT PSYCHOPHYSICAL EXPERIMENTS AND THE DISPLAY SIGNAL-TO-NOISE RATIO CONCEPT

F. A. Rosell and R. H. Willson

5.1. EDITOR'S INTRODUCTION

The objective of this book is to give the reader the ability to specify the parameters of devices for accomplishing a given visual task under given conditions, or to estimate the probability of a viewer accomplishing that task, given the conditions and the parameters of the visual aids.

The earlier chapters were written to give the reader a background and solid acquaintance with past work and the literature documenting that work.

This chapter presents an approach based upon the common foundation presented in Chapters 3 and 4 and then continues to build a more specific foundation leading to, and giving the reader the ability to calculate, the probability of visual detection, recognition, or identification of "real world" scene objects under specified but realistic conditions on electrooptical displays.

To achieve this capability Rosell develops suitable theory from the earlier theories and experiments of Barnes and Czerny, deVries, Rose, Blackwell, Schade, Coltman, and Coltman and Anderson, as well as his own work and that of himself and Willson. Willson tests Rosell's theory first in simple situations and finds excellent predictability; Rosell then includes the criteria put forth by Johnson in 1958 and uses them as an additional, though empirical, input to establish a basis of calculation for more complex imagery. With this combined approach Willson demonstrates prediction of the signal-to-noise ratio required in a given video bandwidth to permit various visual tasks* to be conducted from displayed imagery with various levels of confidence.

This method has given quite useful and accurate results and may be used with good confidence to establish the design parameters of a low-light-level device, given the function desired and the ambient conditions, or conversely, given the design parameters and conditions, will permit good estimation of the probability with which a given function can be accomplished under the specified conditions.

No such computational procedure is the ultimate, and limitations of the present method are pointed out toward the end of the chapter. However, until a more direct form of analysis is developed the present form of computation appears to offer the soundest means for estimating required parameters, given required performance conditions, or conversely, for estimating performance, given parameters.

Basically, Rosell considers a display under nearly ideal conditions but does not preclude computation for the nonideal, i.e., these calculations do not consider factors such as the glare on a display or the effect of a raster in degrading the information transferred to an observer. Ambient illumination effects can be applied to these results by applying the concepts of Chapter 4 or by introducing the glare factor into the modulation in the equations developed below, and similarly the effect of the raster on signal-to-noise ratio developed in Chapters 6 and 7 must be considered with care, especially for reasonably coarse rasters.

Finally, the method of this chapter can be applied empirically with reasonable safety after little study. It is recommended, however, that one start at or near the beginning, building upon the work of Barnes and Czerny (1932) and those who followed, in order to obtain a better understanding of why the system is valid.

5.2. INTRODUCTION

Barnes and Czerny (1932) suggested that the photon detection process of any phototransducer, including the eye's retina, could be subject to statistical fluctuation. This notion was mathematically formu-

---

* A related approach was described by Schnitzler (1971a).

lated by de Vries (1943), who postulated that an image, to be detectable, must have a signal-to-noise ratio (SNR) exceeding some threshold value. The de Vries hypothesis was verified, or at least made reasonable, by Rose (1948a, b) using noisy photographic imagery. Rose also showed that de Vries's simple signal-to-noise ratio concept could be fitted to the Blackwell (1946) disk detection data for the unaided eye over a portion of the range of disk diameters and illuminances used. More recently, Schade (1967), Rosell (1969), and others have applied similar concepts to the detection of images generated by auxiliary sensors such as television cameras.

To increase the generality and applicability of the signal-to-noise ratio concept, Rosell and Willson conducted a large number of controlled psychophysical experiments using noisy televised images. In this chapter we describe a number of these experiments which apply to the detection, recognition, and identification of scene objects along with their use in predicting sensory system performance.

5.3. THE ELEMENTARY MODEL

To begin, we first show the schematic of an electrooptical imaging process in Fig. 5.1. In this figure a rectangular image of area a amid a uniform background has been projected onto a phototransducer by a lens. The phototransducer converts the photon image incident on it to a photoelectron image. The incident photon image can be considered to be noise-free since the existence of a noisy, "coherent" electron emission from a photosurface has yet to be demonstrated. This assumption is made to explain why photon noise is not observed in the phototransduced current (Schade, 1967). The lens forming the image is char-

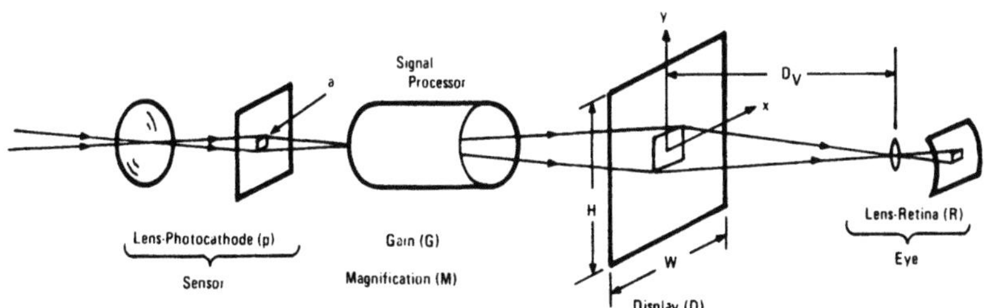

Fig. 5.1. Schematic of the electrooptical imaging process.

acterized by an aperture, or spatial frequency, response whose effect on image signals is analogous to the effect of a filter acting on electrical signals. Isolated images such as squares or rectangles are simply enlarged by the lens, while for more complex images such as periodic patterns the signal amplitudes are reduced. The lens aperture could also filter noise, but as we have noted, the images projected onto the photosurface by the lens are noise-free. From the above discussions it might be inferred that the main degrading effect of the lens is to reduce signal in certain cases, leaving noise unaltered, but it will be seen that the lens can cause an increase in the noise perceived by the observer due to noise subsequently generated within the phototransducer or in the image signal processor. While the lens does not directly generate noise, the photon-to-electron conversion process is noisy and therefore an SNR is established at the output of the phototransducer which will inherently limit the detectability of scene objects.

Continuing the description, the phototransduced image is passed to the signal processor, whose main purpose is to amplify and magnify the image signals and noises alike. If the signals and noises are not further degraded or filtered by the spatial frequency response of the processor or the display and if no further noise is added, then the signal-to-noise ratio of the image appearing on the display will be identical to that on the input photosurface. Furthermore, if the gains and magnifications of the signal processor and display combination are sufficient so that the observer's eye is neither light-level- nor image-size-limited, then the image's signal-to-noise ratio at the output of the eye's retina will be identical to that on the display if due account is taken of the eye–brain combination's ability to integrate in space and time. These conditions can be achieved in practice for a range of image sizes, signal amplifications, apertures, display luminances, image magnifications, and observer-to-display viewing distances. In other cases noise added within the signal processor or even in the observer's retinal photoprocessor may be a factor.

In general, the signal-to-noise ratio which limits the image's detectability will be that generated by a combination of the sensor's primary photoprocess, the signal processor, the display, and the observer. We will begin by considering the sensor's primary photoprocess supposing that the image on it is unaffected by the lens's spatial response. Suppose that the average photosurface irradiance in the image area is $\bar{E}_0$

and that the average irradiance on an equivalent area containing only background is $\bar{E}_b$. The image modulation contrast is specifically defined by

$$M = (\bar{E}_0 - \bar{E}_b)/(\bar{E}_0 + \bar{E}_b) = (\bar{E}_0 - \bar{E}_b)/2\bar{E}_{\mathrm{av}} \qquad (1)$$

where $\bar{E}_{\mathrm{av}} = (\bar{E}_0 + \bar{E}_b)/2$. If the phototransducer is linear, then the image irradiances $\bar{E}_0$ and $\bar{E}_b$ will result in the photoelectron rates $\mathring{n}_0$ and $\mathring{n}_b$, respectively, and the incremental photoelectron signal in the image area a in an integration time t will be equal to

$$\Delta\mathring{n} = (\mathring{n}_0 - \mathring{n}_b)at = 2M\mathring{n}_{\mathrm{av}}at \qquad (2)$$

by use of Eq. (1). The noises associated with the inherent fluctuations in the photoprocess are assumed to follow the Poisson probability distribution law, which states that the fluctuations have a standard deviation, or rms noise, equal to the square root of the number of photoconverted electrons in the image area and integration time. For the case of an object imaged against background the total mean-square noise is assumed to be the average of the object and background photoelectrons summed in quadrature. Thus the mean image signal-to-rms noise ratio, or SNR_I, at the output of the phototransducer becomes

$$\mathrm{SNR}_I = 2M\mathring{n}_{\mathrm{av}}at/[(\mathring{n}_0 + \mathring{n}_b)at/2]^{1/2} = 2M(\mathring{n}_{\mathrm{av}}at)^{1/2} \qquad (3)$$

We observe next that the sensor may be limited not only by photoconversion noise but by noises generated internally within the signal processor or display. If these added noises, referenced to the output of the photosurface, result in a mean-square photoelectron rate of $\mathring{n}_s$, then Eq. (3) becomes

$$\mathrm{SNR}_I = (at)^{1/2}(2M\mathring{n}_{\mathrm{av}})/(\mathring{n}_{\mathrm{av}} + \mathring{n}_s)^{1/2} \qquad (4)$$

Next we write the photoelectron rates in terms of the photocurrent i in amperes, or

$$\mathring{n} = i/eA \qquad (5)$$

where e is the charge of an electron in coulombs and A is the total effective photosurface area in square meters. Combining Eqs. (4) and (5), after noting that $i_{\mathrm{av}} = \mathring{n}_{\mathrm{av}}eA$ and $i_s = \mathring{n}_s eA$, we have

$$\mathrm{SNR}_I = [t(a/A)]^{1/2}(2Mi_{\mathrm{av}})/(ei_{\mathrm{av}} + ei_s)^{1/2} \qquad (6)$$

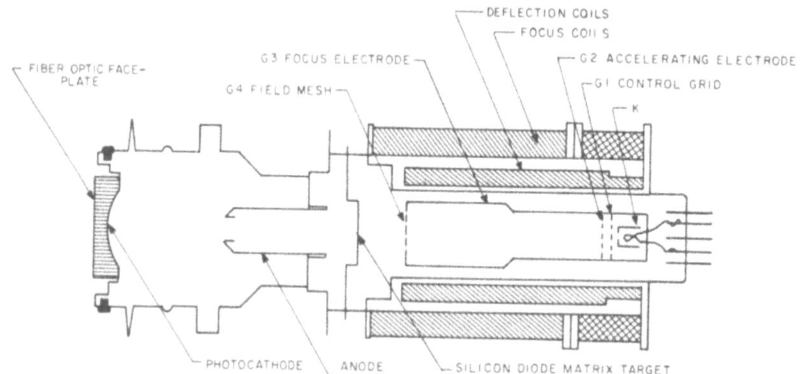

Fig. 5.2. Cross section of the SEBIR camera tube.

Now the numerator and denominator of Eq. (6) are multiplied by $(2\Delta f_V)^{1/2}$ where Δf_V is the video bandwidth in hertz, so that

$$\mathrm{SNR}_I = [2t\,\Delta f_V(a/A)]^{1/2}(2Mi_{\mathrm{av}})/(2ei_{\mathrm{av}}\,\Delta f_V + 2ei_s\,\Delta f_V)^{1/2} \qquad (7)$$

A high-performance television camera tube is shown in Fig. 5.2. The principles of operation of this tube, known as the silicon-electron-bombardment induced response (SEBIR) camera tube, have been described in detail by Rosell *et al.* (1972). A photon image incident on the photocathode is converted to a photoelectron image and greatly accele-

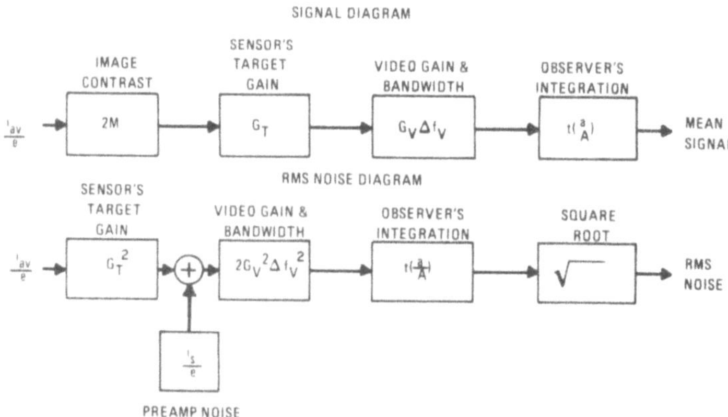

Fig. 5.3. Signal and noise diagram for a high-performance camera tube limited by photoelectron and preamp noise but unlimited by sensor MTF's.

rated to a silicon-diode-matrix target. This target first amplifies the image by a factor G_T and then stores the image for subsequent readout by a scanning electron beam. Assume for the moment that the signal amplification is noise-free and that the image details are not significantly degraded by the sensor's modulation transfer function. After readout the image is passed to a preamplifier which adds noise at its input. This noise may be of a magnitude to substantially reduce the image SNR if the gain G_T is insufficient. The preamplifier is followed by a video amplifier and a display. The signals and noises generated are further amplified in these further elements by a factor G_V. A block diagram of the overall signal and noise processes is shown in Fig. 5.3.

From these diagrams we can write the image signal-to-noise ratio as*

$$\text{SNR}_{DI} = \left[2t\,\Delta f_V\left(\frac{a}{A}\right)\right]^{1/2} \frac{2MG_TG_Vi_{\text{av}}}{(2ei_{\text{av}}G_T^2G_V^2\,\Delta f_V + 2eG_V^2i_s\,\Delta f_V)^{1/2}} \tag{8}$$

Note that we have designated the above equation as SNR_{DI} to indicate that the image SNR is referenced to the display rather than the input photosurface. At this point the display is presumed to be perfect, i.e., noiseless and of unity modulation transfer. Also, since the display brightness and video gain G_V can be arbitrarily large, the observer need not be acuity limited by the display luminance. Observe, in this regard, that the displayed image contrast can exceed the input image contrast by means of a large video gain provided that the display's dynamic luminance range is not exceeded.

Those familiar with television SNR analysis can recognize the quantity on the right side of the equation

$$\text{SNR}_V = 2MG_TG_Vi_{\text{av}}/(2ei_{\text{av}}G_T^2G_V^2\,\Delta f_V + 2eG_V^2i_s\,\Delta f_V)^{1/2} \tag{9}$$

as the video SNR. Thus Eq. (8) may be written as

$$\text{SNR}_{DI} = [2t\,\Delta f_V(a/A)]^{1/2} \cdot \text{SNR}_V \tag{10}$$

This relationship between the image signal-to-noise ratio on a hypothetically perfect display and the video SNR was first suggested by Colt-

---

* In this derivation it is assumed that the spectral density of the preamp noise is uniform with frequency, i.e., "white."

man and Anderson (1960) and provides a very convenient method of generating noisy test imagery for use in the psychophysical experiments discussed below. One caution is in order; the above equation holds only for images unlimited by the sensor's spatial frequency response.

While simple rectangular images and uniform backgrounds are of interest, it is more usual to test sensory systems using periodic test patterns. These test patterns take various forms, from sine wave patterns, to bar pattern wedges, to bursts of bar patterns. Whatever their form, the notion is to project patterns of various spatial frequencies onto the sensor. Then an observer is asked to determine the pattern of highest spatial frequency which can be just barely detected as the pattern's SNR is varied by increasing or decreasing its irradiance. The resolution so measured is called the sensor's "limiting" or "threshold" resolution and is plotted versus the pattern's highlight irradiance. This threshold resolution versus photocathode irradiance characteristic is now used by nearly all major sensor manufacturers to specify and compare the performance of their products with others. While the characteristic curves have become accepted as a means for comparison, they are not necessarily reliable since test procedures are not standardized, different types of test patterns are used by various manufacturers, and the measurements, which are subjective and statistically variable, are usually made by a single experienced observer in a very limited number of trials.

Rosell (1971) found that, given certain data of the type ordinarily supplied by sensor manufacturers, the threshold resolution versus absolute irradiance level can be calculated with good accuracy and has the advantage of avoiding the problem of nonstandard test patterns and test procedures. Furthermore, the calculation technique provides more data at almost no cost relative to the very expensive and time-consuming process of experimentally evaluating a number of sensors or combinations of sensors such as intensifiers cascaded with TV camera tubes. The method of calculation will be described later in this section for a particular case.

Although the equations above are derived for an isolated rectangular image, it is hypothesized that they also apply to the detection of bar and sine wave patterns on the premise that to detect the presence of a bar pattern the observer must detect the presence of a single bar. However, the threshold signal-to-noise ratio required to detect a bar in the presence of a number of bars may differ from that needed to detect an isolated bar on a uniform background but will not differ significantly as we will see. Suppose that a single bar in a pattern is of size $\Delta y/Y$ wide by $\varepsilon\,\Delta y/Y$ long, where ε is the bar length-to-width ratio and Y is the picture height. Note that we use dimensionless size units because of its convenience when a number of image size changes occur within a system. Also, for reasons which will become apparent as we progress, we will prefer to use the reciprocal dimension

$$N = Y/\Delta y \tag{11}$$

Then the bar image area relative to the total effective photocathode area is equal to

$$a/A = \varepsilon(\Delta y^2)/\alpha Y^2 = \varepsilon/\alpha N^2 \tag{12}$$

where α is the displayed horizontal-to-vertical picture aspect ratio and N is designated the pattern's spatial frequency in "lines per picture height." With this result Eq. (10) becomes

$$\mathrm{SNR}_{DI} = (2t\,\Delta f_V/\alpha)^{1/2}(\varepsilon^{1/2}/N) \cdot \mathrm{SNR}_V \tag{13}$$

5.4. EFFECTS OF FINITE APERTURES

In the perfect imaging sensor considered heretofore a point image on the photocathode is assumed to appear as a point image on the display. If this is so, all images are transmitted through the sensor with perfect fidelity. In real sensors the images may be distorted in amplitude, shape, or phase (position) or all three. These distortions are due to finite imaging apertures such as that of the objective lens, fiber-optic faceplates, electron lenses, electron scanning beam, finite phosphor particles, electrical bandwidth limitations, and the like. The effect of these apertures is to smear image detail in a manner analogous to the filtering of electrical signals by electrical filter networks, except that the optical apertures may be two-dimensional.

If the input image to a real lens or sensor is a point, the output image will be a blur denoted as the point spread function $r_0(x, y)$. The Fourier transform of $r_0(x, y)$ is $R_0(N_x, N_y)$ and is the complex steady-state spatial frequency response. $R_0(N_x, N_y)$ has also been designated as the optical transfer function by the International Optics Commission. In this description N_x and N_y are spatial frequencies as defined by Eq.

(11). If the imaging apertures are linear and if either $r_0(x, y)$ or $R_0(N_x, N_y)$ is known, then the aperture's response to any test input image can be determined, i.e., these functions completely specify the aperture. If the input image is a line instead of a point, the output image is designated the line spread function, $r_0(x)$ or $r_0(y)$. The Fourier transform of either $r_0(x)$ or $r_0(y)$ is $R_0(N_x)$ or $R_0(N_y)$. In general,

$$R_0(N) = | R_0(N) | \exp[j\varphi(N)] \tag{14}$$

where $| R_0(N) |$ is the modulation transfer function (MTF) when $| R_0(N) |$ is normalized to unity at zero frequency, and $\varphi(N)$ is designated as the phase transfer function.

In the special case where the variables x and y are independent and separable we can write

$$r_0(x, y) = r_0(x)r_0(y) \tag{15}$$

and

$$R_0(N_x N_y) = R_0(N_x)R_0(N_y) \tag{16}$$

While this situation is usually not the case, it is a reasonable and useful approximation in many instances.

Suppose the input image to an aperture such as a lens is a one-dimensional rectangular pulse as shown by the broken lines in Fig. 5.4.

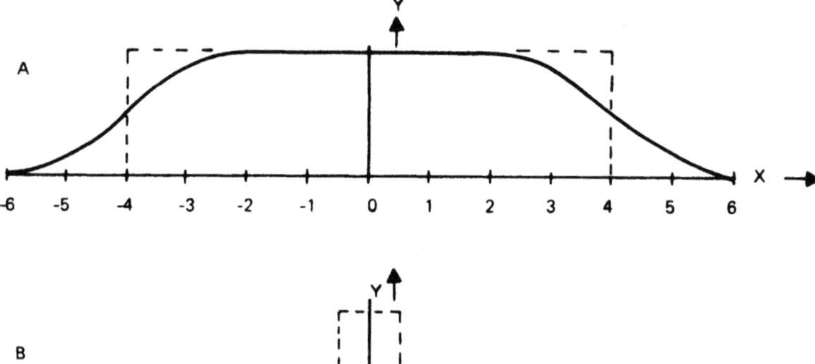

Fig. 5.4. Output pulse (——) after filtering by an MTF as the width of a unit rectangular input pulse (- - -) varies.

With a wide input pulse as in Fig. 5.4A the effect of the aperture is mainly to round its corners, as shown by the solid output pulse curve. With a narrow input pulse as in Fig. 5.4B the aperture both rounds the corners and reduces the pulse amplitude. In both cases the output pulse is spatially enlarged, although the increase is relatively larger for the narrower input pulse. On the other hand, if the aperture is dissipationless, as is usually the case with most optical apertures,* the areas under the output and input pulse curves are identical. Next we will observe, and show later through psychophysical experimentation, that the eye, when confronted with a larger image, simply expands its limits of spatial integration to include all of the image signal (except as noted in the experiments). In this event the aperture has no effect on signal. The tentative conclusion might be that the apertures have no effect on signal-to-noise ratio, but this is not to be.

To account for the effect of finite apertures on simple aperiodic images, Schade (1967) proposes the following. Suppose the image area for a finite aperture is a_L and that the mean signal amplitude is Δi_L. Then SNR_{DI} for a photoelectron-noise-limited sensor will be

$$\mathrm{SNR}_{DI} = (t/A)^{1/2}(\Delta i_L)a_L/(a_L e i_{\mathrm{av}})^{1/2} \qquad (17)$$

If the finite lens aperture is dissipationless, then

$$(\Delta i_0)a_0 = (\Delta i_L)a_L \qquad (18)$$

where Δi_0 and a_0 are the mean signals for an infinite aperture and Δi_L and a_L are the corresponding quantities for a finite aperture. Using Eq. (18) in (17), we find

$$\mathrm{SNR}_{DI} = (t/A)^{1/2}(\Delta i_0)a_0/(a_L e i_{\mathrm{av}})^{1/2} \qquad (19)$$

That is, the signal remains unchanged but the noise is increased because $a_L > a_0$. Observe that the lens did not increase noise; it merely increased the image area. However, the photoconversion process is noisy and the enlarged image area includes more of the noise. The process is illustrated in Fig. 5.5, where an input object of width x_0 becomes an effective width

---

* Except for light collecting efficiencies, transmission losses, and other arbitrary gains which are spatially independent to a first approximation.

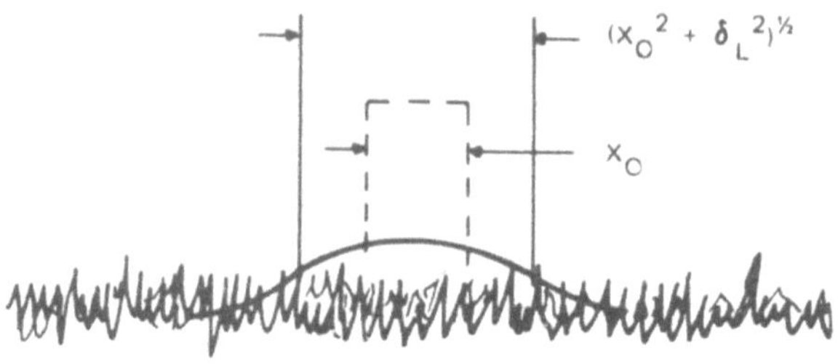

Fig. 5.5 Increase in noise perceived by an observer due to an increase in effective image size due to an MTF preceding a point of noise insertion.

of approximately $(x_0{}^2 + \delta_L{}^2)^{1/2}$ after passing through the lens (the quantity δ_L will be defined below).

To determine the increase in image dimensions, suppose the test object irradiance is described by a function $kf(x, y)$, where k is the amplitude of the function at $x, y = 0, 0$. Let the area before passing through the aperture be $a_0 = \int f(x, y)\, dx\, dy$. Suppose the irradiance function after the lens is $g(x, y)$, which is normalized to its peak amplitude; then the area a_L will be given by

$$a_L = a_0{}^2 \Big/ \int\!\!\int_{-\infty}^{\infty} g^2(x, y)\, dx\, dy \tag{20}$$

The area a_L is designated the noise-equivalent sampling area. Also, note that by the Fourier energy theorem,

$$a_L = a_0{}^2 \Big/ \int\!\!\int_{-\infty}^{\infty} G(N_x, N_y)^2\, dx\, dy \tag{21}$$

where $G(N_x, N_y)$ is the Fourier transform of $g(x, y)$. When the test object is rectangular of dimension $x_0 \cdot y_0$ and of uniform amplitude, Schade gives the following useful approximation:

$$\begin{aligned} a_L &\approx x_0 y_0 [1 + (\delta_L{}^2/x_0{}^2)]^{1/2} [1 + (\delta_L{}^2/y_0{}^2)]^{1/2} \\ &\approx x_0 y_0 (\xi_{x_L} \xi_{y_L}) \end{aligned} \tag{22}$$

where δ_L is the noise-equivalent impulse diameter of the aperture (assuming the aperture's point spread functions are equal, independent, and separable in x and y). Numerically, we have

$$\delta_L = 1/N_{eL} = 1 \bigg/ \int_0^\infty |\, R_{0L}(N)\,|^2 \, dN \tag{23}$$

where R_{0L} is the aperture MTF. The quantity N_e has been designated the noise-equivalent passband by Schade. With these equations SNR_{DI} may be written as

$$\text{SNR}_{DI} = \left(\frac{t}{eA} \, \frac{x_0 y_0}{\xi_{x_L} \xi_{y_L}} \right)^{1/2} 2 M i_{\text{av}}^{1/2} \tag{24}$$

It should be observed that an MTF prior to the point of a noise generation or insertion is more serious than an MTF following such noise insertion since in the latter case the noise is reduced from a white to a finite spectrum.

To account for a finite noise spectrum, we define a function $\Gamma_x \Gamma_y$ which is the factor by which an SNR_{DI} computed for a white noise spectrum must be increased to correct for a finite noise spectrum. To illustrate the processes involved, suppose the image is passed through the lens as above and then passed through a second aperture T of MTF $R_{0T}(N)$ prior to photoconversion. Then SNR_{DI} would be

$$\text{SNR}_{DI} = \left(\frac{t}{eA} \, \frac{x_0 y_0}{\xi_{x_{LT}} \xi_{y_{LT}}} \right)^{1/2} 2 M i_{\text{av}}^{1/2} \tag{25}$$

for

$$\xi_{x_{LT}} \xi_{y_{LT}} = \left[1 + \left(\frac{\delta_L}{x_0} \right)^2 + \left(\frac{\delta_T}{x_0} \right)^2 \right]^{1/2} \left[1 + \left(\frac{\delta_L}{y_0} \right)^2 + \left(\frac{\delta_T}{y_0} \right)^2 \right]^{1/2} \tag{26}$$

where δ_T is the equivalent impulse diameter for the second aperture. If the second MTF followed the phototransducer rather than preceding it, then we would have to correct for the fact that the photoelectron noise is filtered. The correction factors $\Gamma_x \Gamma_y$ are

$$\Gamma_x \Gamma_y = \frac{\xi_{x_{LT}} \xi_{y_{LT}}}{[1 + (\delta_L/x_0)^2 + 2(\delta_T/x_0)^2]^{1/2}[1 + (\delta_L/y_0)^2 + 2(\delta_T/y_0)^2]^{1/2}} \tag{27}$$

Multiplying the right side of Eq. (26) by Eq. (27), we obtain

$$\text{SNR}_{DI} = \left(\frac{t}{eA} \, \frac{x_0 y_0}{\Gamma_x \xi_{x_{LT}} \Gamma_y \xi_{y_{LT}}} \right)^{1/2} 2 M i_{\text{av}}^{1/2} \tag{28}$$

This result can be extended to any number of apertures in cascade. Equation (28) is rather specific in that it applies only to a sensor in which the primary noise is that generated by the input photosurface. For the case where a preamp noise current I_s is added subsequent to the second aperture, we have

$$\mathrm{SNR}_{DI} = \left(\frac{t}{eA}\ \frac{x_0 y_0}{\xi_{x_{LT}} \xi_{y_{LT}}}\right)^{1/2} \frac{2G_T M i_{\mathrm{av}}}{(G_T^2 \Gamma_x \Gamma_y i_{\mathrm{av}} + I_s)^{1/2}} \qquad (29)$$

where G_T is the signal gain between the photosurface and the preamp input. Note that the preamp noise current I_s is referenced to the preamplifier's input and not to the output of the photosurface as in Eq. (7).

We turn next to the effect of apertures on periodic test patterns which are aperiodic in the direction along the individual bars and periodic across the bars. The effect of the apertures along the bars may be taken into account by the methods already described although, for bars of large length-to-width ratio, the effect along the bars can and often will be neglected since it is small relative to the effect in the periodic direction. In the aperiodic case we observed that the apertures increased the distance over which the eye integrates so that the eye includes more noise and leaves the signal unchanged. In the periodic direction the integration width remains unchanged by the apertures while the mean signal amplitude decreases as shown in Fig. 5.6. Schade (1967) gives the new mean amplitude in terms of the square wave flux response defined by

$$R_{\mathrm{SF}}(N) = (8/\pi^2) \sum_{k=1}^{\infty} (1/k^2)\,|\,R_0(kN)\,|, \qquad k = 1, 3, 5, \ldots \qquad (30)$$

where $|\,R_0(kN)\,|$ represents the values of the sensor MTF at frequencies kN, where k are the odd harmonics of a square wave of frequency N. This new measure is made necessary, according to Schade, because the square wave amplitude response bears no fixed relationship to the aver-

Fig. 5.6. Actual image amplitude A_I and wave shape (——) compared to equivalent square wave flux amplitude A_F and wave shape (- - -).

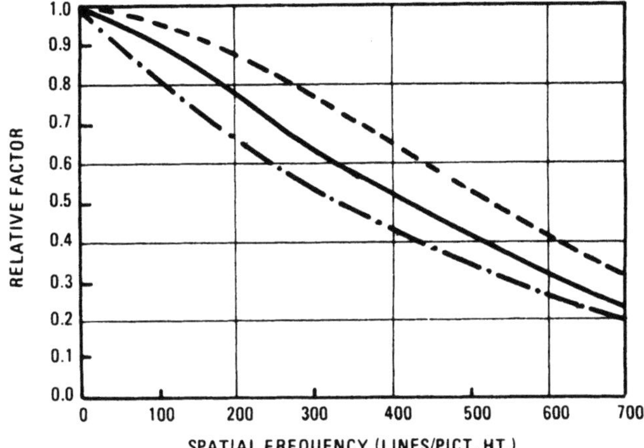

Fig. 5.7. Square wave amplitude (- - -), modulation transfer (———), and square wave flux (- · -) functions for the $1\frac{1}{2}$-in. vidicon used in the bar pattern recognition and identification experiments.

age value of flux as in the case with sine waves but instead depends on the harmonic components of the waveform. The relative magnitudes of the square wave flux and amplitude and the modulation transfer are shown in Fig. 5.7.

In the analysis which follows we will consider the case of a bar pattern which is aperiodic in the y direction and periodic in the x direction. The sensor apertures consist of a lens with response $|\,R_{0L}(N)\,|$ and a target with response $|\,R_{0T}(N)\,|$ as before. For convenience, we make the image-area change of variable as given by Eq. (12). The signal diagram is shown in Fig. 5.8, where the signal due to the bar length is seen

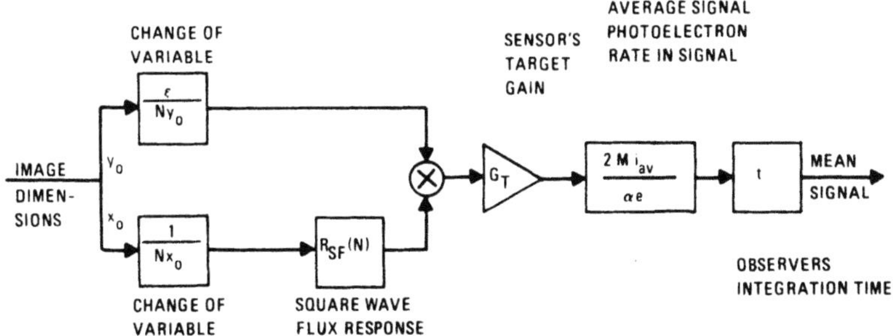

Fig. 5.8. Mean signal diagram.

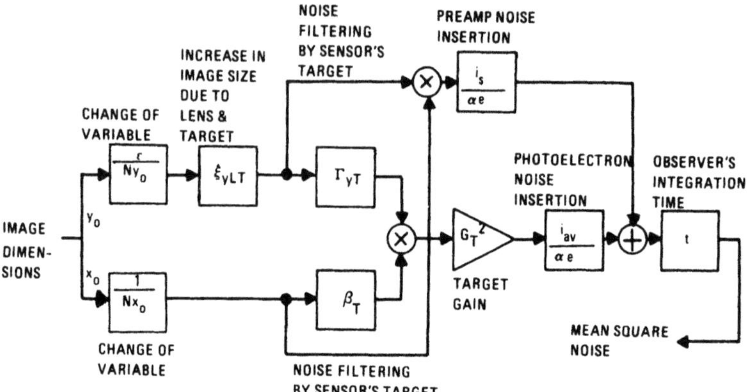

Fig. 5.9. Mean-square noise diagram.

to pass through the sensor to the observer unchanged while the signal due to the bar in the periodic direction is multiplied by the square wave flux response $R_{\mathrm{SF}}(N)$ of the combined lens and target of the sensor. Thus the signal becomes

$$S_{DI} = \frac{t\varepsilon}{\alpha} \; \frac{R_{\mathrm{SF}}(N)}{N^2} \; \frac{2MG_T i_{\mathrm{av}}}{e} \tag{31}$$

The mean-square noise diagram is shown in Fig. 5.9. Here it is seen that the photoelectron noise in the periodic direction is reduced by the filtering effect β_T of the sensor target, where

$$\beta_T = \left[\int_0^N | R_{0T}(N) |^2 \, dN \right] \Big/ N \tag{32}$$

In the aperiodic direction the image is enlarged by the lens and target by an amount*

$$\xi_{y_{LT}} = [1 + (N/\varepsilon N_{eL})^2 + (N/\varepsilon N_{eT})^2]^{1/2} \tag{33}$$

where N_e is defined as in Eq. (23). The photoelectron noise is increased by the lens but filtered by the target by a factor Γ_{yT} as given by Eq. (28). The mean-square preamp noise is increased by the factor $\xi_{y_{LT}}$ but is not filtered. The result is that the rms noise expression becomes

$$N_{DI} = [(t\varepsilon \xi_{y_{LT}}/\alpha N^2 e)(G_T^2 \Gamma_{yT} \beta_T i_{\mathrm{av}} + I_s)]^{1/2} \tag{34}$$

---

* Equation (33) is equivalent to $(\xi_{y_L})^{1/2}$ of Eq. (26).

and SNR_{DI} becomes

$$\mathrm{SNR}_{DI} = \left(\frac{t\varepsilon}{\alpha e \xi_{y_{LT}}}\right)^{1/2} \frac{R_{\mathrm{SF}}(N)}{N} \frac{2MG_T i_{\mathrm{av}}}{(G_T^2 \beta_T \Gamma_{yT} i_{\mathrm{av}} + I_s)^{1/2}} \qquad (35)$$

It should be observed that under certain conditions aperture correction or compensation may be of benefit. The loss in signal amplitude due to the lens cannot be compensated since it occurred prior to noise insertion, but the losses due to the apertures, such as that of the gain-storage target which follows, can be partially compensated through appropriate aperture-correcting networks. The extent of the correction which can be made depends upon the phase shift generated by the correcting network and upon the magnitude of any noises inserted between the aperture and the correcting network. Thus aperture correction will be much more effective in the photoelectron-noise-limited case of Eq. (28) than in the case of Eq. (29), where preamp noise is a factor. In particular, the preamp noise generally is an increasing function of frequency such that aperture correction may increase noise at a faster rate than the signal improves.

5.5. LEVELS OF DISCRIMINATION

The purpose of much electrooptical equipment is to augment an observer's capability so that he can discern a scene object at longer range than he could with his unaided eye. These objects may be seen with greater or lesser clarity depending on the object size, distance, contrast, and radiance and the sensor resolving power. In this chapter we are concerned with a condition where a scene object needs to be discerned with sufficient clarity to serve some intended function and where the observer is highly motivated to do so. The words "sufficient clarity" should be stressed. In one case it may be sufficient to merely detect a blob such as a channel buoy, while in other cases much higher acuity is needed. For example, it is of no use to televise and record a burglary if the recording acuity is insufficient to identify the burglar in a court of law.

The acuity with which an object is seen depends upon range. When we speak of range, we generally mean the maximum, or threshold, range at which the object can be barely discerned with the needed acuity; the

TABLE 5.1

Levels of Object Discrimination

Classification of discrimination level	Meaning
Detection	An object is present
Orientation	The object is approximately symmetric or asymmetric and its orientation may be discerned
Recognition	The class to which the object belongs may be discerned (e.g., house, truck, man, etc.)
Identification	The target can be described to the limit of the observer's knowledge (e.g., motel, pickup truck, policeman, etc.)

needed acuity depends upon the level of discrimination desired, whether mere detection of an object, its recognition, or its positive identification. The process is, typically, as follows. At very long range a scene object appears first as a blob. Moving ever closer, the object begins to take some shape such as a rectangle. Closer yet, the observer becomes able to classify the object as to type and finally to identify the object positively. Johnson (1958) has arbitrarily divided these levels of object discrimination into four categories as given in Table 5.1.

It is readily evident that a higher degree of visual acuity is needed to identify an object as opposed to just detecting it. To obtain a quantitative feel for the problem, Johnson performed a series of experiments using electrooptical sensors. In these experiments an attempt was made to correlate the detectability of a bar pattern of a given spatial frequency with the level of object discrimination. The procedure was to increase the object range until it was just barely detected (or recognized, etc.). Then a bar pattern was placed in the field of view and its spatial frequency was increased until it could barely be resolved at the same range. The spatial frequency of the pattern was specified in terms of the number of lines in the pattern subtended by the object's minimum dimension as illustrated in Fig. 5.10, where the object, for the recognition case, subtends eight lines.

Johnson's results, as tabulated in Table 5.2, are not unexpected. If the observer could only just resolve a coarse pattern corresponding to

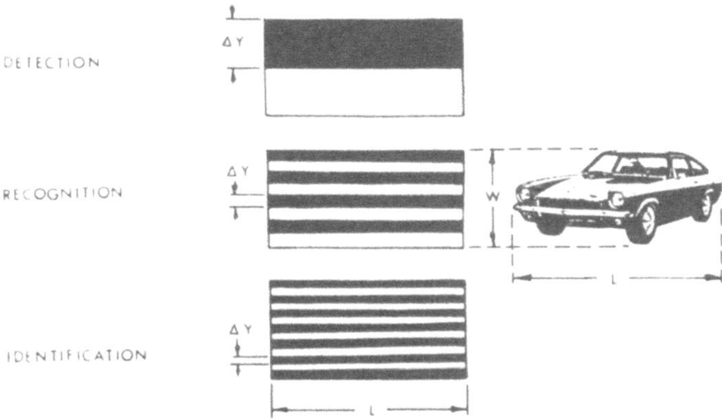

Fig. 5.10. Resolution required per minimum object dimension to achieve a given level of object discrimination expressed in terms of an equivalent bar pattern.

two bars per minimum object dimension, the level of object discrimination was limited to detection. With higher acuity a bar pattern of higher spatial frequency could be discerned and the level of object discrimination increased in turn.

Table 5.2 has been widely used and misused by systems designers from the time of its publication to the present. The misuse stems from the neglect of additional requirements imposed by Johnson, to wit, that

TABLE 5.2

Johnson's Criteria for the Resolution Required per Minimum Object Dimension versus Discrimination Level

Discrimination level	Resolution per minimum object dimension, TV lines
Detection	$2^{+1.0}_{-0.5}$
Orientation	$2.8^{+0.8}_{-0.4}$
Recognition	$8.0^{+1.6}_{-0.4}$
Identification	$12.8^{+3.2}_{-2.8}$

the "signal-to-noise" ratio and image contrast must also be sufficient. However, it was not too clear how these quantities were to be measured and calculated and thus the further requirements were neglected in many cases. However, many competent designers did use the sensor threshold resolution versus scene irradiance curves in estimating the level of discrimination. Since the threshold curves do contain image signal-to-noise ratio as a factor in their measurement, estimates made on this basis turn out to be reasonable if not precise.

Most sensors are characterized by an absolute limiting resolution. If the sensor sensitivity at a given scene irradiance level and object contrast is sufficient to realize the limiting resolution but is not sufficient to allow the performance of the desired discrimination task, further increases in scene irradiance level will be to no avail. The only solution is to move closer. On the other hand, if the resolution were sufficient at a given irradiance level, a decrease in scene irradiance level could cause the acuity of a sensor/observer combination to fall below the level required for the wanted level of object discrimination.

In the above we have implied that image signal-to-noise ratio, image contrast, and sensor/observer resolution are independent quantities, whereas in our view these quantities are functionally related, i.e., the image signal-to-noise ratio is proportional to the image size, contrast, irradiance level, sensor sensitivity, etc. In this case Johnson's requirements reduce to one; namely that an object should be discriminated at the desired level if its signal-to-noise ratio at the output of the observer's retina after processing and interpretation by the brain is large enough. Obviously, the signal-to-noise ratio as defined in this manner is not directly measurable, but as will be seen, can be indirectly measured through psychophysical experimentation.

The quantitative models developed in this chapter are based on simple test images such as rectangles or bar patterns for which an image size or "resolution" can be precisely defined. Through psychophysical experiments the threshold signal-to-noise ratios, as calculated on the basis of image geometry, measured electrical quantities, and estimated psychophysical parameters are determined. While the test images are of simple geometry, it is hypothesized that these images and the requirements for their discrimination can be correlated with the discrimination of more complex imagery as encountered in a real world scene. Such correlation does appear to exist, as will be discussed.

Detection is the lowest level of object discrimination since it usually implies only that an object of undeterminable shape has been sighted in the field of view. Recognition usually requires shape information but in some cases shape need not be known if other clues are available. For example, a series of regularly spaced and moving blobs on a road may be interpreted as vehicular traffic. On the other hand, a single stationary blob on a road may be the shadow of a tree, a puddle, a truck, or any number of other objects. While a blob on a road has a reasonable probability of being a vehicle, the same blob in a field or in a sparse forest can be almost anything. Thus there are obviously many degrees of discrimination even within a discrimination level. A single criterion such as that based on resolution and signal-to-noise ratio is unlikely to be sufficient to cover every case. Rather, a number of cases must be considered and subclasses formed.

However, Johnson's hypothesis and definitions are reasonable and have some basis in experimental fact. Thus we elected to adopt his premises and to further test their validity. For this purpose we assume the image to be detected or otherwise discriminated to be of size $L \times W$ as shown in Fig. 5.10 on a focal plane of dimensions $X \cdot Y$. To correlate the detectability of a bar pattern with a level of object discrimination, we divide the minimum dimension W of the image by a factor k_d which is numerically equal to the lines per minimum object dimension given in Table 5.2, e.g., for recognition $k_d = 8$. To add these notions to the SNR_D equation, we let $\Delta y = W/k_d$ and by Eq. (12),

$$\frac{a}{A} = \frac{LW}{XY} = \frac{L\,\Delta y}{k_d \alpha Y^2} = \frac{\varepsilon}{k_d \alpha N^2} \tag{36}$$

which can then be used in the various SNR_{DI} equations (7)–(9) or (13). *When the above area ratio is applied together with Johnson's criteria it is assumed that the object is in a comparatively cluttered scene.* For the specific case of detection of an object amid a uniform background we will assume $k_d = 1$ and we will also treat the object analytically as aperiodic rather than as periodic. Recall that SNR_{DI} for an aperiodic object has the form of Eq. (29) while SNR_{DI} for a periodic object has the form of Eq. (35). In general, an object amid a uniform background will be much easier to detect than an object in a cluttered background, as one would expect.

5.6. PSYCHOPHYSICAL EXPERIMENTATION— APERIODIC AND PERIODIC IMAGES

In the preceding we have associated a signal-to-noise ratio with an image as generated by an electrooptical sensor. This signal-to-noise ratio is a function of image size or a quantity inversely related to it, the image resolution, as we observed in deriving Eq. (12). In the previous section it was noted that the higher the level of object discrimination desired, the higher the resolution required. Now suppose that the signal-to-noise ratio obtainable from a sensor for a given image size or resolution is known either through measurement or calculation. Then if we knew the observer's signal-to-noise ratio needs, we could predict the overall sensory system performance by comparing the image signal-to-noise ratio provided by the sensor to that needed by the observer for the level of object discrimination wanted.

In this section the observer's needs will be investigated through psychophysical experimentation. The test images used will be simple geometric shapes such as squares and rectangles and periodic images such as bar patterns.

The aperiodic images used were squares and rectangles for which the display signal-to-noise ratio has been derived in the form of Eq. (10), which holds if the test images are large relative to the point spread functions of the sensory system and if the noise is of substantially uniform spectral density over the video bandwidth Δf_V. The experimental setup we used to perform the experiments is shown in Fig. 5.11. In this experiment a signal pulse of rectangular wave shape but variable duration is electronically generated and mixed with band-limited white noise of Gaussian distribution. The spatial image displayed on the cathode ray

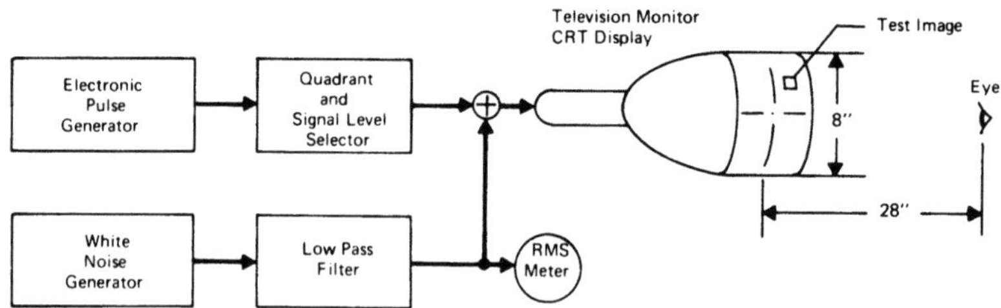

Fig. 5.11. The display signal-to-noise ratio experiment.

tube (CRT) display is a rectangle which can appear in any of four quadrants (but always in the same position in the quadrant selected). The observer is asked to specify the quadrant in which the image is located as the video signal-to-noise ratio and the image locations are randomly located. The observer is asked to specify the image location whether he could see it or not. The probability of detection determined in this manner was then corrected for chance using the formula

$$P_d = (P_0 - P_c)/(1 - P_c) \qquad (37)$$

where P_d is the corrected probability, P_0 is the raw probability data, and P_c is the probability due to chance (0.25 in the case cited). Two noise bandwidths were used, 7.1 and 12.5 MHz, and the observation times per trial were usually 10 sec. The observer distance from the 8-in.-high picture displayed was 28 in. unless otherwise specified and the display background luminance was either 0.2–0.3 or 1 ft-L. The television monitor was operated at 30 frames/sec with a 525-line scan in the vertical. The precise experimental conditions for each experiment are given in Table 5.3.

For the rectangle experiments it was found to be convenient to define the image size in terms of the dimensions of a single scan line. Thus we define the quantities L_x and L_y as

$$L_x L_y = (490)^2 \alpha(a/A) \qquad (38)$$

where 490 is the number of active lines in a conventional 525-line television display and α is the width-to-height picture aspect ratio of the total effective picture on the CRT. Combining Eqs. (10) and (38), we have for the aperiodic images

$$\mathrm{SNR}_{DI} = (1/490)(2L_x L_y t\, \Delta f_V/\alpha)^{1/2} \cdot \mathrm{SNR}_V \qquad (39)$$

This is the equation used to calculate SNR_{DI} for the rectangular images used in the experimental program reported below. The numerical values used were $t = 0.1$ sec and $\alpha = 4/3$.

For the first experiment of Table 5.3 we hypothesized that the SNR_{DI} required to liminally* detect a stationary rectangular image of

---

* By liminal detection we mean a 50%, or threshold, probability of detection.

TABLE 5.3

Conditions for the Aperiodic and Periodic Televised Image Experiments*

Expt. No.	Test image	D_V/D_H	L_D, ft-L	Δf_V, MHz	Number of trials	Number of observers	N_s	Frames sec^{-1}	Figure number
1	Long, thin rectangles	3.5	0.2–0.3	7.1	800	5	525	30	5.12, 5.13
2	Large rectangles	3.5	1.0	12.5	800	5	525	30	5.14, 5.15, 5.17
3	Small to large squares	3.5	1.0	12.5	1200	5	525	30	5.16, 5.17, 5.20
4	Small to large squares	7.0	1.0	12.5	700	5	525	30	5.20
5	Bar patterns, variable ε, N	3.5	1.0	12.5	2350	5	875	25	5.22–5.25
6	Bar patterns, variable N	3.5	1.0	12.5	460	5	875	25	5.26, 5.27
7	Horizontal bar patterns	3.5	1.0	12.5	280	1	875	25	5.28
8	Bar patterns, variable D_V/D_H	1.75 } 3.5 } 7.0	1.0	12.5	2476	14	875	25	5.29
9	Bar patterns, optimum D_V/D_H	Opt.	1.0	12.5	1210	5	875	25	5.30
10	Moving squares, 5 and 20 sec/picture width	3.5	0.2	7.1	1200	10	525	30	5.31
11	Moving bar patterns, 20 sec/picture width	3.5	1.0	12.5	280	1	875	25	5.32, 5.33
12	4 × 4 squares, no added noise	3.5	0.2–10	7.1	600	5	525	30	5.34, 5.36
13	8 × 8 squares, no added noise	3.5	0.2–10	7.1	1000	5	525	30	5.35, 5.36
14	Bar patterns, 200 mV added noise	3.5	10	12.5	2732	4	875	25	5.37, 5.38, 5.39

* D_V/D_H is the display viewing distance-to-height ratio; L_D is the average monitor luminance, ft-L; Δf_V is the video bandwidth, MHz; N_s is the number of scan lines per picture height, interlaced 2 : 1.

variable length would be a constant independent of the image area. To test this notion, we measured probability of detection versus SNR_V and SNR_{DI} for rectangles of size 4×4, 4×64, 4×128, and 4×180 scan lines. The results, plotted in the form of probability of detection versus video SNR, are shown in Fig. 5.12, where it is seen that the larger the image length,* the smaller the SNR_V required for a given level of probability. When the probability of detection is plotted versus the display signal-to-noise ratio as calculated using Eq. (39) and as shown in Fig. 5.13, the data collapse to a single curve, confirming the original hypothesis. Observe that the angular extent of the test images relative to the observer's eye varied from 0.13×0.13 deg for the smallest rectangle to 0.13×6.02 deg for the largest rectangle. This experiment implies that the eye can integrate over very large angles in space, angles which are much larger than were previously thought to be the case. A review of the literature indicates that the eye's ability to spatially integrate with high effectiveness is limited to considerably smaller angles—perhaps $\frac{1}{2}$–1 deg. To explain this discrepancy, it was postulated that the eye is more sensitive to edges than to areas (the Mach effect) and that the test image is nearly "all edge" since it is a long, thin rectangle.

A second experiment was performed using a rectangle of length 96 scan lines or angular subtense 3.2 deg and of variable widths of 4, 8, 16, and 32 scan lines, corresponding to angular subtenses of 0.13, 0.267, 0.534, and 1.07 deg relative to the observer. The corrected probability of detection for this case is shown for the various rectangles in Fig. 5.14 and a plot of the thresholds as a function of image size is shown in Fig. 5.15. For long, narrow rectangles the same threshold value of SNR_{DI} is obtained as was obtained for narrow rectangles of various lengths (Fig. 5.13) and we conclude that for narrow widths (angular subtense of up to about 0.5 deg) the eye fully integrates the whole area of the rectangle but for wider rectangles of angular subtense larger than 0.5 deg the eye is apparently less efficient in utilizing image area.

To further investigate, experiments were performed using squares of larger angular extent with the results shown in Figs. 5.16 and 5.17. These data were taken using a wider noise spectrum (12.5 MHz as opposed to 7.1 MHz previously) and with a brighter display background (1 ft-L versus 0.2 ft-L previously). The increase in video noise bandwidth

---

* In this experiment the longitudinal axis of the rectangle was horizontal.

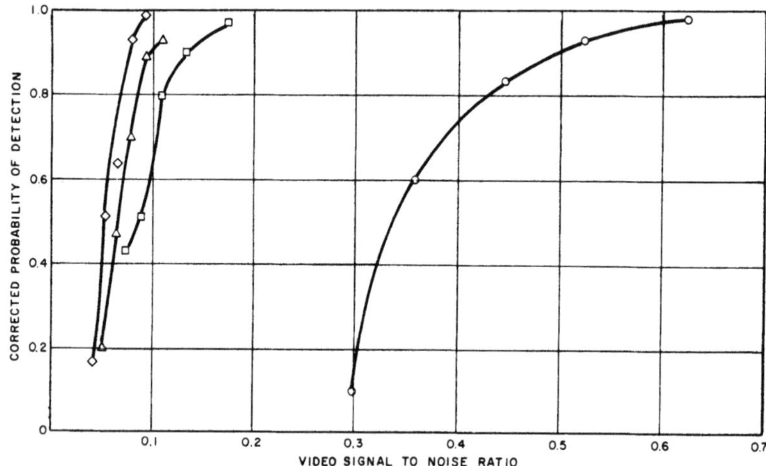

Fig. 5.12. Experiment 1. Corrected probability of detection versus video signal-to-noise ratio for rectangular images of size (○) 4 × 4, (□) 4 × 64, (△) 4 × 128, and (◇) 4 × 180 scan lines.

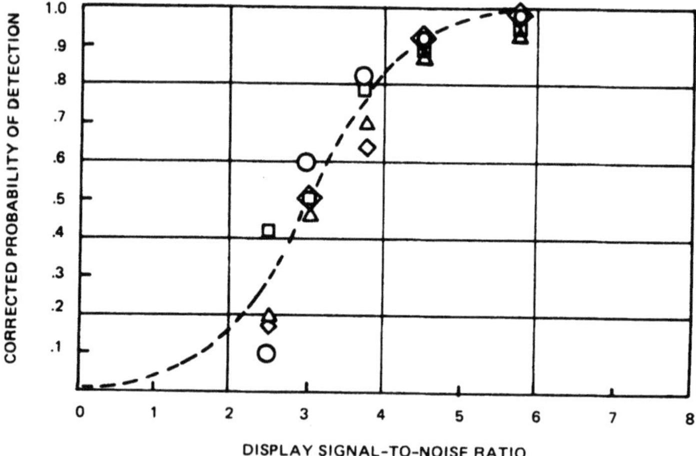

Fig. 5.13. Experiment 1. Corrected probability of detection versus SNR_{DI} required for rectangular images of size (○) 4 × 4, (□) 4 × 64, (△) 4 × 128, and (◇) 4 × 180 scan lines.

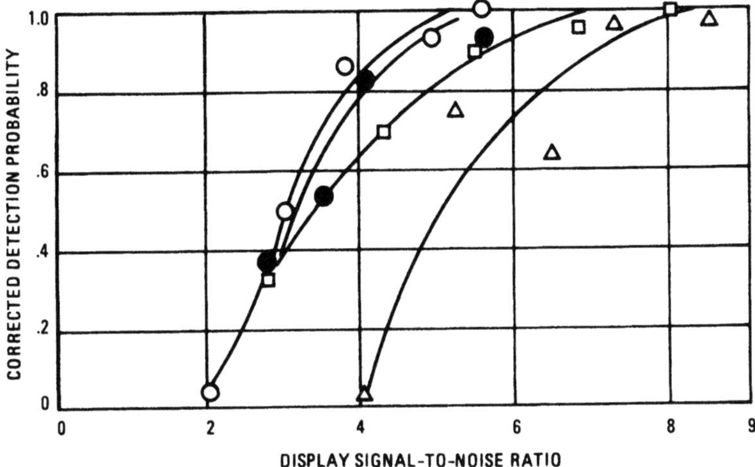

Fig. 5.14. Experiment 2. Corrected probability of detection versus SNR_{DI} for a rectangle of height 96 scan lines and width of (●) 4, (○) 8, (□) 16, and (△) 32 scan lines.

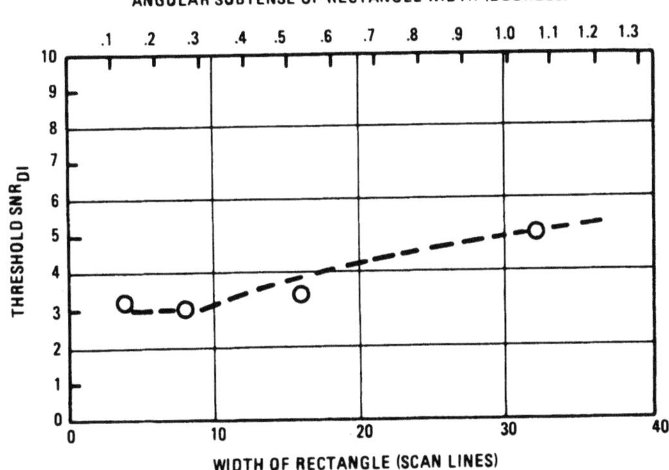

Fig. 5.15. Experiment 2. Threshold SNR_{DI} as a function of the linear and angular extent of a rectangle of height 96 scan lines and variable width 4, 8, 16, and 32 scan lines. Broken curve is theoretical.

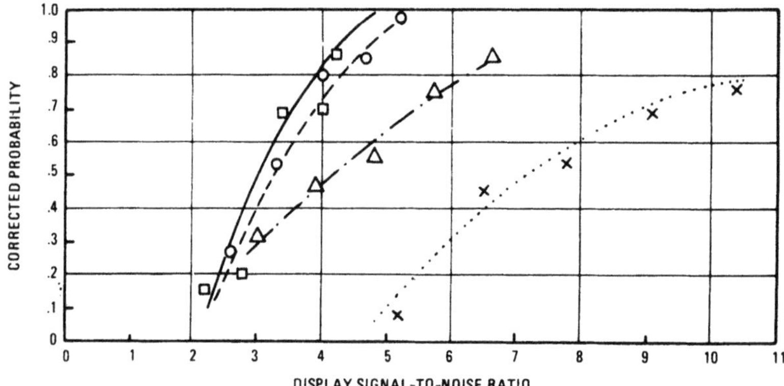

Fig. 5.16. Experiment 3. Corrected probability of detection versus SNR_{DI} required for square images of size ($\square$) 8 × 8, ($\bigcirc$) 16 × 16, ($\triangle$) 32 × 32, and ($\times$) 64 × 64 scan lines.

and display brightness was not expected to change the SNR_{DI} thresholds and did not. The thresholds did increase for the larger angular sizes, however, being about 2.5 times larger for the squares of 2 deg angular subtense than for squares of subtense less than 0.5 deg.

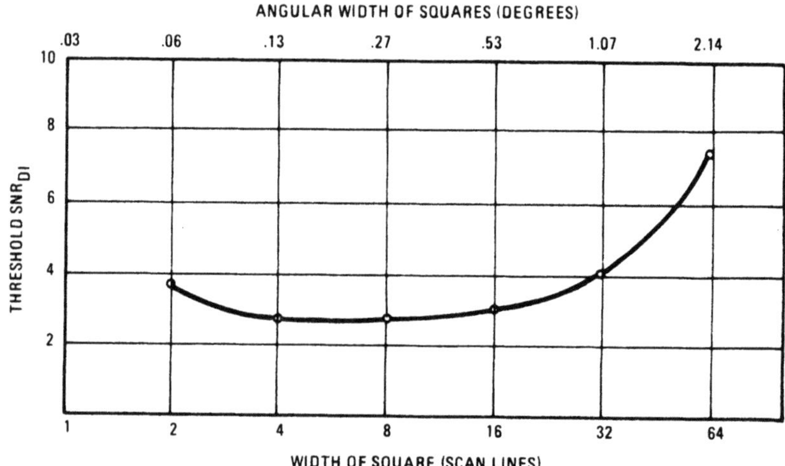

Fig. 5.17. Experiments 2 and 3. Threshold SNR_{DI} required to detect square images of various sizes and angular extent relative to the observer.

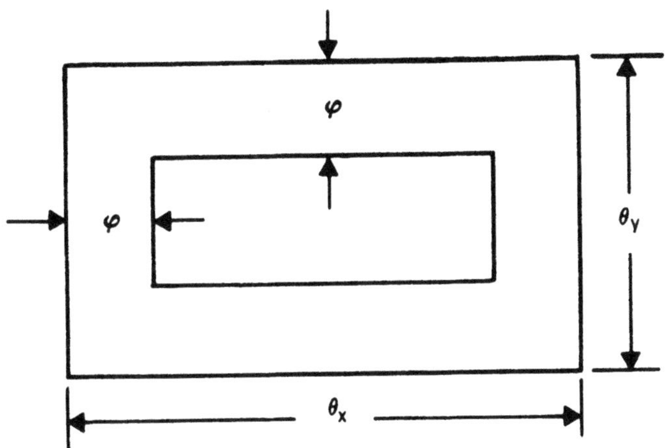

Fig. 5.18. Geometry for the display signal-to-noise ratio analysis for
rectangles of large angular extent.

Thus for both rectangles and squares we found as a result that the
eye is efficient in integrating the full image area only up to angular subtenses of about 0.5 deg.*

As a possible explanation for the observed increases in threshold,
we hypothesized that the eye actually integrates signal from an area
around the perimeter of the area rather than the total area. This is
indicated by the geometry of Fig. 5.18. Let the angular extent of the
test image rectangle be θ_x by θ_y relative to the observer's eye. Assume
that the eye integrates the total area of the image if both θ_x and θ_y are
less than 2φ, i.e.,

$$\text{SNR}_{DI} \approx (\theta_x \theta_y)^{1/2} \tag{40}$$

and for larger rectangles, where both θ_x and θ_y are both larger than 2φ,

$$\text{SNR}_{DI} \approx [\theta_x \theta_y - (\theta_x - 2\varphi)(\theta_y - 2\varphi)]^{1/2}$$
$$\approx [2\varphi(\theta_x + \theta_y) - 4\varphi^2]^{1/2} \tag{41}$$

For squares

$$\text{SNR}_{DI} \approx \theta, \qquad \theta \leq 2\varphi \approx 2(\varphi\theta - \varphi^2)^{1/2} \tag{42}$$

* When the angular subtense is greater than 0.5 deg in two dimensions.

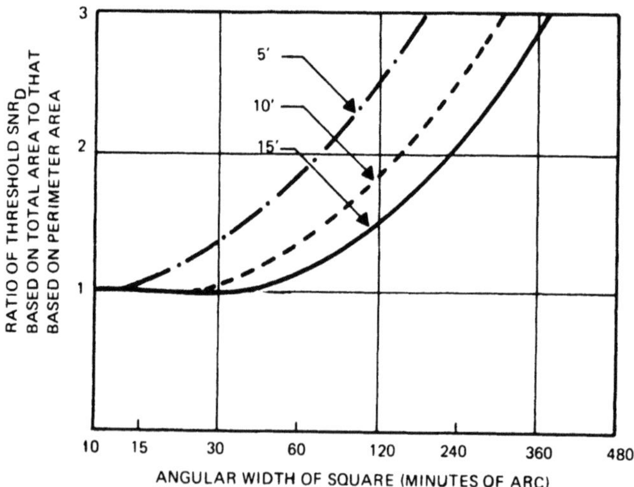

Fig. 5.19. Ratio of threshold display signal-to-noise ratio computed for squares on the basis of total area to that calculated on the basis of perimeter area, versus the square's angular width relative to the observer's eye for three values of perimeter area integration angles.

For comparison purposes, we form the ratio of thresholds

$$\frac{\text{SNR}_{DI}(\text{threshold calculated on total area bases})}{\text{SNR}_{DI}(\text{threshold calculated on perimeter area basis})} \tag{43}$$

These ratios are plotted in Fig. 5.19 for three values of φ: 5, 10, and 15 min of arc. It is seen by comparing Figs. 5.17 and 5.19 that the shapes of the curves are similar and that the computed 10-min curve of Fig. 5.19 would give a good fit to the measured data of Fig. 5.17. The predicted curve of Fig. 5.17 was also calculated on the same basis and is seen to fit the experimental curve.

If the perimeter area concept is to have any merit, then we should be able to offset the increase in threshold that was noted by increase in the object's angular extent by simply increasing the observer-to-display viewing distance. Hence an experiment was performed in which the observer's viewing distance was increased from 28 to 56 in. As shown in Fig. 5.20, the expected effect did occur. Thus the premise that the eye integrates only around the perimeter, though not proven, is at least made plausible.

Before proceeding we observe that the probability model used to fit the experimental points of Fig. 5.13 is based on a model originally suggested by Legault (1971). In this model it is assumed that the mean number of photoelectrons within the sampling area and period has become sufficiently large so that the Gaussian or normal probability distribution law given by

$$f_Z(z) = [\exp(-z^2/2)]/(2\pi)^{1/2} \tag{44}$$

becomes a good approximation to the Poisson distribution law, which actually represents the signal and noise processes. In the above Z is a random variable which is numerically equal to

$$Z = \text{SNR}_{DI} - \text{SNR}_{DT} \tag{45}$$

where SNR_{DT} is the threshold display signal-to-noise ratio, which is generally regarded to be that needed to obtain a detection probability of 0.5. The random variable Z is of unit mean and variance. Other values of probability are obtained from the formula

$$P_d(-\infty < Z < z_2) = (1/2\pi^2) \int_{-\infty}^{z_2} \exp(-z^2/2) \, dz \tag{46}$$

which cannot be integrated in closed form but is widely available in standard mathematical tables.

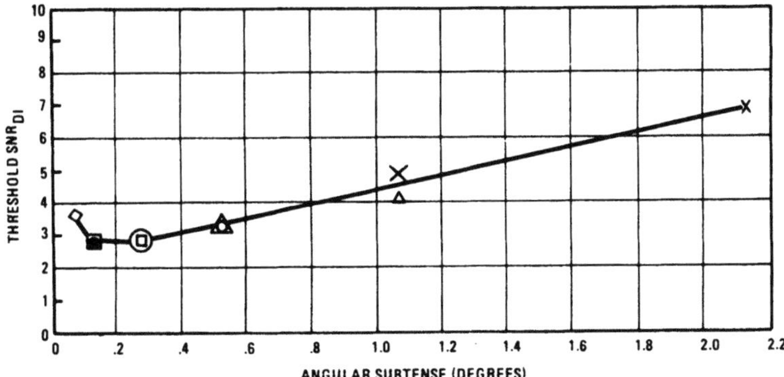

Fig. 5.20. Experiment 4. Threshold SNR_{DI} required to detect square images as a function of their angular size for two viewing distances, 28 and 56 in. Image size on display is ($\diamond$) 2, ($\bullet$) 4, ($\square$) 8, ($\bigcirc$) 16, ($\triangle$) 32, and ($\times$) 64 scan lines on a side. Large symbols at 56, small at 28 in.

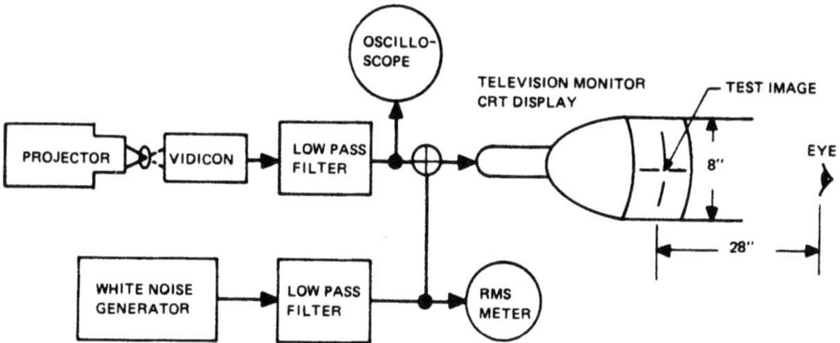

Fig. 5.21. Experimental setup for the television-camera-generated imagery.

The periodic test images used were primarily bar patterns of various height-to-width ratios, spatial frequencies, and numbers of bars. The experimental setup of Fig. 5.21 was used to perform the necessary psychophysical experiments. Test images are projected on the faceplate of a high-resolution $1\frac{1}{2}$-in. vidicon operated at highlight video signal-to-noise ratios of 50:1 or better. The camera and TV monitor were operated at 25 frames/sec with 875 scanning lines (825 active). Band-limited white noise of Gaussian distribution was mixed with the camera-generated signal. Both the signals and noise were passed through identical filters

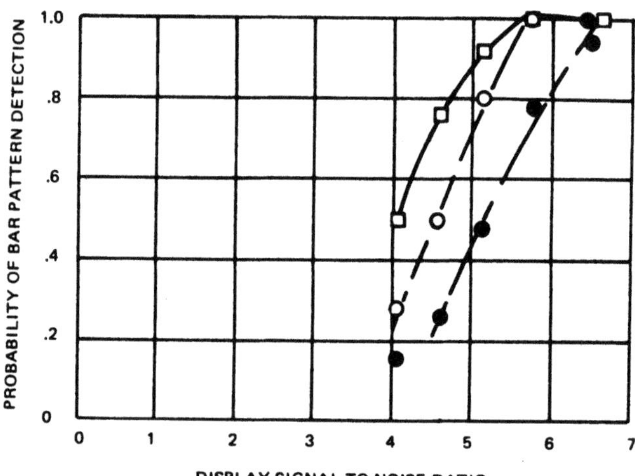

Fig. 5.22. Experiment 5. Probability of bar pattern detection versus SNR_{DI} for a 104-line bar pattern of bar height-to-width ratio ($\square$) 5:1, ($\bigcirc$) 10:1, and ($\bullet$) 20:1.

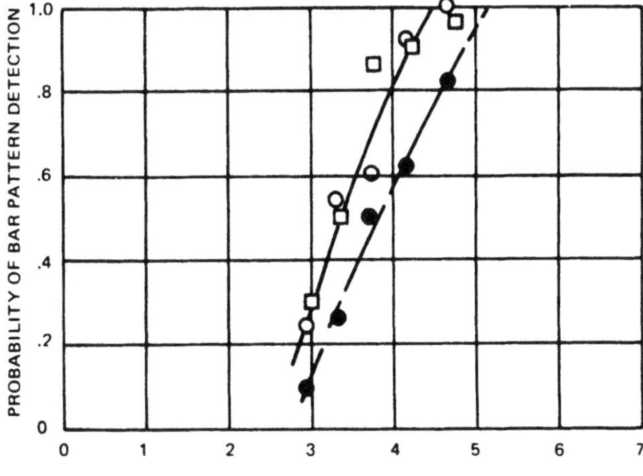

Fig. 5.23. Experiment 5. Probability of bar pattern detection versus SNR_{DI} for a 200-line bar pattern of bar height-to-width ratio (□) 5:1, (○) 10:1, and (●) 20:1.

(noise equivalent bandwidth of 12.5 MHz) prior to mixing in the monitor. The monitor luminance was approximately 1 ft-L unless otherwise specified. The displayed picture height was 8 in. and, unless otherwise stated, the observer–display distance was 28 in.

For the first series of experiments the test images were bar patterns of various spatial frequencies and bar height-to-width ratios. The observer was required to state whether or not the pattern displayed was resolvable as the image SNR_{DI} was randomly varied. Chance was not involved since the patterns were always present on the display. The experimental constants for the various experiments are given in Table 5.3 as before. The purpose of the first series of experiments was to determine the effect of bar height-to-width ratio on the bar pattern detectability, with the results shown in Figs. 5.22–5.24. At low spatial frequencies of 104 lines/picture height the threshold signal-to-noise ratio SNR_{DT} is seen to increase slightly with bar height-to-width ratio (Fig. 5.22), while at the highest spatial frequency of 396 lines/picture height the SNR_{DT} required is very nearly independent of bar height-to-width ratio, as shown in Fig. 5.24. By threshold SNR_{DI} it is implied that 50% of the total number of patterns displayed are resolved (or "barely discerned"). The SNR_{DT} values are summarized in Fig. 5.25 as a function of spatial

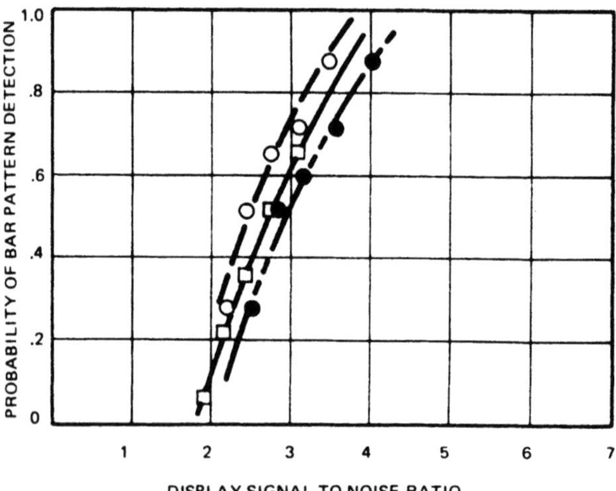

Fig. 5.24. Experiment 5. Probability of bar pattern detection versus SNR_{DI} for a 396-line bar pattern of bar height-to-width ratio (□) 5:1, (○) 10:1, and (●) 20:1.

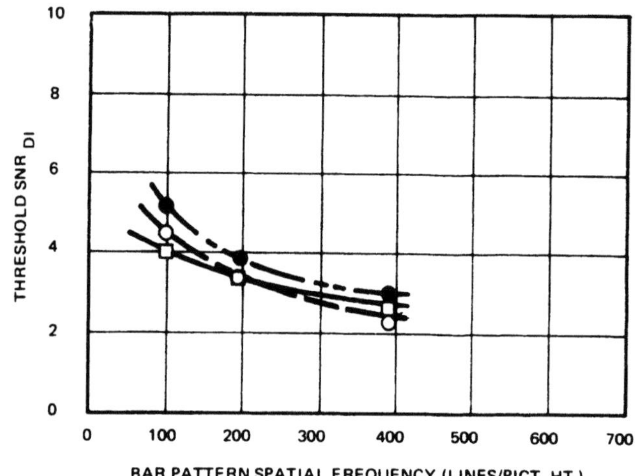

Fig. 5.25. Experiment 5. Threshold SNR_{DI} versus bar pattern spatial frequency for three bar height-to-width ratios of (□) 5:1, (○) 10:1, and (●) 20:1.

frequency and are seen to decrease slowly with line number for all of the patterns. We note that SNR_{DI} is calculated on the basis of the total area of a bar. Specifically, the equation

$$\text{SNR}_{DI} = \left(\frac{2tn_V\,\Delta f_V}{\alpha}\right)^{1/2}\frac{R_{\text{SF}}(N)}{N}\frac{\Delta i}{I_n} \qquad (47)$$

was used. Here Δi is the peak-to-peak signal current for a broad-area pattern (unity modulation transfer function) and I_n is the rms noise that is added to the camera-generated image. Real cameras, of course, have a response that is a function of spatial frequency and the value of Δi in the video channel for square wave inputs becomes $\Delta i_{p\text{-}p}$, the peak-to-peak value of the video signal when the spatial frequency effects are included, that is, we have

$$\Delta i_{p\text{-}p} = (\Delta i)R_{\text{SQ}}(N) \qquad (48)$$

and

$$\text{SNR}_{DI} = \left(\frac{2tn_V\,\Delta f_V}{\alpha}\right)^{1/2}\frac{1}{N}\frac{R_{\text{SF}}(N)}{R_{\text{SQ}}(N)}\frac{\Delta i_{p\text{-}p}}{I_n} \qquad (49)$$

where $R_{\text{SF}}(N)$ is the value of the flux factor at N, $R_{\text{SQ}}(N)$ is the value of the square wave response at N, and $\Delta i_{p\text{-}p}$ is the value of the peak-to-peak signal corresponding to N as measured in the output of the video channel. For calculation purposes t, the integration time of the eye, is taken to be 0.1 sec and α, the picture aspect ratio, is 4/3. At low spatial frequencies the displayed images approach a square wave while at high spatial frequencies, above about 500 lines/picture height, the displayed images were nearly pure sine waves. In Fig. 5.7, the values of modulation transfer function, square wave response, and flux factor are shown for the vidicon. The main point of the above discussion is to note that the SNR_{DI} are calculated on the basis of mean signal-to-rms-noise ratio since we have corrected the measured signal using the square wave amplitude and square wave flux factors.

Before proceeding we note that in the bar pattern experiments above the number of bars in the various patterns varied. For the 5:1 pattern five bars were used, for the 10:1 pattern nine bars, and for the 20:1 pattern 17 bars. This should make no difference if our premise that the eye uses only a single bar in judging the pattern's resolvability is correct, as we believe it to be on the basis of the experimental evidence shown

TABLE 5.4

**Angular Subtense (deg) of a Bar in Each Experiment Relative to the Observer
As a Function of the Bar Height-to-Width Ratio**

Bar pattern spatial frequency, lines/picture height	Angular subtense in the horizontal	Angular subtense in the vertical for bar height-to-width ratios of		
		5	10	20
104	0.157	0.785	1.57	3.14
200	0.0818	0.409	0.818	1.636
396	0.0413	0.2065	0.413	0.826

below.* We observe further that the signal-to-noise ratios and height-to-width ratios in the above experiments were randomly varied. The angular subtense of the various bars relative to the observer's eye are given in Table 5.4.

To further confirm the falloff in thresholds at the higher spatial frequencies, a second series of experiments was performed using bar patterns with bars of 5:1 height-to-width ratio. A number of bar patterns of various spatial frequencies were displayed at the same time and the observer was asked to indicate the pattern of highest frequency that was barely visible as the signal-to-noise ratio was systematically varied from high levels to low levels and the reverse in steps of 1 dB. The fraction of patterns resolved by this "method of limits" is plotted in Fig. 5.26 as a function of display signal-to-noise ratio and the thresholds determined are plotted in Fig. 5.27. Also plotted are the results determined in the previous experiment. As can be seen, the "method of limits" gives the same values for the thresholds and does confirm the falloff in threshold with spatial frequency. The method of limits is used in measuring the limiting resolution of real camera tubes and this method of test thus appears to be appropriate.

In the above experiments the bar patterns were vertically oriented with their longitudinal axes perpendicular to the direction of scan. The

---

* We believe this premise to be valid for live televised images where the eye obtains many samples of any given bar as opposed to a single sample as for noisy photographic imagery. In the latter case more bars should result in higher detectability.

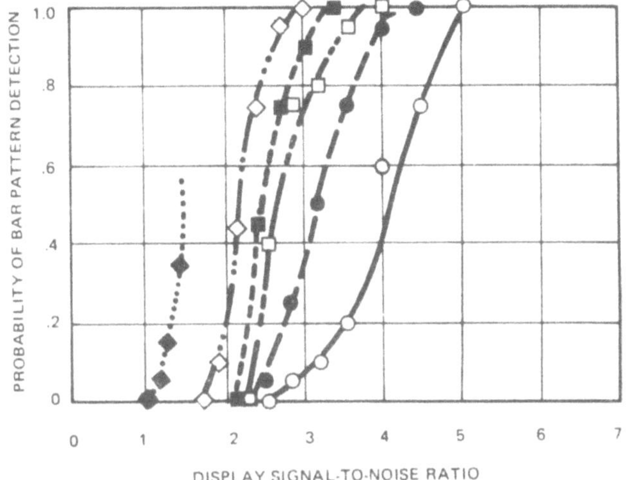

Fig. 5.26. Experiment 6. Probability of bar pattern detection versus SNR_{DI} for bar patterns of spatial frequency (○) 104, (●) 200, (□) 329, (■) 396, (◇) 482, and (◆) 635 lines per picture height. Bar height-to-width ratio was five in all cases.

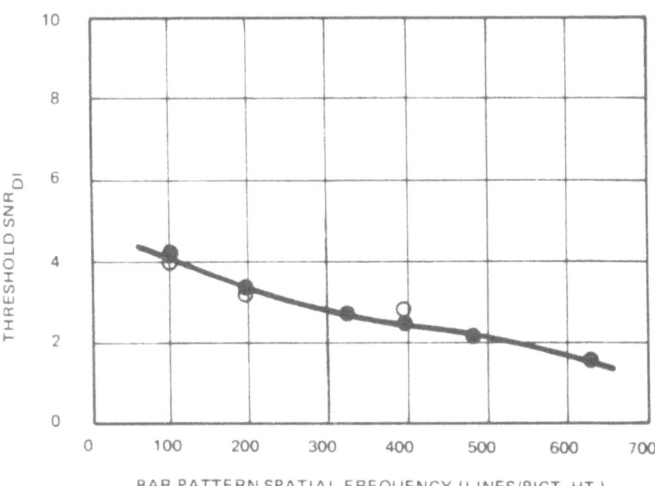

Fig. 5.27. Experiment 6. Threshold SNR_{DI} versus bar pattern spatial frequency obtained using two different experimental techniques: (●) method of limits and (○) method of random SNR variation.

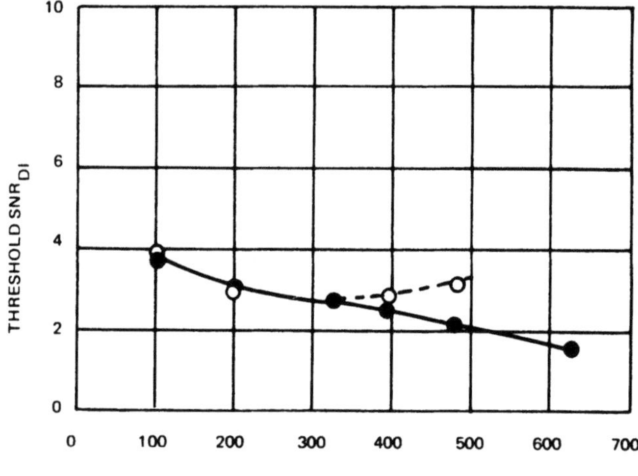

Fig. 5.28. Experiment 7. Threshold SNR_{DI} required to recognize the presence of (●) horizontally oriented and (○) vertically oriented bar pattern versus bar pattern spatial frequency.

results with the axes horizontal are shown in Fig. 5.28. One observer was used for the experiment. Also shown in Fig. 5.28 are the results for vertical bars for the same observer (the same results as those shown in Fig. 5.27). As can be seen, the orientation is immaterial at the low spatial frequencies but the thresholds for the high spatial frequencies increase with horizontal* bars. Indeed, the 635-line pattern could only be seen with no noise added to the signal.

In the next bar pattern experiment viewing distance was varied in discrete steps from 14 to 28 to 56 in. The bars were vertically oriented. The results are as shown in Fig. 5.29. At the short distances the low pattern frequencies become less detectable, while the reverse was true at the long viewing distances. It is clear from Fig. 5.29 that for a given line number there is a viewing distance that minimizes threshold SNR_{DI}. In the final bar pattern experiment the observer adjusted his position at each SNR_{DI} value to enable him to see the highest line number possible for the SNR_{DI} value; the threshold values of SNR_{DI} are shown in Fig. 5.30. A comparison of Figs. 5.29 and 5.30 shows that the optimum distances for low line numbers are greater than 56 in. whereas for high

* This topic is the subject of Chapters 6 and 7.

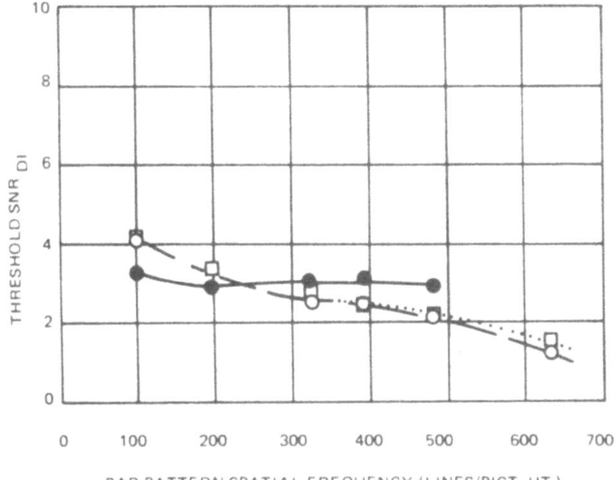

Fig. 5.29. Experiment 8. Threshold SNR_{DI} versus bar pattern spatial frequency for display-to-observer viewing distances of ($\bigcirc$) 14, ($\square$) 28, and ($\bullet$) 56 in.

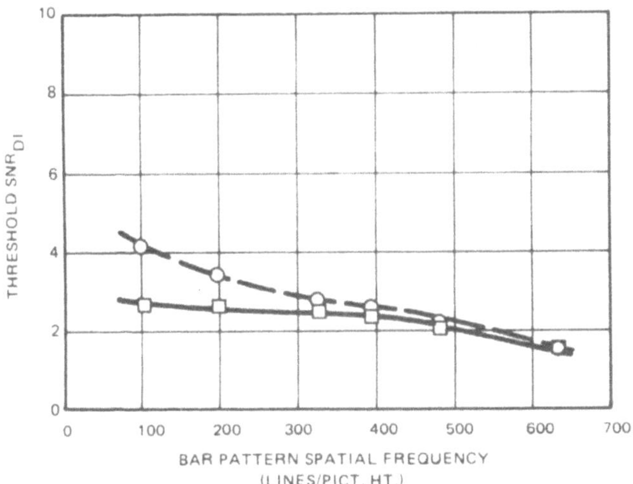

Fig. 5.30. Experiment 9. Threshold SNR_{DI} versus bar pattern spatial frequency for ($\square$) optimum viewing distance and ($\bigcirc$) 28 in. viewing distance for one observer.

line numbers optimum distances are approximately 14 in. from the 8-in.-high display.

In the above the images were held stationary during the period of observation. It has been observed that images in motion, when viewed on the display of an electrooptical sensor, are degraded. This degradation can be due to the combination of image motion and TV camera frame time, to sensor time constants, and to psychophysical effects. In the next series of experiments attempts are made to determine the image degradations due to psychophysical effects alone.

It is common experience that objects can be tracked visually at very high angular rates. Quantitatively, it has been found that objects moving at rates up to 20 deg/sec are nearly as detectable as stationary objects, while at 120 deg/sec small objects can still be detected but must be increased in size threefold over their threshold size at a given light level when stationary. For our most usual display viewing distance-to-height ratio of 3.5 the display width subtends about 21.6 deg at the observer's eye. Thus a 20 deg/sec speed corresponds to one picture width/sec, a rate which is very high for most TV viewing. In testing the effects of image motion on TV cameras, it is more customary to employ rates of 10 and 20 sec/picture width, which are very slow from a psychophysical point of view. Because of these considerations, we did not expect any appreciable image detectability degradations due to image motion and this premise is confirmed in the results reported below.*

In the first motion experiment we electronically generated square images which could be set in motion across a display. With electronic generation the lag and exposure time effects normally associated with camera tube imagery are absent. The display itself has some lag but these lags are negligible for the image sizes and pattern speeds used. The test squares could appear in either the top, middle, or lower one-third of the displayed picture and the motion was from left to right at speeds of 20 and 5 sec/picture width (only 93% of the actual picture width was used). The effects of these speeds on the threshold signal-to-noise ratio is shown in Fig. 5.31 for various image sizes. It is seen that at 20 sec/picture width, motion has almost no effect on the required SNR_{DT} except for the smallest object, which increased in detectability since

---

* In a cluttered background, scene objects in motion may actually increase their detectability.

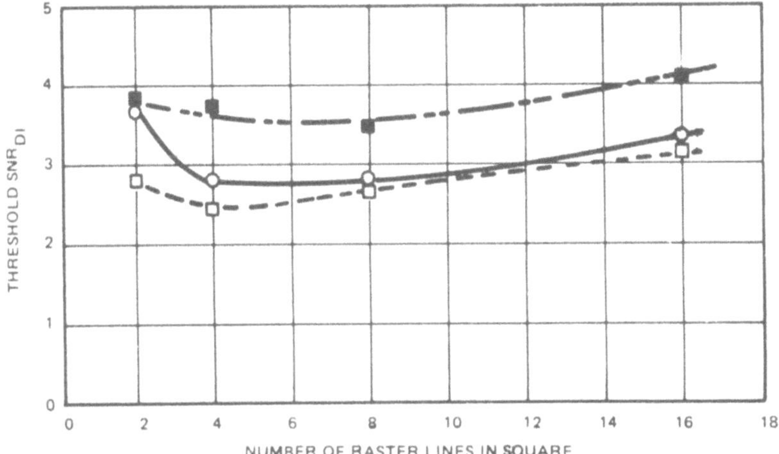

Fig. 5.31. Experiment 10. Threshold SNR$_{DI}$ versus number of raster lines in square for (○) stationary patterns, (□) 20 sec/picture width motion, and (■) 5 sec/picture width motion.

the SNR$_{DT}$ required was 24% lower than for the static case. With a rate of 5 sec/picture width the threshold SNR$_{DI}$ is the same as for the static case for the smallest square but is 26% higher for the larger squares. The conditions for these motion experiments are summarized in Table 5.3.

In the next series of experiments bar patterns generated by a vidicon camera were employed. The vidicon camera significantly degrades the bar pattern modulation as can be seen in the A-scope trace of Fig. 5.32. As is seen, the peak-to-peak signal in the first bar is significantly less than that in the subsequent two bars. For calculation purposes the average

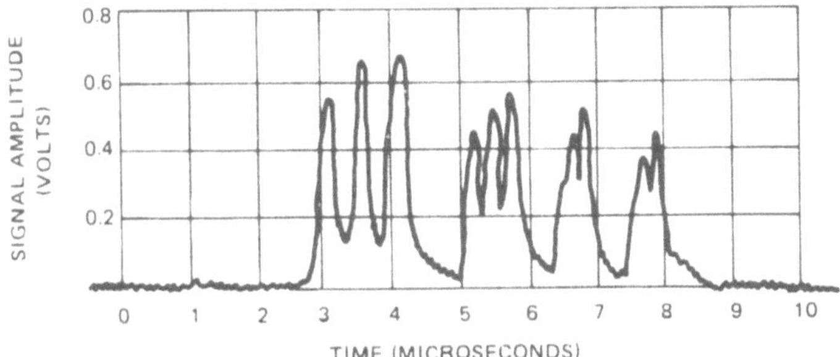

Fig. 5.32. Experiment 11. Trace of the video signals for bar pattern motion of 20 sec/picture width.

peak-to-peak signal in the three bars was used. The results of this experiment and the corresponding result with no motion for the same observer are shown in Fig. 5.33. No appreciable difference in detectability is discerned. The above results suggest that the purely psychophysical effects of image motion are quite small for the image motion rates expected in real imaging situations.

In the analysis of sensors we have normally assumed that the displayed images have been sufficiently amplified and magnified so that the observer is limited only by the displayed-image signal-to-noise ratio. The assumption has been that the displayed image is large enough and bright enough so that the observer is not acuity-limited by the image size or luminance. As we will see, it is also necessary that the image incremental luminance above background, or its contrast, be sufficient.

We have previously observed that the displayed image contrast need not be the same as the contrast of the input image. In the usual cathode ray tube (CRT) display two controls are available to the observer. One is the brightness control which establishes the average display luminance level. The second is usually referred to as the contrast control but is actually a video gain control. This is not a complete misnomer, however, since this control determines the swing of the display luminance

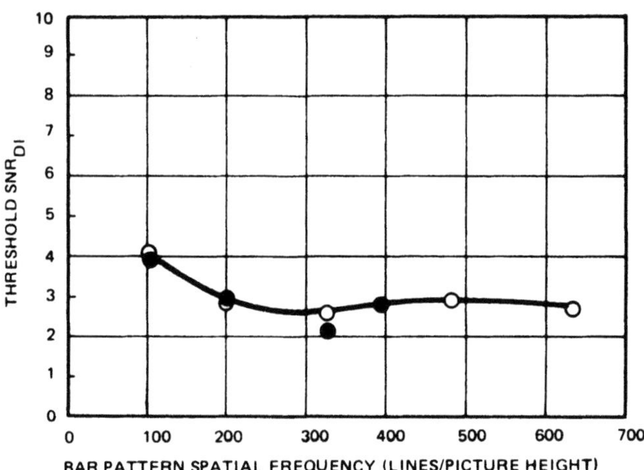

Fig. 5.33. Experiment 11. Threshold SNR$_{DI}$ versus bar pattern spatial frequency with (●) 20 sec/picture width motion and (○) no motion for the same observer in both cases.

about the average and therefore increases or decreases the contrast of the displayed image.

In viewing the display a fluctuation noise is generated in the observer's retinal photoprocess which is proportional to the display's average luminance. Suppose the displayed image is essentially noise-free. As the video gain is reduced, the image signal level will eventually drop below the retinal fluctuation noise. Thus, the image detectability becomes limited by the SNR_I* developed in the observer's retina. Obviously, the observer will not intentionally reduce the image contrast until it becomes undetectable but as a practical matter he may be forced to it. This situation arises when the intrascene light level range, i.e., the ratio of image highlight to low-light irradiance range, becomes very large. In this case the dc display brightness may need to be increased to accommodate large signal swings. Simultaneously, the video gain may need to be reduced in order to keep the total signal swing within the display's dynamic range limitations. The net result may be that small signals in the scene are so reduced by video gain reduction that they cannot be discerned in the presence of the retinal fluctuation noise associated with the average monitor luminance.

To investigate the possibility of retinal fluctuation noise limitations to image detection, a number of psychophysical experiments were performed. In the first series of experiments we used electronically generated square images of height four and eight scan lines. The test conditions for all of the display luminance experiments are given in Table 5.3. In experiment 12 the test image was a 4×4 scan line square. The monitor gain was set at the arbitrary level of 15.5. For the 8×8 scan line square the video gain level was 5.5. In both cases the display luminance values were systematically varied over the values 0.2, 1.0, 5.0, and 10.0 ft-L. At each luminance level the video signal levels were randomly varied over five values. No noise was added in the video channel prior to display.

The results for the 4×4 scan line square are shown in Fig. 5.34 and the results for the 8×8 scan line square are shown in Fig. 5.35. In both cases the higher the monitor luminance, the higher the video signal required for a given level of detection probability.† The shift in

---

* SNR_I is the SNR of the image at the retina.

† Note that in the SNR_D experiments previously discussed the fixed system noise levels injected in the video prior to the display were much larger than the retinal fluctuation noise levels encountered here, i.e., the fixed system noise dominated.

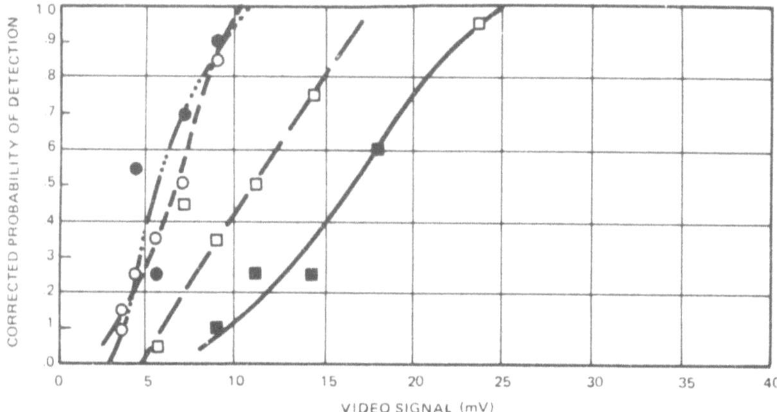

Fig. 5.34. Experiment 12. Influence of monitor luminance for a square of size 4 × 4 raster lines for monitor luminance values of (○) 0.2, (●) 1, (□) 5, and (■) 10 ft-L. No noise added at video input.

video signal level for the 8×8 scan line square as opposed to the 4×4 scan line square is due to the image size and video gain changes. When the data are compensated for these factors the data from one fall directly over the data for the other, as shown in Fig. 5.36. To obtain this figure, we assume that the threshold SNR_{DI} required to liminally detect the squares is the same for the retinal-fluctuation-noise-limited eye as it is

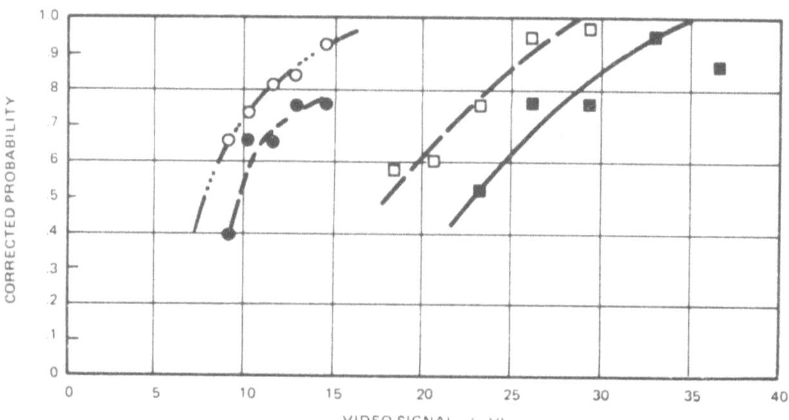

Fig. 5.35. Experiment 13. Influence of monitor luminance for a square of size 8 × 8 raster lines for monitor luminance values of (○) 0.2, (●) 1, (□) 5, and (■) 10 ft-L. No noise added at video input.

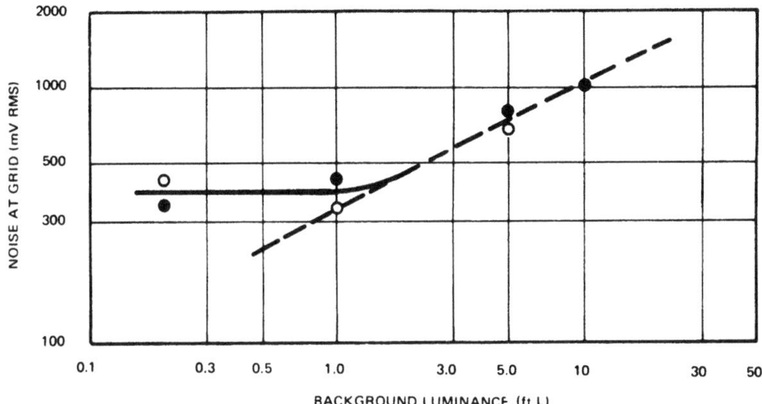

Fig. 5.36. Experiment 13. Equivalent noise at the display grid as a function of average display luminance for display gains of ($\bigcirc$) $G = 15.5$ with 4×4 scan line image and ($\bullet$) $G = 5.5$ with 8×8 scan line image. No noise added at video input.

for a sensory-system-noise-limited eye. Then we calculate the amount of system noise which would have to be present at the output of the monitor's video amplifier (or input to the grid of the CRT) to obtain the same threshold SNR_{DI}. For lack of a better name, we call this the system noise equivalent of the retinal-fluctuation-noise V_{ND}.

From Fig. 5.36 we observe that V_{ND} is nearly constant for display luminances from 0.2 to 1.0 ft-L but increases at a rate which is approximately equal to the display luminance raised to the one-half power for display luminances in the range from 1 to 10 ft-L, as can be inferred from the theoretical dashed line. The cause of the constant value of V_{ND} below 1.0 ft-L has not been determined.

Vidicon-generated bar patterns were used in experiment 14 using the conditions given in Table 5.3. The display luminance was fixed at 10 ft-L and two monitor gain settings of 2.3 and 11 were used. In this series 200 mV of white noise in a 12.5-MHz bandwidth was added prior to the monitor amplifier. The results of the experiments with a monitor gain of 2.3 are shown in Fig. 5.37 and the results with a monitor gain of 11 are shown in Fig. 5.38. From the experiments with squares we would expect that the mean-square system noise equivalent of the retinal fluctuation noise V_{ND}^2, should be approximately equal to 10^6 (mV)^2 for our setup at 10 ft-L. With a monitor gain of 2.3 the mean-square system noise should be about 0.212×10^6 (mV)^2. Thus the retinal fluctuation

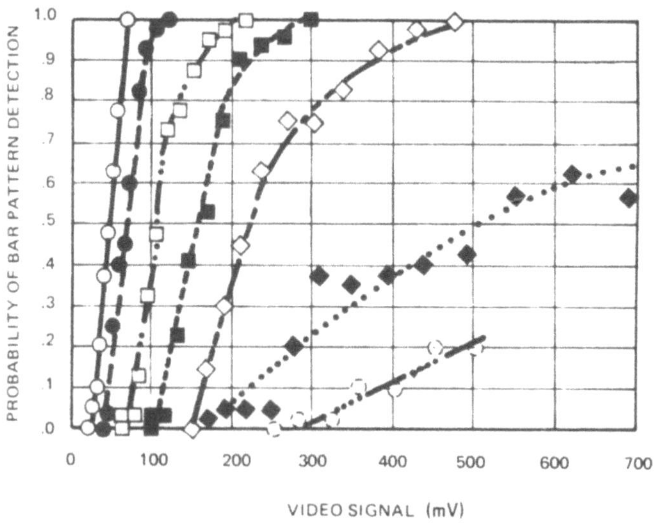

Fig. 5.37. Experiment 14. Probability of bar pattern detection versus video signal. Monitor gain of 2.3 and bar pattern spatial frequencies of (○) 104, (●) 200, (□) 239, (■) 396, (◇) 482, (◆) 635, and (broken circle) 729 lines/picture height.

noise should dominate. With a monitor gain of 11 the mean-square system noise should be about 4.84×10^6 (mV)^2 and it should dominate.

With the assumption of a V_{ND} of 1050 mV we calculate the threshold SNR_{DI} for the data of Figs. 5.37 and 5.38. These data are plotted in Fig. 5.39. With a monitor gain of 11 the threshold SNR_{DI} values obtained are consistent with those values obtained when the bar pattern discernibility was limited by system noise alone, while with a monitor gain of 2.3 the threshold SNR_{DI} shows a sudden and dramatic increase for the 635-line pattern. In both cases we ignored the thresholds for the 729-line pattern, for which the data were insufficient. The increase in threshold for the 635-line pattern with a low monitor gain is not explained. At first it was thought that the monitor MTF could be the cause but this would affect the experimental results for both experiments equally. At present we favor an explanation based on some lower contrast limit for the observer's eye.

In summary, it was hypothesized that a retinal fluctuation noise proportional to the average display luminance could be limiting to image discernibility. It is felt that this hypothesis has been confirmed.

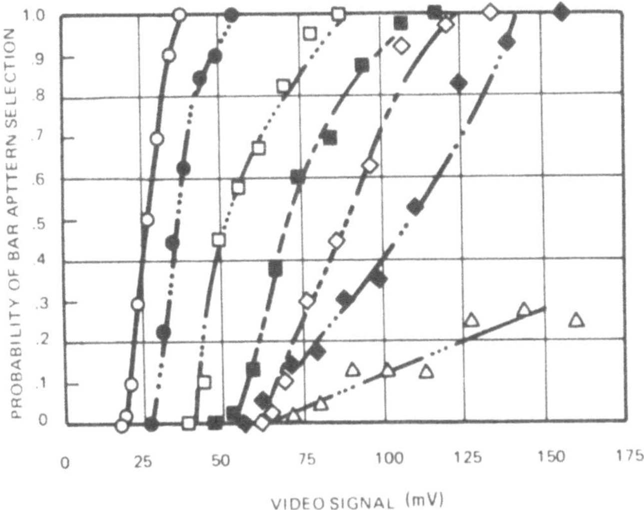

Fig. 5.38. Experiment 14. Probability of bar pattern detection versus video signal. Monitor gain of 11 and bar pattern spatial frequencies of (○) 104, (●) 200, (□) 329, (■) 396, (◇) 482, (◆) 635, and (broken circle) 729 lines/picture height.

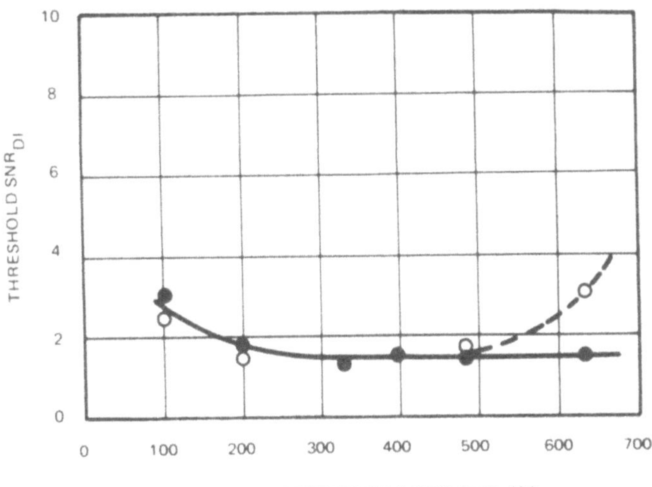

Fig. 5.39. Experiment 14. Threshold SNR_{DI} versus bar pattern spatial frequency obtained with 200 mV additive noise and video gains of (○) 2.3 and (●) 11.

An unexplained noise source was noted with low monitor luminances and an unexplained noise source or contrast threshold was noted with low monitor gains with high monitor luminance.

5.7. PSYCHOPHYSICAL EXPERIMENTS; RECOGNITION AND IDENTIFICATION

In the previous sections various aspects of target detection were discussed and equations were derived for calculating the display signal-to-noise ratio detecting aperiodic laboratory targets such as squares and rectangles as well as periodic patterns such as bar patterns. It was shown that for a given detection probability a specific value of display signal-to-noise ratio was required and that this value was independent of target size for aperiodic targets and was a slowly varying function of line number for periodic patterns. To perform higher levels of object discrimination than detection, such as target recognition or identification, we would intuitively expect that a greater object resolution would be needed since greater target detail must be discerned. As we noted previously, this notion was the cornerstone of Johnson's experiments and theories, which we will attempt to extend in the following.

Two different types of backgrounds were used for the recognition experiments. The first background was a uniform white background and the second was a transparency of a real background. The experimental setup is the same as that shown in Fig. 5.21 in which the transparency on a light box is imaged onto the photosurface of a high-resolution $1\frac{1}{2}$-in. vidicon. Experimental conditions are detailed in Table 5.5.

The transparencies used in the first experiment were high-quality photographs of vehicles amid a uniform white background. The photographs were taken at a depression angle of 45 deg from the horizontal and perpendicular to the vehicle's longitudinal axis, i.e., the sides and tops of the vehicles were imaged as shown in Fig. 5.40. The vehicles included a tank, a van truck, a half-track with top-mounted radar antenna, and a tracked bulldozer with derrick. The areas of the various vehicles were approximately 0.057 in.^{2} on the 8×10.7 in. display and subtended angles of about 0.34 by 0.68 deg at the observer's eye. The vehicle types and video SNR were randomly varied and the probabilities of recognition, corrected for chance, were determined. The SNR_{DI} for the various images were calculated on the basis of the area of a bar

TABLE 5.5

Conditions for the Recognition and Identification Experiments*

Expt. No.	Test image	Background	D_V/D_H	L_D, ft-L	Δf_V, MHz	Number of trials	Number of observers	N_s	Frames sec^{-1}	Figure number
15	Vehicle recognition	Uniform	3.5	1.0	12.5	1000	5	875	25	5.41, 5.42
16	Bar patterns	Uniform	3.5	1.0	12.5	872	5	875	25	5.43
17	Vehicle recognition	Road	3.5	1.0	12.5	800	8	875	25	5.45
18	Vehicle recognition	Grass	3.5	1.0	12.5	800	8	875	25	5.46
19	Vehicle recognition	Grass–trees	3.5	1.0	12.5	800	8	875	25	5.47
18	Tank identification	Uniform	3.5	1.0	12.5	1250	5	875	25	5.49
19	Tank identification	Uniform	3.5	1.0	12.5	1250	5	875	25	5.50

* D_V/D_H is the display viewing distance-to-height ratio; L_D is the average monitor luminance, ft-L; Δf_V is the video bandwidth, MHz; N_s is the number of scan lines per picture height, interlaced 2 : 1.

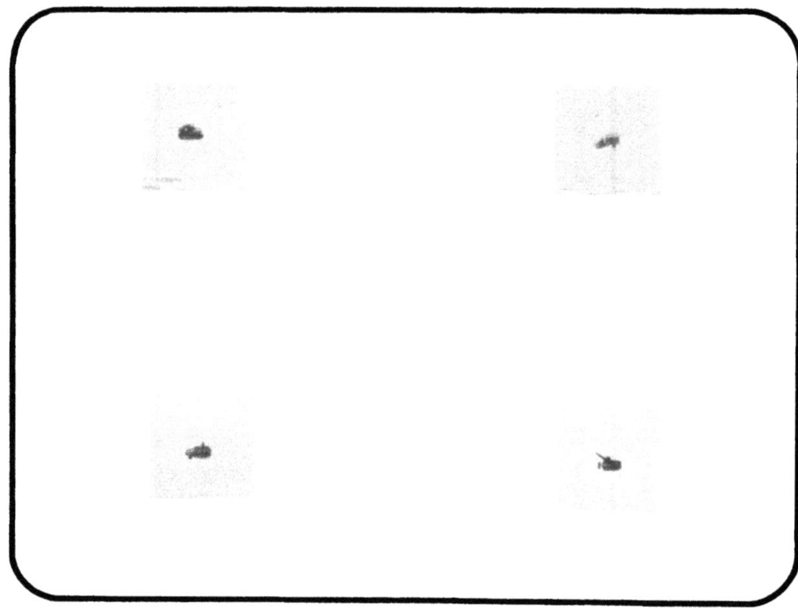

Fig. 5.40. Photographs of models used for recognition experiments. Upper left, tank; upper right, van truck; lower left, half-track with antenna; and lower right, derrick half-track.

whose length and width are equal to the length of the vehicle's image and the width of the vehicle's image divided by eight. This is in accord with the equivalent bar pattern concept discussed above and illustrated in Fig. 5.10. We note, however, one difference between the calculations for the bar pattern and the vehicular image SNR_{DI}. In the case of the vehicular image the signal amplitude was measured from the background signal level, which was approximately constant, to the peak object signal level. For the "equivalent bar patterns" the signal levels were measured in terms of the mean signal excursion within the bar pattern area in the periodic direction. Had the peak-to-peak excursions about the average signal within the vehicle area been used (when the object is imaged against a uniform background) the threshold SNR_{DI} would have been somewhat lower.

These difficulties result from the necessity of defining an image area and a signal excursion in order to calculate an SNR_{DI} threshold. In this connection we observe that the criterion for bar pattern "recognition" is that the observer must be able to discern a modulation within the bar pattern, whereas for vehicle image recognition the vehicle outline must be discerned. This outline may have periodic features but is more likely to be aperiodic. While we recognize and even emphasize these differences, we do not now have better criteria to suggest. It is felt that recognition predictions based on the equivalent bar pattern concept will tend to be somewhat pessimistic because the effect of an MTF is generally more severe on periodic objects than on aperiodic objects. It is interesting to observe that the thresholds measured for recognition of various vehicle types are nearly constant but that the scatter in the data increases as we move from the lowest level of discrimination (detection) to the highest (identification). An alternative concept for recognition and identification might be to define some subarea on the image and treat it as an aperiodic object. Intuitively, we would expect that range predictions made on this basis would be optimistic because sensor MTF's will tend to reduce the modulation of image detail features that are not well separated from one another.

Each object used in the recognition experiments did have a feature which approached the idea of an isolated object feature such as a boom arm, a turret, etc. It may be that a single feature is characteristic of a specific object and to recognize the object the characteristic feature must be discerned. In each case the maximum signal excursion in the "characteristic feature" was the same as the largest found for the whole target, or nearly so. Also, the width of the feature was about one-eighth the total minimum width of the object. A similar situation was found in the identification experiments, where the features needed were closer to 1/13 the object's minimum width.

In Fig. 5.41 we show the signal excursions for the tracked bulldozer on selected horizontal lines from the top to the bottom of the vehicle with line 1 being just above the object and line 17 just below. These traces were taken on every other line from a line selector oscilloscope. The dominant features of the bulldozer shown pictorially in Fig. 5.40 can be located in the traces of Fig. 5.41. The boom arm is in traces 3–6, the cab in 6–8, and the bulldozer blade in 9–13. The boom appeared to be the most characteristic feature and the width was of the order of one-eighth the width of the bulldozer.

In the identification experiments the photographic patterns were larger and the signal modulations were more pronounced. The camera

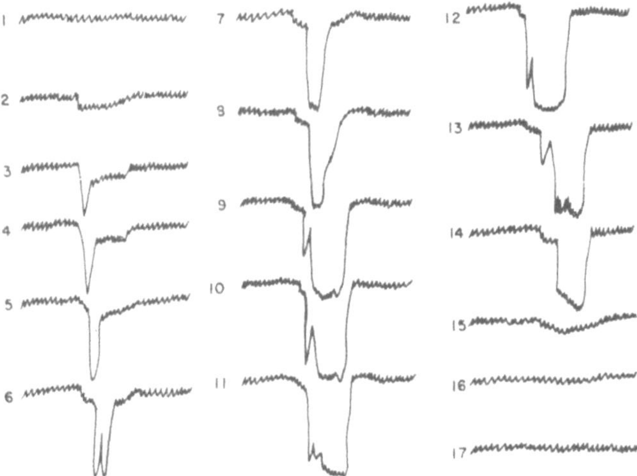

Fig. 5.41. Waveform of derrick half-track along the horizontal
as a function of vertical position in scan lines.

MTF had a small but noticeable effect on the recognition experiment
images but had nearly no effect on the identification experiment images.
In the latter case the equivalent bar pattern had a spatial frequency of
only about 100 TV lines per picture height.

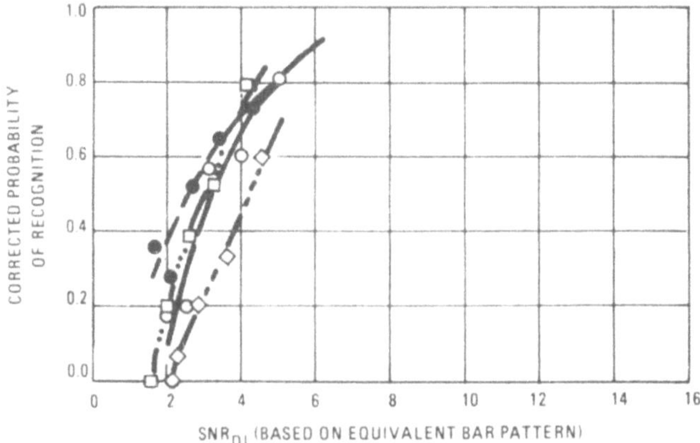

Fig. 5.42. Experiment 15. Probability of recognition versus SNR_{DI}
for a ($\bigcirc$) tank, ($\Diamond$) radar half-track, ($\square$) van truck, and ($\bullet$)
derrick bulldozer. Background was uniform.

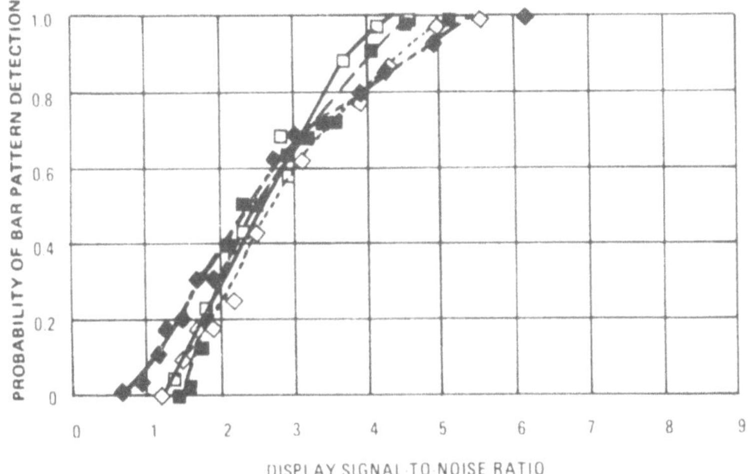

Fig. 5.43. Experiment 16. Probability of bar pattern detection versus SNR_{DI} for pattern of area equal to average real object area of $N =$ ($\square$) 329, seven bars; ($\blacksquare$) 396, seven bars; ($\lozenge$) 482, nine bars; ($\blacklozenge$) 635, 11 bars.

The results of the vehicle recognition experiment are shown in Fig. 5.42. It is seen that for 50% probability of correct recognition a threshold SNR_{DI} of 3.3 is needed. For purposes of correlation a bar pattern experiment was performed using a bar pattern of length and width approximately equal to the length and width of the vehicles, i.e., the vehicles are of 2:1 length-to-width aspect and the bar pattern length was about equal to the length of the vehicle and the width across all the bars was about equal to the height of the vehicles. The length-to-width ratio of a single bar in the bar pattern was 16 in each case. The bar pattern spatial frequencies were 329, 396, 482, and 635 lines/picture height and there were seven, seven, nine, and 11 bars in each pattern, respectively.* The background was black. From the results shown in Fig. 5.43, it is seen that the threshold SNR_{DI} is 2.9 $\pm$ 15%. The threshold SNR_{DI} obtained in the vehicle recognition experiment was 3.3, which is only 14% more than 2.9.

The second recognition experiment was performed by superimposing a background transparency over the same vehicle transparencies as were

---

* The number of bars in each pattern is actually immaterial so long as transient effects due to finite video bandwidth are avoided.

Fig. 5.44. Photograph of real background used for recognition experiments #1 grass,
#2 road, and #3 grass–trees.

used above. The background transparency as shown in Fig. 5.44 has
regions containing a road, a grassy field, and grass among trees. In the
experiment a vehicle was randomly chosen and located within one of
the randomly chosen background regions noted above. In all, 12 vehicle-
background combinations were shown at five randomly selected SNR_{DI}
values. The SNR_{DI} was calculated as in the first recognition experiment.
The results using a road background are shown in Fig. 5.45. The spread
in the data was small and a threshold value of SNR_{DI} is 3.8, which is
only 15% higher than that noted for a uniform background. The results
with grass background are shown in Fig. 5.46 and give a threshold value
of SNR_{DI} of 4.1; a value that is 24% higher than for a uniform background. In Fig. 5.47 a threshold value of 5.0 was obtained for the recognition of vehicles in a grass and tree background, which is 52%
higher than the uniform background case, 31% higher than the road
background case, and 22% higher than the grass background case. As
one would expect, increases in background complexity increase the
SNR_{DI} thresholds observed but the increases are not large.

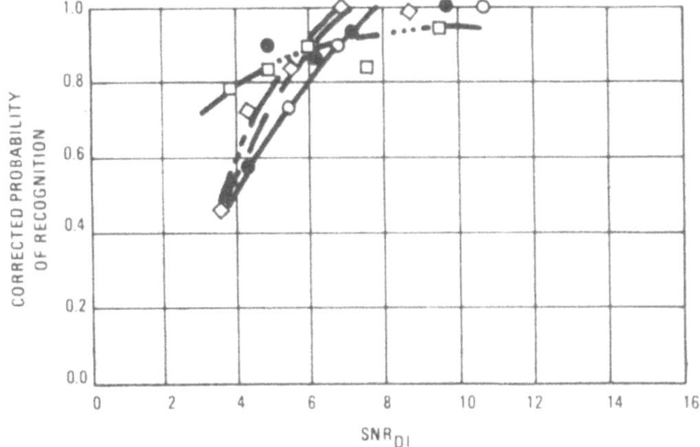

Fig. 5.45. Experiment 17. Probability of recognition versus SNR_{DI} for a (○) tank, (◇) radar half-track, (□) van truck, and (●) derrick bulldozer against a road background.

Considering Figs. 5.42 and 5.45–5.47, it is seen that, on the average, the derrick bulldozer and van truck have an SNR_{DI} threshold that is about 26% lower than that required for the tank and radar half-track. By visually comparing these objects in Fig. 5.41, it appears reasonable that less resolution would be needed to recognize the derrick bulldozer

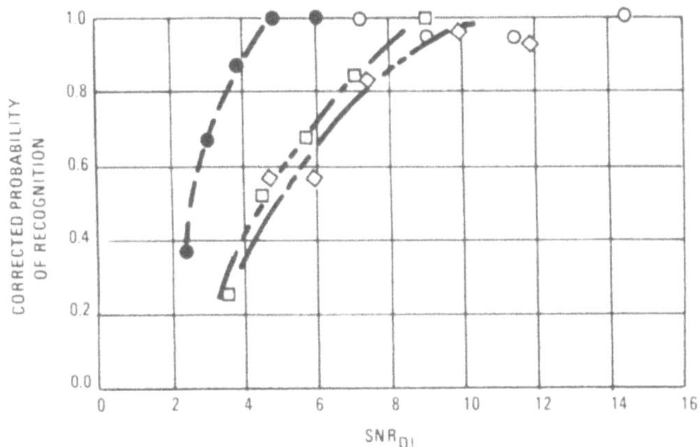

Fig. 5.46. Experiment 18. Probability of recognition versus SNR_{DI} for a (○) tank, (◇) radar half-track, (□) van truck, and (●) derrick bulldozer against a grass background.

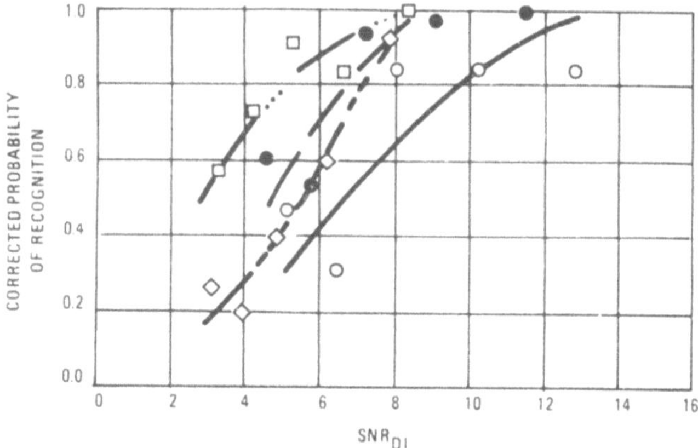

Fig. 5.47. Experiment 19. Probability of recognition versus SNR_{DI} for a ($\bigcirc$) tank, ($\diamondsuit$) radar half-track, ($\square$) van truck, and ($\bullet$) derrick bulldozer against a grass–tree background.

Fig. 5.48. Photographs of tank models used in identification experiments. Upper left, M47; upper right, M48; center left, Stalin; center right, Panther; lower center, Centurion.

and van truck. Assume that, in general, a different resolution is required to recognize each object and that if this resolution were used in the SNR_D calculation (instead of eight) that the value of SNR_D would be the same for the various images. The spread in the measured data, using the value of eight, should correspond to the spread in the required resolution and from the spread we have that the required resolution for target recognition is 8 ± 2 lines.

Next we progressed to vehicle identification experiments. The images for this experiment were transparencies of the M47 Patton, the M48 Centurion, the Panther, and the Stalin tanks as shown in Fig. 5.48. Again the photographs were taken to show the sides and tops of the vehicles from an angle of 45 deg from the horizontal. Two image sizes were used. In the first experiment the average image on the 8×10.7 in. display was 0.9 in.^{2} and subtended 1.3×2.6 in. at the observer's eye; and in the second experiment the image was about 2.2 in.^{2} and the angular extent was about 2×4 in. The objects and SNR values were randomly varied and the probabilities of correct identification were corrected for chance. For the calculation of SNR_{DI} the image area was assumed to be a bar whose length and width are equal to the length of the vehicle image and the width of the vehicle image divided by 13 using

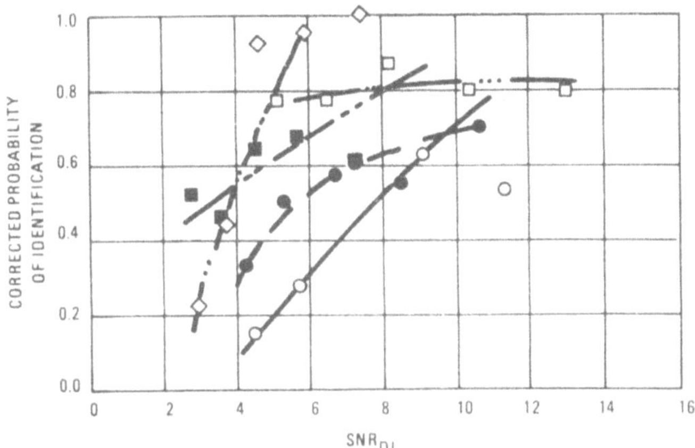

Fig. 5.49. Experiment 18. Probability of identification versus SNR_{DI} for tanks of angular extent 1.3×2.6 deg relative to the eye against a uniform background. Tanks were (O) M47, (□) Centurion, (■) Panther, and (◇) Stalin.

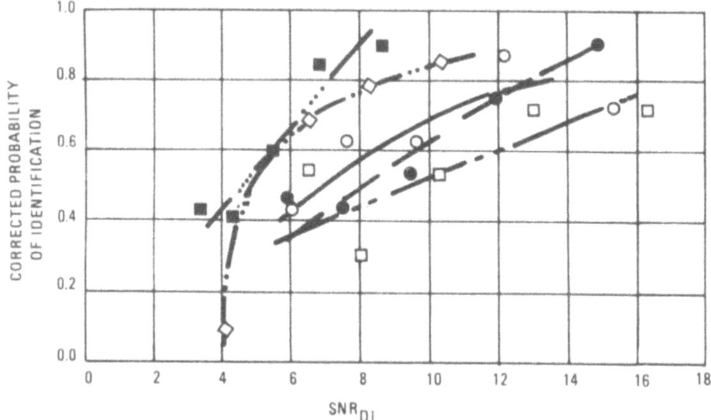

Fig. 5.50. Experiment 19. Probability of identification versus SNR$_{DI}$ for tanks of angular extent 2 × 4 deg relative to the eye against a uniform background. Tanks were (○) M47, (□) Centurion, (■) Panther, and (◇) Stalin.

the equivalent bar pattern concept discussed and illustrated in connection with Fig. 5.10. The results of the first identification experiment are shown in Fig. 5.49, where we obtain a threshold SNR$_{DI}$ of 5.2. This is nearly identical to the value obtained for the equivalent bar pattern, which had 13 bars with a length-to-width ratio of 26 and a spatial frequency of 137 lines/picture height, as can be inferred from Fig. 5.25.

For the second identification experiment we obtain a threshold value of 6.8 as shown in Fig. 5.50. An estimate of 6.2 is obtained from Fig. 5.25 for the equivalent bar pattern of 26:1 length-to-width ratio and a spatial frequency of 101 lines/picture height. From Figs. 5.49 and 5.50 it is seen that, on the average, the Panther and Stalin tanks are seen with an SNR$_{DI}$ at threshold which is about 42% smaller than that required for the M48 and M47. From the photographs of the tanks in Fig. 5.48 it is seen that the Stalin and Panther tanks appear more distinctive than the M47 and M48 so that the lower threshold SNR$_{DI}$ is not entirely unexpected.

In summary, it was hypothesized that the recognizability and identifiability of "real world" objects could be correlated with the discernability of an "equivalent bar pattern." The "equivalent bar pattern" was defined (in image space) in terms of a single bar in the pattern of length equal to the length of the image of the "real world" object and width equal to the "real world" image's width divided by k_d. The factor k_d is a number associated with a given discrimination level, e.g., eight for recognition. When the bar pattern was defined in the above manner it was found experimentally that the SNR_{DI} required to liminally discern the equivalent bar pattern was very nearly equal to the SNR_{DI} required to recognize the "real world" image when its SNR_{DI} is calculated on the basis of its area divided by k_d. It might be thought that defining the area of the bar in the equivalent bar pattern as one-eighth the area of the "real world" image's area and then calculating the SNR_{DI} for the "real world" image on the basis of one-eighth its area is redundant, but this is not so. The key reason for defining the bar pattern in the manner described is to define the required level of sensor resolution in terms of a spatial frequency. This permits us to take into account the effects of finite sensor apertures. As we noted previously, Johnson specified that an image, to be discerned at a given level of discrimination, must both be resolved and have a sufficient SNR_{DI}. In the above formulation SNR_{DI} is a function of resolution so that if the SNR_{DI} is sufficient, the resolution will be sufficient as well. The threshold SNR_{DI} required for the various levels of discrimination are estimated in Table 5.6. These

TABLE 5.6

Best Estimate of Threshold SNR_{DI} for Detection, Recognition, and Identification of Images

Discrimination level	Background	k_d, TV lines per minimum dimension	Threshold SNR_{DI} for a single bar of spatial frequency (in lines/picture height) equal to			
			100	300	500	700
Detection	Uniform*	1	2.8	2.8	2.8	2.8
Detection	Clutter	2	4.8	2.9	2.5	2.5
Recognition	Uniform	8	4.8	2.9	2.5	2.5
Recognition	Clutter	8	6.4	3.9	3.4	3.4
Identification	Uniform	13	5.8	3.6	3.0	3.0

* Treated as an aperiodic object.

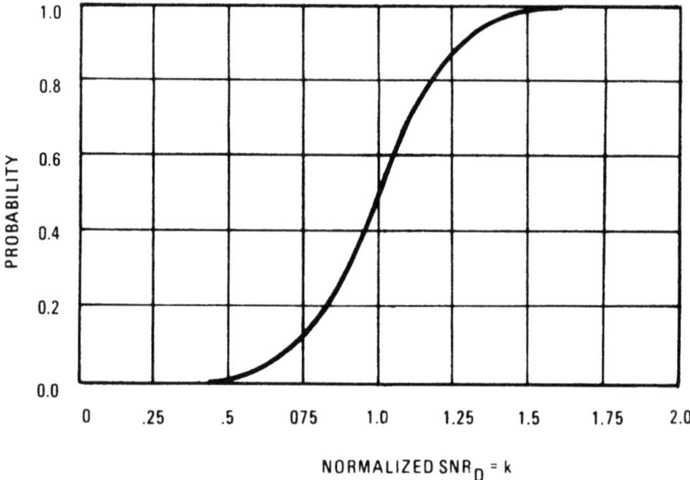

Fig. 5.51. Probability versus normalized SNR_{DI}. For any probability value, obtain SNR_{DI} from Table 5.6 for 50% probability. Find value of k for desired probability and multiply value of SNR_{DI} by k to obtain new value of SNR_{DI} required.

thresholds apply to a 50% probability but can be adjusted for any other level of probability by use of Fig. 5.51.

We noted an important caution in connection with the method of calculation. The "real world" objects are not periodic in general but rather assemblages of aperiodic objects of different sizes. These aperiodic objects which make up the object may be relatively isolated as in the case of the derrick on the tracked bulldozer or more clutterlike as in the case of the half-track. Thus the equivalent bar pattern approach weights cluttered scenes more heavily and may tend to be somewhat pessimistic, particularly at high line numbers.

5.8. PREDICTION OF ELECTROOPTICAL SENSOR RESOLUTION

The primary purpose of the analysis and experimentation discussed in this chapter is to provide methods of estimating the ability of a sensor-augmented observer to resolve scene detail. This capability is required to design a system with a reasonable expectation of actually serving its

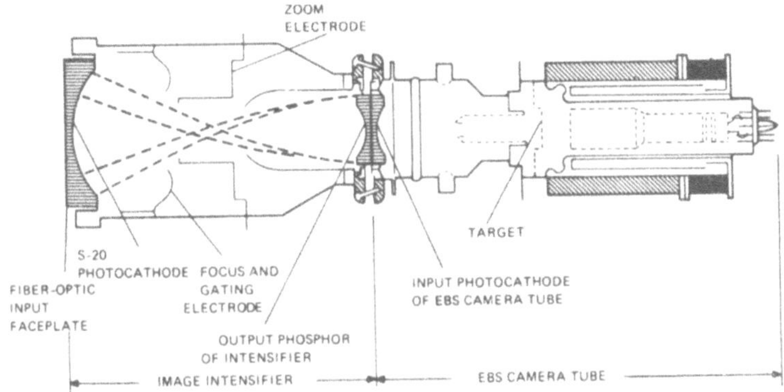

Fig. 5.52. Schematic of the intensified SEBIR camera tube with electronic viewfield zoom.

intended purpose. We noted previously that the threshold resolution versus photosurface irradiance characteristic is widely used to compare various sensors. Also, we noted that while these characteristic curves have become accepted, they are costly and time-consuming to obtain, particularly when a large number of sensor candidates must be evaluated. As will be shown, these characteristics can be readily calculated with the additional advantage of eliminating factors due to nonstandard measurement techniques.

In the usual laboratory resolution measurement bar patterns of high spatial fidelity are projected onto the photosurface, i.e., the projection lens is of high enough quality that its MTF can be ignored. Thus, the only MTF's of concern are those of the sensor itself. For threshold calculations we will assume that the input test image is the Air Force pattern consisting of three white and two black bars with each bar having a length-to-width ratio ε of 5. The sensor is assumed to be an intensified SEBIR camera as shown in Fig. 5.52.* The intensifier shown incorporates electronic viewfield zoom by means of varying the input photosurface diameter from 80 to 40 mm. Further viewfield zoom may be obtained in the SEBIR camera tube if desired.

For generality, we will perform our analysis in terms of the intensifier photocurrent i, which is related to the image irradiance E_I through

---

* In Fig. 5.52 the SEBIR camera tube is designated the EBS Camera Tube, which is the name used by the Westinghouse Electric Corp. for its version. NATO calls tubes of this class EBSICON.

the equation

$$i = \sigma_p A \int_0^\infty R(\lambda) E_I(\lambda)\, d\lambda \tag{50}$$

where σ_p is the photosurface responsivity at its spectral peak, $R(\lambda)$ is its relative spectral response, A is the effective photosurface area, and λ is the wavelength of the image irradiance. For specific test sources such as a tungsten lamp operated at 2854°K Eq. (50) simplifies to

$$i = \sigma_T A E_T \tag{51}$$

where σ_T is the specific response to the tungsten source irradiance E_T. Typically σ_T is 3–8 mA/W in the current state of the art for an S-25 photosurface. Equation (35) is to be used in making the calculations and the constants assumed are $t = 0.1$ sec, $\varepsilon = 5$, $\alpha = 4/3$, $G_T = 2 \times 10^4$, $e = 1.6 \times 10^{-19}$ C, and $eI_s = 8 \times 10^{-25}$ C²/sec. The quantity eI_s is the mean-square preamp noise per half-cycle of bandwidth and is obtained from

$$2eI_s\, \Delta f_V = I_p{}^2 \tag{52}$$

where $I_p{}^2$ is the mean-square preamp noise referred to as the preamp noise input. Equation (52) assumes that the preamp noise is white. The value of 8×10^{-25} C²/sec for eI_s assumes an rms preamp noise of 4×10^{-9} A in a bandwidth of 10^7 Hz. For the test pattern of 5:1 length-

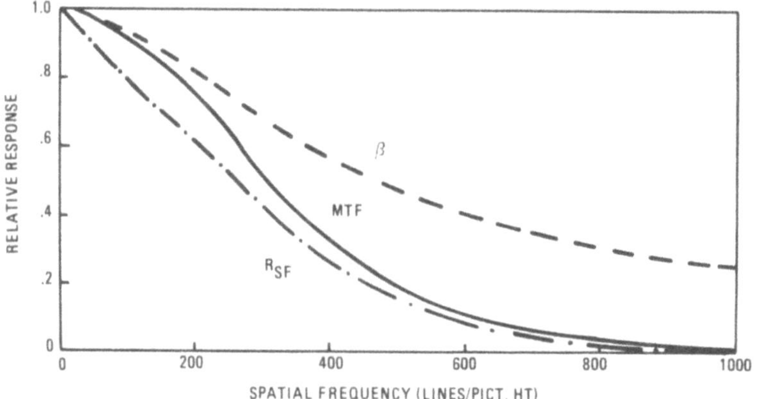

Fig. 5.53. Modulation transfer function MTF, noise filtering factor β, and square wave flux response of the I-SEBIR camera tube.

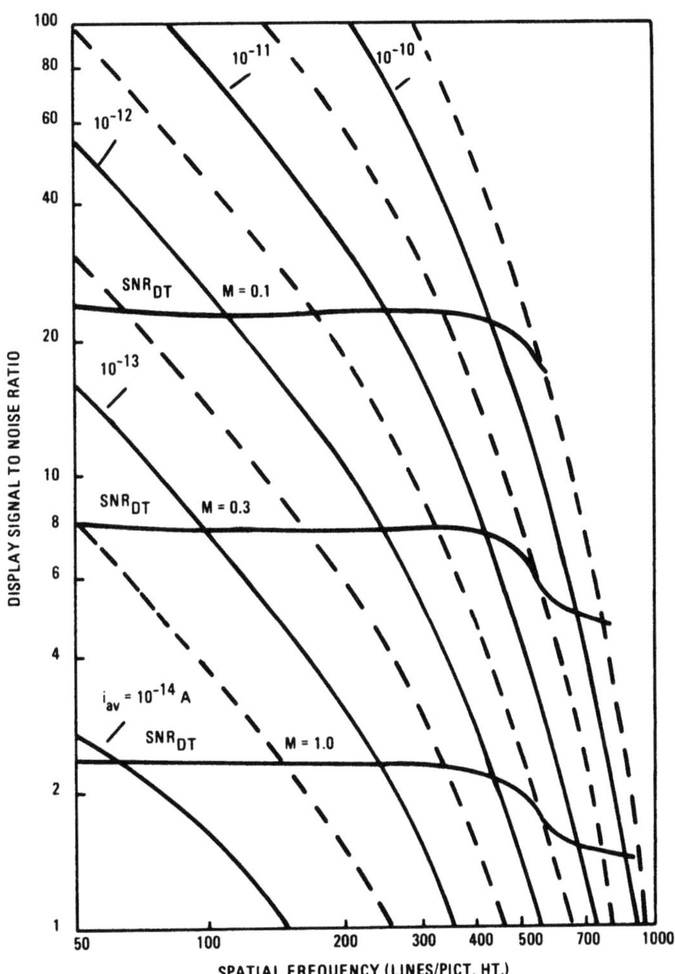

Fig. 5.54. The SNR_{DI} obtainable and threshold, SNR_{DT}, for various average input photocurrents and image contrasts as a function of bar pattern spatial frequency. Dashed average current curves are $\sqrt{10}$ times larger than the value of the current curves below them.

to-width ratio ξ_{yT} and Γ_{yT} are approximately unity and Eq. (35) reduces to

$$\text{SNR}_{DI} = \left(\frac{t\varepsilon}{\alpha e}\right)^{1/2} \frac{R_{\text{SF}}(N)}{N} \frac{2MG_T i_{\text{av}}}{(G_T^2 \beta_T i_{\text{av}} + I_s)^{1/2}} \tag{53}$$

The MTF, $R_{\text{SF}}(N)$, and β_T quantities are plotted in Fig. 5.53.

Using the above quantities and functions, the SNR_{DI} obtainable from the sensor are calculated for various input photocurrents in Fig. 5.54. In this figure the input image contrast is assumed to be unity. The threshold SNR_{DI} required by the observer for 50% probability of detection are taken from Fig. 5.30 for the optimum viewing case and plotted in Fig. 5.53. Note that as the input image contrasts M are reduced, the SNR_{DI} obtainable from the sensor are reduced in turn. To minimize the number of curves needed, it is convenient to assume that the observer's threshold SNR_{DI} requirement increases as image contrast decreases using the formula

$$\text{SNR}_{DI}(\text{threshold}) = \frac{\text{SNR}_{DI}(\text{threshold for } M = 1.0)}{M \text{ actual}} \tag{54}$$

Again note that the observer's threshold SNR_{DI} actually remains constant with M and that Eq. (54) merely represents a mathematical convenience.

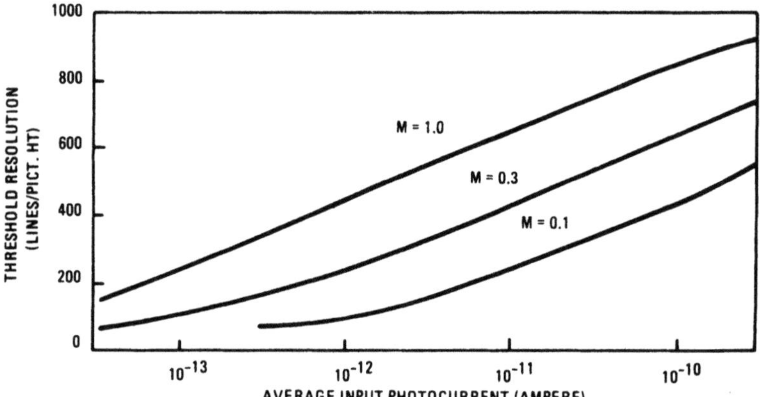

Fig. 5.55. Threshold resolution versus average input photocurrents for various image contrasts for the assumed I-SEBIR camera tube (Fig. 5.52).

The intersection of the SNR_{DI} obtainable from the sensor curves and the threshold SNR_{DI} required by the observer curves gives the threshold resolution versus input photocurrent characteristic as shown in Fig. 5.54. The threshold resolution versus photosurface irradiance curves can also be computed using Eq. (50) or (51) (Fig. 5.55). In the above calculation the average photocurrent was used as a parameter and thus the conversion to irradiance will be in terms of average irradiance. Ordinarily, tube manufacturers specify their limiting resolution versus irradiance characteristic in terms of the test pattern highlight irradiance rather than the average. To convert from average to highlight photocurrent (or irradiance), use the relation

$$i_{\max} = (M + 1)i_{av} \tag{55}$$

which is seen to be primarily significant for the higher image contrasts.

5.9. EDITOR'S POSTSCRIPT

Section 5.8 above and previous material have been flavored strongly with the point of view of television systems designers and analysts.

Infrared systems such as FLIRs are quite similar in their analysis. If one goes back to the equations showing the various relations between SNR_{DI} and SNR_V it becomes quite clear that there is no distinction about the sensor that provides SNR_V.

Now we may note that the usual texts and reports on IR system design—for example, see Sendall and Rosell (1972) for an excellent TV *versus* IR comparison—show that

$$(S/N)_V = \Delta T \frac{\delta_x \delta_y \eta_s D}{2(\Delta f)^{1/2}} \int_\lambda \tau_0(\lambda) D^{**}(\lambda) M'(\lambda, T)\, d\lambda$$

For a FLIR sensor, the noise-equivalent input of interest is the noise-equivalent ΔT (NET), the temperature difference between adjacent large objects which would provide an SNR equal to 1. The usefulness of NET is that for a well-designed system the video SNR for an input of ΔT can be predicted directly. That is,

$$(SNR)_V = \frac{\Delta T}{NET}$$

Thus one can compute the SNR_V for various inputs to be considered, and from SNR_V one can calculate SNR_{DI}. From that point on, the estimation of observer performance is identical. For details about the equations and computational procedures which are outside the prime interest of this book, the reader is again referred to Sendall and Rosell (1972) or to the texts by Jamieson *et al.*, *Infrared Physics and Engineering*, McGraw-Hill (1963), or by Richard Hudson, *Infrared System Engineering*, John Wiley & Sons (1969).

IMAGE REPRODUCTION BY A LINE RASTER PROCESS

Otto H. Schade, Sr.

6.1. EDITOR'S INTRODUCTION

The last part of Chapter 2 reviewed a number of psychophysical experiments concerned with sampled imagery, i.e., imagery with a raster as an integral part of its composition. At the end of that section we quoted from F. T. Thompson's experiment. He was concerned with removing the line raster features from a television display in order that the viewer could (and also chose to) move closer to see more detail without seeing the objectionable lines.

We are not so much interested, as was Thompson, in improving the display for the sake of bringing larger displays into the livingroom; rather we are concerned with suppressing the raster to permit the observation of more detail usually not discernable because it is "swamped" by the raster frequency signal level.

Three symposia were held on this topic in 1971, one in the spring, sponsored by IDA, and two in the fall, by Perkin-Elmer.

The Perkin-Elmer symposium contained papers by R. Scott, E. D. Brown, R. J. Arguello, H. Sellner, T. W. Barnard, B. M. Boyce, S. J. Kishner, K. Preston, J. F. Cunniff, and R. M. Vesper which were published in a document simply entitled *A Symposium on Sampled Images*.

The IDA symposium (Aug. 1971) included papers by L. M. Biberman, R. L. Legault, F. A. Rosell, O. H. Schade, A. D. Schnitzler, and H. L. Snyder, much of which has been rewritten to form Chapter 6 and 7 of this book.

In writing the synopsis to the IDA spring meeting, several things became quite clear (if not quantitative) to me and in the following material I abridge those remarks as an introduction to the quantitative treatments by Schade and then in Chapter 7 by Legault.

Since I quote myself, I take some editorial liberties, among them neglecting quotation marks.

Periodically a need arises for sampled data systems to transmit photographs, maps, or other images. Each time the problem gives rise to questions such as: "What does sampling do to picture quality?" and "What must the sampling frequency and format be to minimize image deterioration?"

The use of sampled data systems affects the signal-to-noise ratio of the imagery and, in addition, introduces spurious signals caused by a heterodyne-like process. The need to maintain desirably large (near unity) modulation transfer function (MTF) values over the entire spatial frequency of interest must be weighed against the need for prefiltering before sampling so as to minimize aliasing. This is the dominant tradeoff.

In photographic imagery the actual detectors are silver halide grains. A fairly large number of grains usually define the image. Since more than a planar layer of grains is involved, because of the combination of depth of focus and the third dimension (depth) of the emulsion, a significant number of grains are involved in the reproduction of even a point image.

The spread function of the optics used is usually large compared to the dimensions of an individual grain, so that in photography the problems involved in sampling the imagery with a finite number of detectors often become insignificant in relation to other considerations.

In the case of image formation by a matrix of detectors with electrical outputs the problems associated with fabrication, cost, interconnections of wiring, number of amplifiers, and so forth serve to constrain the number of detectors used. Thus there tend to be many fewer detector elements per image in such a matrix than in film, and the limitations associated with sampling the imagery must be considered.

It is important to estimate accurately how few individual detectors one can use in the matrix without significantly degrading the image.

Basically, there are three factors affecting this problem of finite sampling and image quality: (1) The number of samples per image.

(2) The signal-to-noise ratio per sample. (3) The generation of spurious signals by the sampling process.

Earlier we discussed Johnson's criteria and showed the number of resolvable lines on an output display necessary per linear "critical" dimension of a target image for various visual tasks, such as detection or recognition. In Chapter 3 we discussed the number of errors of a photointerpreter versus a quantity known as modulation transfer function area (MTFA). It is clearly shown that a high value of MTF is desirable relative to the modulation threshold of the observer to ensure few interpreter errors.

Unfortunately, it is also true that a spatial frequency passband larger than half the sampling (or raster) frequency permits the transmission of spurious "sideband" frequencies that are reflected back into the passband. That is, with a sampling system it is possible for image transformation effects to occur that cannot be simply described by an MTF. These effects are called aliasing, and they occur because, although the sampling process can be linear, it is not translationally invariant. If aliasing occurs, an input of one spatial frequency can cause a spurious output at another spatial frequency. Under such circumstances an analysis of sampling theory is required to understand the output. The central result of sampling theory (Nyquist's theorem) demonstrates that it is quite impossible to obtain useful, unambiguous information from a sampled system concerning input spatial frequencies that are greater than $1/2L$, where L is the appropriate "detector" or "display" sample spacing. Higher spatial frequencies could be observed in the output, but these would be impossible to interpret, i.e., they would be effectively spurious. In the absence of other constraints the sampling process can thus set the resolution of the system at $1/2L$. If a space-filling detector or display element configuration (abutting detectors) is used, this of course does not mean that bright target images which at the detector plane have a dimension smaller than L cannot be seen, but rather that these target images cannot be localized any better than $\pm 1/2L$.

Besides imposing a resolution limit on the system, the sampling process can cause higher spatial frequencies to be translated back into the useful passband of the system. For the most obvious situation where detector and spot sizes just fill the plane (the detector part of this is necessary for efficient detection) aliasing can convert spatially periodic images with frequencies between $1/2L$ and $1/L$ back into the 0 to $1/2L$

frequency passband. When dealing with periodic inputs this could cause a false signal. However, as Schade shows in the material that follows, if a lens system is designed so that its MTF is zero for frequencies greater than $1/2L$, no frequencies greater than $1/2L$ will fall on the detector plane and no aliasing problem will occur; i.e., no outside frequencies will be translated into the operational passband of the system.

The only problem here is that any realizable lens filter that has an MTF of zero for frequencies greater than $1/2L$ is likely to have an MTF much less than unity for frequencies approaching $1/2L$. Thus there is a tradeoff between the desire to eliminate aliasing and a desire not to degrade the MTF for spatial frequencies less than $1/2L$. Schade suggests for commercial TV situations that the use of a lens MTF that drops to 35% at a spatial frequency of $1/2L$ is tolerable.

Where the prefiltering tradeoff should fall depends critically on the type of images being viewed. When dealing with periodic inputs aliasing can cause a false signal. Without prefiltering, when counting a row of small objects where the fundamental frequency of the row is higher than $1/2L$ it would be possible to count a smaller number than is actually present. If there were no post-sampling filtering, another aliasing effect would be present, and a naive photointerpreter might neglect the sampling process and count a larger number of objects than is actually present. However, if the overlap of the output display were adjusted to give a flat picture for a uniform input, the necessary post-filtering to eliminate transformations to higher spatial frequencies would be automatically provided. Since no amount of post-filtering can eliminate the down conversion of spatial frequencies caused by aliasing, this is the principal design consideration. It should be remembered that in real space the effects of aliasing are sharply localized.

When one is dealing with localized signals one must consider both frequency and phase relations. The physical situation of the sampling process is such that a signal derived from an x to $x + L$ sampling interval cannot be translated by the sampling process out of that interval. This means that the high-frequency part of a sharp edge cannot produce a ghost at a distance more than the detector or display element spacing L away from that edge. Thus for an ordinary aperiodic scene no false signals can be produced, and prefiltering is not an important design goal. The sampling process may produce some structure near sharp edges, but unless several edges are close together, no effects will occur that cannot be explained by the loss of information concerning input frequencies greater than $1/2L$.

If periodic images are likely to be viewed, it would be prudent to experimentally investigate the photointerpretation problem caused by aliasing (clutter formation) in order to help the system designer make the necessary tradeoff. If a low-frequency target were to be viewed against a high-frequency periodic background, aliasing could only cause a significant clutter problem if the amplitude difference between the target signal and the high-frequency part of the background were less than a factor of three.

At a minimum, photointerpreters should be made aware of the aliasing phenomenon when using sampled systems.

Now having drawn a number of conclusions mostly judgmental, almost subjective and qualitative, it is time to turn to the quantitative and rigorous analysis of the sampling effects of the raster form of imagery that follows, first in this chapter by Schade and then in Chapter 7 by Legault.

6.2. NOTATION

a_e	Noise equivalent sampling area
B_C	Critical luminance for flicker threshold
CRT	Cathode ray tube
d	Viewing distance
f	Spatial frequency, cycles per unit length
f_m	Modulation frequency
f_r	Raster frequency
f_e	Noise equivalent passband expressed in cycles/unit length
f_d	Difference frequency $(f_r - f_m)$
I	Intensity
$I(y)$	Intensity distribution in y
L	Luminance, foot-lamberts
$N = 2f$	Television line number (half-cycles/frame dimension V)
$N_\delta = 1/\delta$	TV line number at which the length of one half-cycle equals the aperture diameter δ

N_e	Noise equivalent passband expressed in TV lines/frame dimension V
$n_{(1)}$	Particle density
n_r	Raster constant $=$ raster line number $= f_r$
MTF	Modulation transfer function $\tilde{r}(f)$
$\tilde{r}$	Sine wave response factor
$\tilde{r}_c$	Sine wave response factor of camera
$\tilde{r}_d$	Sine wave response factor of display
$\tilde{r}_{\text{sp}}$	Spurious frequency response factor
r_0	Radius of Gaussian spot at amplitude $1/\varepsilon$
S	Spread function
V	Vertical frame dimension
$W_{L(0.5)}$	Middle of line profile between one-half amplitude values
$W_{S(0.5)}$	Middle of spot profile between one-half amplitude values
W_{FF}	Linewidth determined by the shrinking raster method
Y	Vertical image coordinate
X	Horizontal image coordinate
α	Angle between raster line and pattern line
θ	Angle between direction of modulating frequency f_m and x coordinate
ψ	Total aperture flux

6.3. RASTER PROCESSES

6.3.1. Raster Constant and Frequency

The formation of images by lenses or optical systems is continuous in both coordinates of the image format. Every point in the image area undergoes an aperture process and the aperture shape becomes indistinguishable in areas of constant luminance. In an image the mean flux can be considered as a "carrier" flux of constant luminance L "modulated" by a "signal," the intensity variations of the image.

Printing, facsimile, and television are sampling processes in which the number of aperture positions is finite in one or both coordinates of the image frame. The image flux is no longer continuous in two coordinates but contains periodic components. An arrangement limiting aperture positions to a fixed number of uniformly spaced points in the image frame is termed a point raster; an arrangement providing continuous aperture positions along uniformly spaced parallel lines is termed a line raster. The raster constant n_r specifies the number of aperture positions in the length unit of a geometric arrangement of points or lines; it does not specify the dimensions of the "points" or "lines" themselves, which are determined by the geometry of the sampling apertures used with the raster process.

The conversion of continuous luminance functions $L(x, y)$ into one-dimensional electrical time functions $I(t)$ and reconversion into continuous two-dimensional luminance functions in a television system involves scanning of the format area with an "aperture" along the uniformly spaced parallel lines of a line raster. The scanning process in the camera and display yields a set of continuous electrical intensity functions $I_x(t)$ corresponding to luminance functions $L(x)$ along the lines, whereas luminance functions $L(y)$ are transmitted as discrete amplitude samples $I_y(t)$ taken at intervals Δy determined by the raster line spacing. It follows at once that continuity in the displayed image requires a display aperture having a particular spread $S(y)$ to fill the interline spaces of the raster and establish continuity in y, whereas the aperture spread $S(x)$ could be very much smaller, making the resolution in the image anisotropic. Similarly, the effective spread $S(y)$ of the sampling aperture in the camera must have a particular value to prevent loss of information contained in the interline spaces of the raster. Continuity in y can be obtained by selecting a raster line density to provide a large overlap of aperture positions in successive line scans as illustrated in Fig. 6.1. A high raster line density, however, is wasteful in terms of the electrical frequency channel and raises two questions: Is a "flat," i.e., structureless, field necessary in both camera and display? What is the optimum relation of raster line density and aperture size and shape needed in order for a true image reproduction to be achieved?

Most television displays have a visible line structure on the screen, and increased viewing distances are required to effect integration by the eye into a structureless field. An image containing a visible line structure may appear to be sharper, but more detail becomes visible when the line structure is removed by integration. This can be demonstrated convincingly by modulating a CRT producing a pronounced

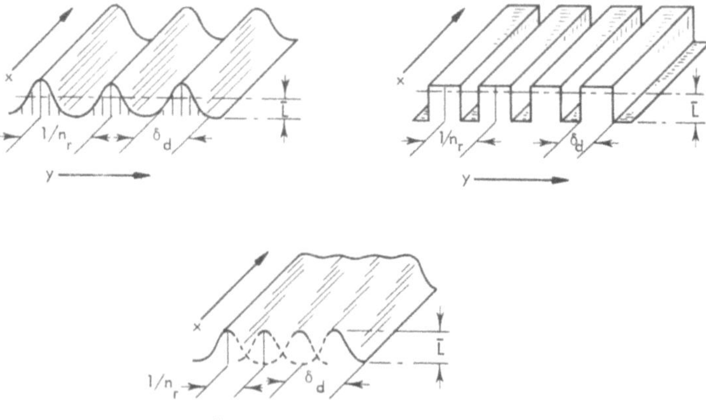

Fig. 6.1. Luminance distribution and "carrier" waves in the y coordinate of line rasters.

line structure with wideband noise. It will be observed that the noise is much more visible when the interfering line structure is removed by defocusing the CRT spot or by increasing the viewing distance. The line structure is an interfering signal which, like noise, prevents detection of small detail.

Various other effects occur when the effective sampling apertures of the camera and display are too small relative to the raster line spacing. Diagonal lines become staircases, spurious diamond-shaped patterns appear in horizontal line wedges, low-frequency beat patterns occur in "vertical" resolution charts of parallel lines at higher frequencies, and the reproduction of significant detail depends on position relative to the raster lines.

Quantitative specifications can be derived by convolution of luminance and time functions in the space and time domains. An analysis in the frequency domain, however, is more convenient.

6.3.2. Carrier Wave and Line Structure

Areas of constant luminance are reproduced in a line raster process by the sampling aperture δ_d of the display system as a luminous flux pattern in which the luminance is constant in the direction x parallel to the raster lines but contains a more or less pronounced periodic component in the y coordinate defined as the coordinate perpendicular to the raster lines as illustrated by Fig. 6.1. The following analysis of conditions in the y coordinate of a line raster applies to optical as well as television processes* and also to point rasters which cause periodic components in both x and y coordinates.

The periodic component can be regarded as a constant carrier wave added by the raster to the continuous carrier flux of a normal aperture process. The mean signal flux from the analyzing aperture δ_c in the camera determines the video signal intensity I_y and the scale factor of the image flux in the display. It is seen by inspection of Fig. 6.1 that the length of the carrier wave is the reciprocal of the raster constant and that the raster constant n_r is the frequency f_r of the carrier wave, $n_r = f_r = 1/\Delta y$, while waveform and relative amplitude of the carrier wave are determined by the raster line density and the geometry of the synthesizing aperture δ_d of the display system.

A Fourier analysis of this "pulse carrier wave" shows that the luminance distribution L_y contains the constant signal term $\bar{L}$ and a series of harmonic cosine waves:

$$L_y = \bar{L}(1 + 2 \sum \tilde{r}_{d,pf_r} \cos(p \cdot 2\pi f_r y)), \qquad p = 1, 2, 3, \ldots \qquad (1)$$

The cosine terms specify the harmonic components of the carrier wave, which are integral multiples of the fundamental frequency f_r. Their relative luminances are specified by the sine wave response factors $\tilde{r}_d$ of the modulation transfer function (MTF) of the particular aperture δ_d in the y coordinate at the frequencies pf_r of corresponding carrier harmonics. The cosine wave components are in phase at the aperture center (on the raster line) when the aperture has axial symmetry[†] and its reponse decreases asymptotically to zero. When the response characteristic has an oscillatory form the phase may reverse at each zero response point. Examples of a numerical synthesis of the luminance distribution expressed by Eq. (1) are illustrated by Figs. 6.2a and 6.2b for the MTF shown in Fig. 6.3. The cross section of an isolated line (Fig. 6.2a) is constructed by selecting a low raster frequency ($f_r = 100$

---

* A two-dimensional Fourier analysis of the television picture was presented in an early paper by Mertz and Gray (1934).

† Apertures with asymmetric cross sections introduce phase shifts between cosine terms.

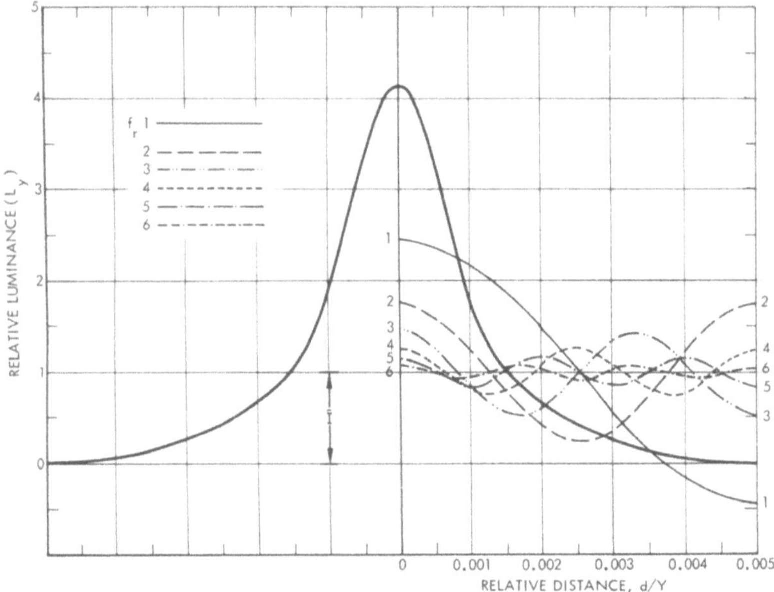

Fig. 6.2a. Synthesis of line profile from its frequency components [Eq. (1)]; $f_r = 100$ cycles in Fig. 6.3.

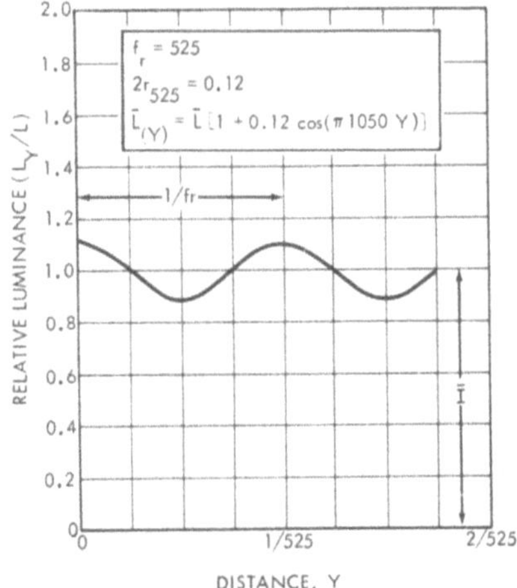

Fig. 6.2b. Luminance distribution of display with an aperture $\delta_d > 1/n_r$ passing only one cosine term of carrier wave.

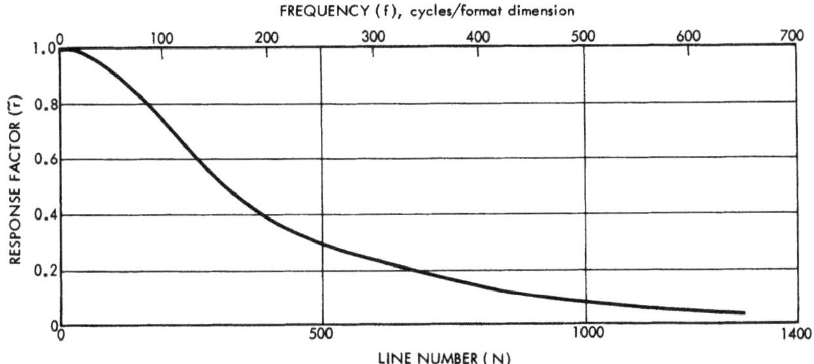

Fig. 6.3. MTF_d of a display tube.

cycles in Fig. 6.3). When the raster line number is increased to 525 lines ($f_r = 525$ cycles) the MTF (Fig. 6.3) gives a value only for the fundamental component which results in a large overlap of the line images and the field intensity function shown in Fig. 6.2b.

The presence of a pronounced line structure in the image is highly undesirable. Perfect continuity is restored when none of the carrier-wave components are reproduced by the aperture δ_d, i.e., when the aperture response is zero at frequencies $\geq f_r$. Practical display devices usually have an aperiodic response characteristic. In some cases the response is oscillatory but is usually low beyond the first zero. A substantially continuous of "flat" field* is therefore obtained when the sine wave response factor at f_r has a value

$$\tilde{r}_{(f_r)} \lesssim 0.005 \tag{2}$$

This response factor causes a ripple amplitude of 1%, i.e., a peak-to-peak luminance variation of 2%. The aperture process in the display device is followed by other imaging processes, for example, by the process of vision or by a photographic process. It is therefore unneces-

---

* The "flat field" requirement refers to a structure-free reproduction of a continuous field of uniform intensity by a line raster process. In the camera it specifies a uniform charge readout leaving no interline charges on the storage surface, and is satisfied when the sum of the effective line image cross sections of the scanning aperture spaced at raster line distances yields a constant intensity function $I(y)$.

sary to restrict the response of the display device alone by Eq. (2) but rather the overall sine wave response of the aperture system following the raster process.* It is obvious that line pairing (which may occur in an interlaced raster) introduces a lower raster frequency component ($f_r/2$) which destroys the continuity of the field and increases the visibility of the raster lines.

6.3.3. System Response to Sine Wave Test Patterns

A line raster has no effect on the sine wave response or modulation transfer function (MTF) of the apertures δ_c and δ_d in the x coordinate (parallel to the raster lines), in which the aperture process is continuous. The discrete aperture positions in the y coordinate affect the total y response of the two apertures in a different manner.

The analyzing aperture δ_c of the camera "samples" the flux of a test pattern in the y direction at the raster points only; all other aperture positions are "blocked" by the raster. What is left of the normally continuous aperture signal is a series of samples of its response at regularly spaced distances $\Delta y = 1/n_r$ as indicated by Fig. 6.4(a). The reader may visualize the raster as an opaque plate with *very fine* slits (holes for a point raster) through which the observer or a photoelectric device views the test pattern from a fixed distance. He can control δ_c by varying the spacing between the raster plate and test object. When the modulation frequency f_m of the test pattern is varied the sample amplitudes depend on the degree of integration by δ_c and vary in direct proportion to the normal sine wave response of δ_c.

The discontinuous electrical amplitude samples $I(y)$ obtained by the line raster process in the camera represent a pulse carrier wave with infinitesimal pulse width of spatial frequency f_r, amplitude modulated by the spatial frequencies f_m contained in the image, which are limited by the MTF$_c$ of the camera aperture δ_c. The fundamental frequency f_r of the spatial carrier wave is equal to the number of raster lines per unit length. The spatial luminance functions $L(y)$ are converted by the sequential scanning process into electrical time functions $I_y(t)$ contained in the video signal, which can be displayed with an electrical sampling

---

* For the case of a photographic recording the luminance values L_y and L in Eq. (1) are intensitites or densities.

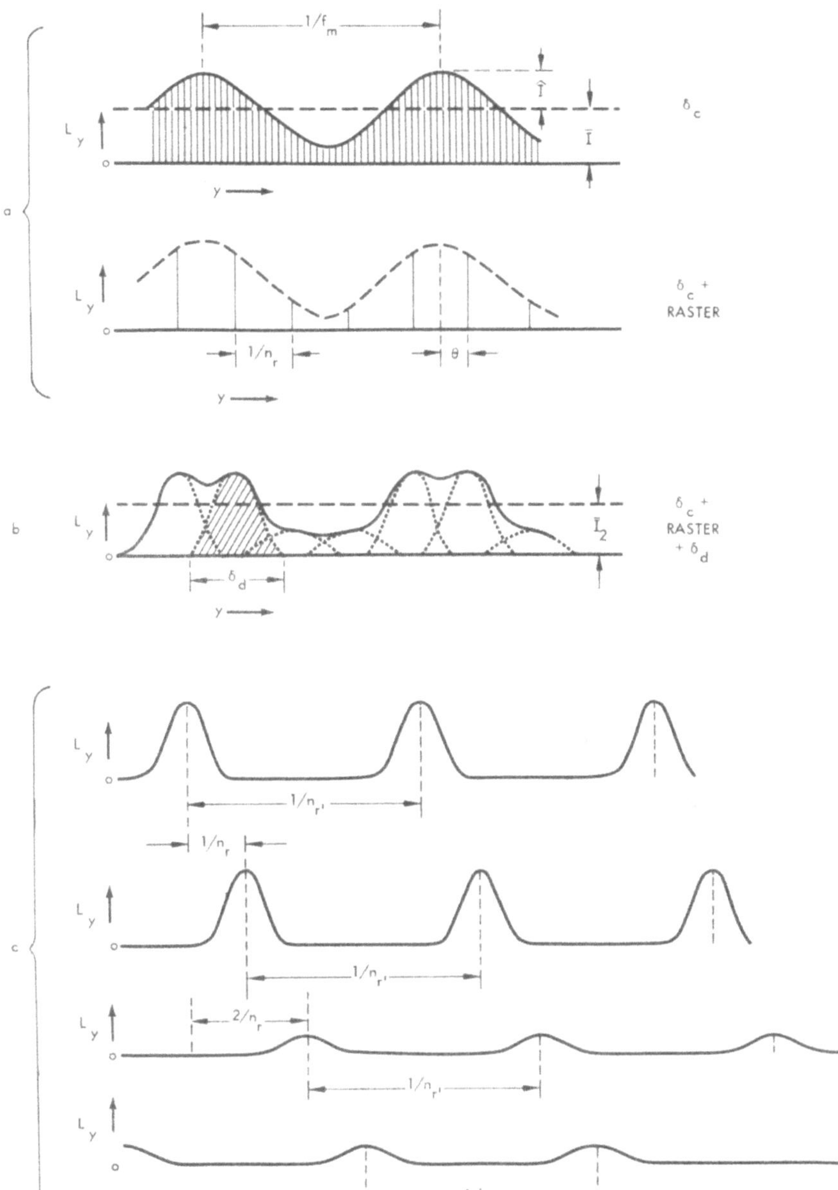

Fig. 6.4. Development of raster equation. (a) Input function and samples taken by small aperture δ_c; (b) luminance function synthesized with δ_d; (c) analysis of displayed function by interlaced carrier waves.

circuit on an oscilloscope.* The time signals are converted back into spatial modulated carrier waves in the display system by a synchronized scanning process where the sample amplitudes are replaced by the spread function of the display aperture δ_d. In terms of modulation theory, the pulse carrier is "demodulated" by a low-pass filter, the MTF_d of the display aperture δ_d, which should have a particular form to restore a continuous undistorted modulation envelope from the transmitted samples.

For a linear system the luminance of the light flux from the synthesizing aperture δ_d is proportional to the signal amplitude $I(y)$ delivered by δ_c at corresponding raster points. The reproduced waveform, however, may be only an artifical approximation of the test pattern waveform determined by the raster constant and the geometry of the aperture δ_d as illustrated by Fig. 6.4(b). The fundamental sinewave response and the waveform distortion can be evaluated by a Fourier analysis. For this purpose the periodic wave may be regarded as the sum of a series of interlaced carrier waves, each having a constant amplitude and a wavelength $1/n_r'$ which is longer than the normal raster period (see Fig. 6.4c). These component carrier waves are displaced in phase by distances $1/n_r$, $2/n_r$, etc. with respect to one another and can be expressed by Fourier series [Eq. (1)] differing only in amplitude and phase of the terms. A vectorial addition of corresponding terms yields an expression for the waveform. For the conditions that the average intensity $\bar{L}_2$ in the image of the test pattern has the same numerical value as the test pattern intensity $\bar{L}_1$ and the transfer ratio of signals (gamma) is unity, the general expression for the intensity function $I(y)$ resulting from a carrier modulation by a frequency f_m is the following equation:

$$
\begin{aligned}
L(y) = \bar{L}\Big(1 + 2 \sum \tilde{r}_{d,pf_r} \cos(p \cdot 2\pi f_r y) \qquad &(C)\\
+ \hat{L}_{f_m} \tilde{r}_c \tilde{r}_d \cos(2\pi f_m y + \theta) \qquad &(f_m)\\
+ \hat{L}_{f_m} \tilde{r}_c \sum^p \{\tilde{r}_{d,(pf_r+f_m)} \cos[2\pi(pf_r+f_m)y+\theta]\} \qquad &(S)\\
+ \hat{L}_{f_m} \tilde{r}_c \sum^p \{\tilde{r}_{d,(pf_r-f_m)} \cos[2\pi(pf_r-f_m)y+\theta]\} \qquad &(D)
\end{aligned}
\tag{2}
$$

---

* The sampling functions $I_y(t)$ in a standard U. S. television system contain 490 sampling pulses of 0.1176 μsec duration, occurring at line frequency intervals ($\Delta t = 1/15{,}750$ sec), and repeat in the total frame time $T_f = 1/30$ sec.

where $p = 1, 2, 3, \ldots$; f_m is the modulation frequency, cycles/unit length; f_r is the raster frequency, number of sampling points/unit length; y is the distance along the y coordinate; $\bar{L}$ is the mean luminance of the test object waveform; $\hat{L}_{f_m}$ is the crest luminance of the sine wave test object; $\tilde{r}_c$ is the sine wave response factor of the camera at f_m; $\tilde{r}_d$ is the sine wave response factor of the display system at f_m; $\tilde{r}_{d,\text{(index)}}$ is the sine wave response factor of the display at the index frequency; and θ is the phase displacement between $\hat{L}_{f_m}$ and the raster lines.

The first term (C) contains the dc level $\bar{L}$ and an infinite number of steady carrier frequency components pf_r (p is an integer) with amplitudes $2\bar{L}\tilde{r}_d$ depending only on the MTF_d of the display system. The second term (f_m) is the normal MTF product $\tilde{r}_c\tilde{r}_d$ of the system as obtained without raster process at any modulation frequency.

The third and fourth terms (S and D) are modulation products (sidebands) generated by sum and difference frequencies with the carrier components.

The entire spatial frequency transfer characteristic for the y coordinate of the television process is shown by the graphic representation in Fig. 6.5. The MTF_c (of the camera) under the input frequency scale of the raster characteristic is the product of the MTF's of *all* two-dimensional aperture processes for the y direction preceding the raster process and including the scanning beam in the camera. It represents the "input filter."

The MTF of the video system is unity for the y samples and need not be shown. The transfer functions of the raster itself comprise a network of diagonal lines with constant transfer factors ($\tilde{r} = 1$) for the frequency f_m and the sum and difference frequencies (D, S). The carrier frequencies ($C_1, C_2, \ldots$) are represented by horizontal lines because their existence depends only on the dc term and on the attenuation by the "output filter" which is the MTF_d of the display system.

The MTF_d of the display system is drawn parallel to the output frequency scale of the raster characteristic. It is the product of the MTF's of *all* two-dimensional aperture processes following the raster process and includes, therefore, the MTF's of copying systems and the eye. Unless eliminated by adequate magnification, the MTF of the eye obviously must be considered in MTF specifications of display systems designed for a specific viewing distance.

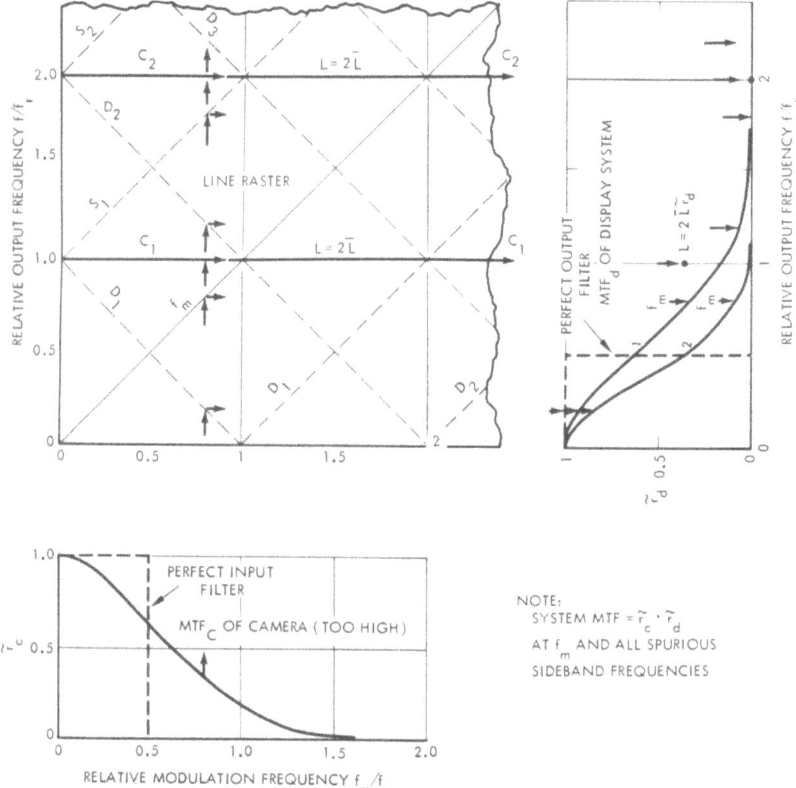

Fig. 6.5. Frequency and modulation transfer characteristic in y coordinate of television systems with line raster process.

The use of the diagram is simple. A vertical projection of an input frequency f_m (see arrows) locates the output frequencies of the raster process at the intersections with the various transfer lines. Horizontal projections from these points onto the output MTF_d indicate the attenuation $2\bar{L}\tilde{r}_d$ of the carrier frequency components and the response factors $\tilde{r}_d$ for determining the relative amplitudes $\tilde{r}_c\tilde{r}_d$ of the modulation products. The example illustrates that the higher MTF_d (curve 1) reproduces the carrier C_1 with a modulation amplitude of 36%, representing a 72% peak-to-peak variation in a uniform field $\bar{L}$. The lower MTF_d reproduces a substantially flat field, but the raster generates a low difference frequency $f_{D1} = 0.2f_r$ of amplitude $\tilde{r}_c\tilde{r}_d = 0.27$ from a modulation frequency $f_m = 0.8f_r$ of 32% amplitude. It is seen at a glance that a complete elimination of all spurious modulation frequencies restricts the MTF's to the spatial frequency bands indicated by the broken-line rectangles, i.e., to frequencies $f_m \leq 0.5f_r$. In other words, a minimum of one sample per half-cycle is necessary to transmit a continuous sine wave by a sampling process. The solution for optimum low-pass MTF's is known from modulation theory and states that the MTF of both input and output filters must be limited to frequencies $f_m \leq 0.5f_r$ to eliminate all raster carrier components and unwanted modulation products. The MTF's should be constant for maximum utilization of the frequency channel. This optimum solution may not be realizable in a practical system.

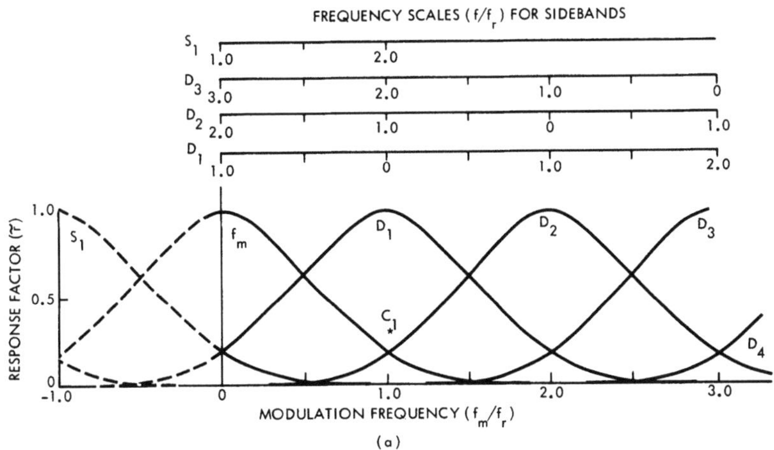

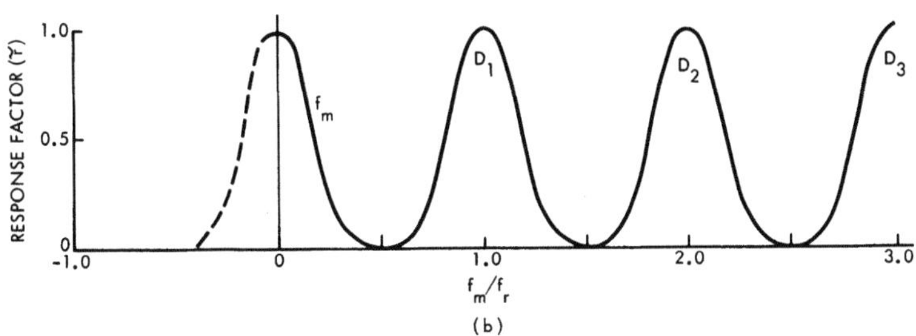

Fig. 6.6. Modulation products (sidebands, "aliasing") of raster process with constant input modulation (MTF$_c = 1$ with unrestricted frequency spectrum) for (a) a small display aperture δ_d; (b) a large aperture δ_d.

A restriction of input frequencies to $f_m < f_r$ is particularly important to prevent generation of repetitive modulation products, electrically known as "sidebands" and more recently termed "aliasing." The modulation products *for a constant input modulation* ($\tilde{r}_c = 1$) without frequency limit are illustrated by Fig. 6.6(a) for the MTF_d, curve 1 of Fig. 6.5. Appropriate scales permit a direct reading of the frequencies and response factors of all associated terms in the y coordinate of the image. The response factor $2\tilde{r}_d$ of the single constant carrier term C_1 is indicated. The normal response MTF_d of the aperture δ_d appears symmetrically repeated at each carrier frequency location. The response pattern MTF_d of the output filter repeats indefinitely. A large aperture, for example, has zero response at $f_m = 0.5f_r$; its response nevertheless repeats up to infinity, periodically going to zero as shown in Fig. 6.6(b).

The fact that the passband of an aperture δ_d is repeated by addition of a raster process when the input modulation is not restricted by a filter is demonstrated in Fig. 6.7(a–d). Figure 6.7(a) is a photograph of a test pattern having a variable line number. A sharp photograph of

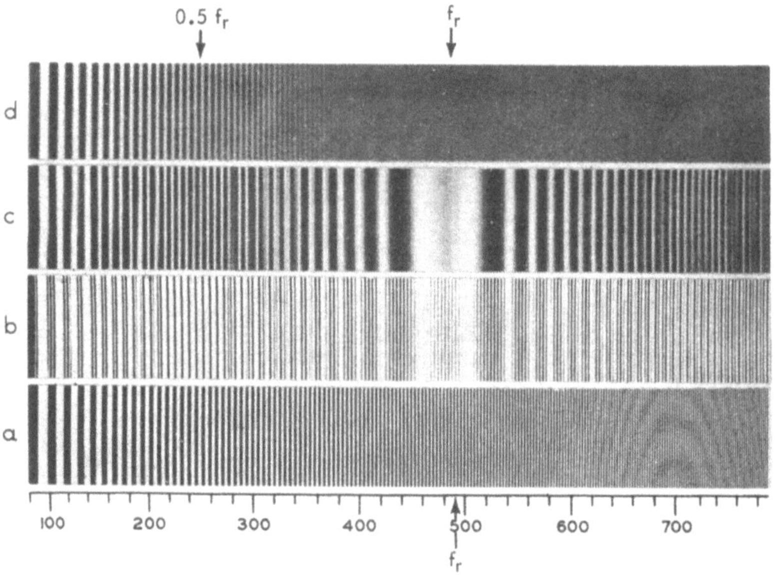

Fig. 6.7. Photographic proof of repeating line-number spectra ("sidebands") obtained by a line raster process ($f_r = 490$).

the pattern through a raster plate having very fine lines (small apertures δ_c and δ_d) is shown in Fig. 6.7(b). A photograph made with a larger aperture δ_d giving a flat field is shown in Fig. 6.7(c) which may be compared with the image Fig. 6.7(d) made without raster and the same aperture δ_d. In all practical cases the infinitely repetitive spectrum MTF_d is limited by the finite response MTF_c of apertures preceding the raster because, excepting carrier components, the overall response of the system goes to zero when the MTF of the camera is zero.

6.4. RASTER LINE FREQUENCIES AND MTF COMBINATIONS FOR LOW SPURIOUS RESPONSE

The inverse transform of the optimum rectangular frequency spectrum is a $(\sin x)/x$ impulse function or line image which can be realized with the coherent light of laser-beam image reproducers by using a rectangular lens stop. Similar functions can be synthesized from Gaussian-type impulses by vertical aperture correction with tapped delay lines for noninterlaced or interlaced scanning (O. H. Schade, Sr., Patent 3,030,440, Vertical Aperture Correction). Such corrections may not be feasible in many cases, which are then restricted to the MTF's of normal cameras and diplay systems, and which are approximated in the illustrations by Gaussian functions.

A substantially structureless field is obtained when the MTF_d of the display is 2.5% or less at f_r. The carrier amplitude for the upper limit is $2\bar{L}r_d = 0.05\bar{L}$, producing a peak-to-peak ripple of 10%. The numerical evaluation of cross products is illustrated by Fig. 6.8. Curves 1–5 represent the MTF_c of various cameras. The MTF of the display system repeats for the sum and difference frequencies; the sidebands MTF_{D1} and MTF_{D2} are shown in Fig. 6.8. The spurious modulation products $\tilde{r}_c\tilde{r}_{D1}$ and $\tilde{r}_c\tilde{r}_{D2}$ are easily evaluated* and shown by curves 1–5 in Fig. 6.9(a) for the five camera MTF's of Fig. 6.8. Note that the zero frequency of the cross products occurs at the modulation frequency $f_m = f_r$ and that the spurious modulation frequencies are higher than f_m for $f_m < 0.5f_r$ and lower than f_m for $f_m > 0.5f_r$. The maximum values of the spurious response are plotted in Fig. 6.9(b) as a function of the

---

* The products $\tilde{r}_c\tilde{r}_{D2}$ are very small or zero and can be neglected.

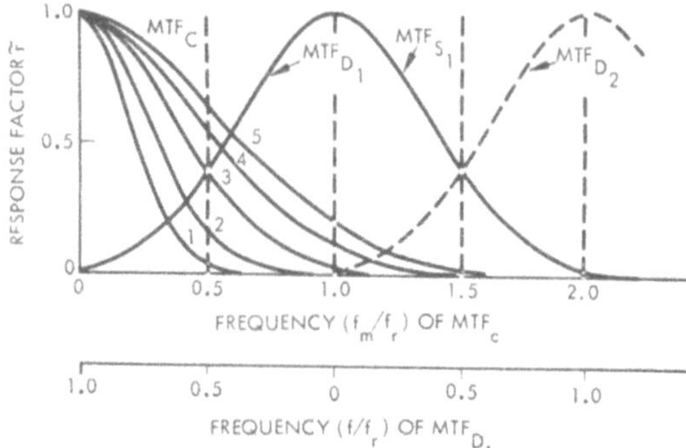

Fig. 6.8. Evaluation of cross products for an MTF$_d$
producing a flat field.

camera response $\tilde{r}_c$ at the theoretical frequency limit $f_m = 0.5f_r$. The
straight line shows the overall MTF ($\tilde{r}_c\tilde{r}_d$) of the system.

A spurious response $\tilde{r}_{sp}$ of 15% may be considered an upper limit
for good system design. This value is a worst case and occurs occasionally
for 100% contrast in the scene. Spurious frequencies occur in the range
indicated by curve 3 in Fig. 6.9a and include weak low-frequency beats

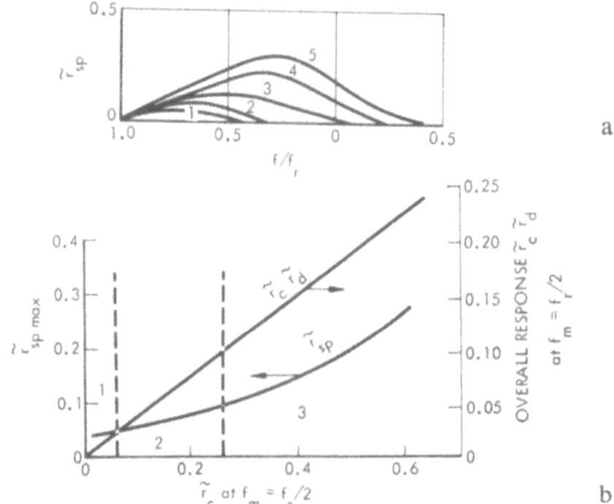

Fig. 6.9. (a) Cross products $\tilde{r}_{sp}$ for MTF$_d$ = MTF
No. 3 and various MTF$_c$ curves 1–5. (b) Maximum
spurious response $\tilde{r}_{sp}$ and overall response $\tilde{r}_c\tilde{r}_d$ for
MTF$_d$ = MTF No. 3 and various camera MTF's
at $f_m = f_r/2$.

which are most visible. It follows that the raster frequency (number of raster lines) should be

$$f_r \geq 2f_{m(0.4)} \tag{3}$$

where $f_{m(0.4)}$ is the spatial frequency at which the camera has a sine-wave response factor $\tilde{r}_c = 0.4$. An overall system response $\tilde{r}_c\tilde{r}_d \simeq 0.15$ is then obtained with a flat-field display system having a sine wave response $\tilde{r}_d = 0.38$ at the theoretical frequency limit $f = f_r/2$. There is, of course, an upper limit for the raster frequency f_r because a frequency $f_m = f_r/2$ with negligible amplitude need not be sampled as expressed by

$$2f_{m(0.05)} \geq f_r \geq 2f_{m(0.4)} \tag{4}$$

The index numbers specify the camera response $\tilde{r}_c$ at f_m. The upper limit $f_r = 2f_{m(0.05)}$ provides a very low level of spurious signals but requires a wide video passband.

The preceding analysis has treated the modulation transfer in the y direction. Sine wave test fields for this direction give constant-amplitude signals in the x direction for each raster line, varying from line to line in accordance with the changing amplitude of the sine wave field. This is no longer true when the sine wave field is rotated relative to the line raster. The scanning aperture in the camera then passes at an angle α over the sine wave field and the signal amplitude varies sinusoidally along each line. The frequency of this spurious x component increases from zero to increasingly higher values when the angle α is increased. A plot of the crest amplitudes on all raster lines reveals a beat frequency pattern which has a higher frequency than for $\alpha = 0$ and a different direction. The normal beat frequency for $\alpha = 0$ is the difference frequency $f_D = f_r - f_m$ and is zero for $f_m = f_r$. A rotation of the raster with respect to the sine wave field for this particular case results in a beat frequency $f_{D(\alpha)} = f_r \sin \alpha$ as illustrated in Fig. 6.10(a). The direction of the beat frequency pattern rotates suddenly from y to x for a very small angle α and forms an angle equal to $\alpha/2$ with the x coordinate. For modulation frequencies $f_m < f_r$ the frequency of the beat pattern is the normal difference frequency at $\alpha = 0$. The rotation toward the x-coordinate is slower and the frequency $f_{D,\alpha}$ increases at a different rate with α as shown in Fig. 6.10(b, c) for the frequencies $f_m = 0.9f_r$ and $f_m = 0.8f_r$, respectively.

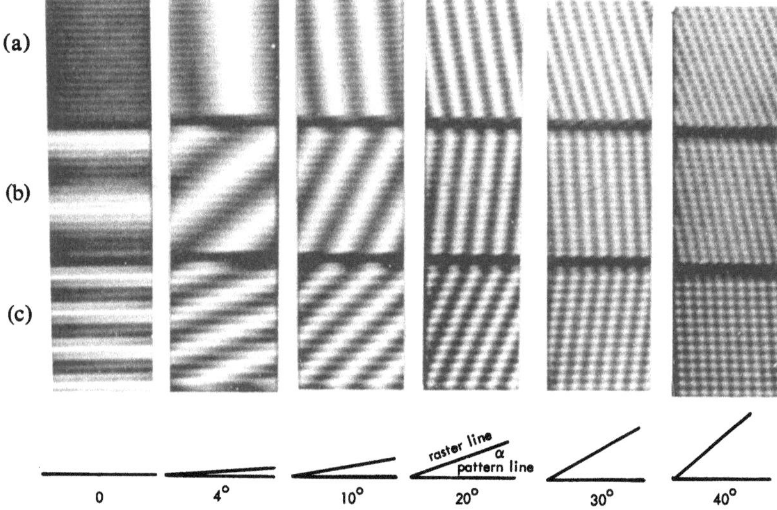

Fig. 6.10. Beat frequences $f_D(\alpha)$ generated when the raster is rotated by an angle α relative to a line pattern. (a) $f_m = f_r$, (b) $f_m = 0.9f_r$, (c) $f_m = 0.8f_r$.

The photographs were made with square wave gratings and are defocused to attenuate the raster carrier f_r for better visibility of the beat patterns. The test wave strip is stationary and the raster is rotated counterclockwise, indicated by the position of the raster lines which are faintly visible in the prints for $\alpha = 40$ deg. Similar conditions occur for $f_m > f_r$. The beat patterns rotate in opposite directions when the raster is rotated because the order of the frequencies is reversed.

Common properties for all cases are the increase of the beat frequency $(f_{D,\alpha})$ from the initial difference frequency to higher frequencies for $\alpha > 0$, and the fact that the amplitude of the beats remains constant for a given set of frequencies f_r and f_m.*

The amplitude of the spurious (beat) frequencies in the camera output (video signal) is therefore determined by the response factor $\tilde{r}_c$ of the camera, whereas the amplitude of the modulation products in the display depend on the frequency $(f_{D,\alpha})$ of the beats, determined by the angle α, and is given by the product $\tilde{r}_{sp} = \tilde{r}_{c(f_m)} \tilde{r}_{d(f_{D,\alpha})}$.

---

* This fact is used to generate variable-frequency beat patterns of constant amplitude with two constant-frequency gratings for MTF tests.

It follows that the evaluation of spurious signal amplitudes and frequencies given for $\alpha = 0$ is truly a worst case, because the frequencies of the spurious signals increase for $\alpha > 0$, and MTF_d reduces their amplitude and visibility.

The visual appearance of the line raster structure and of spurious signals in a two-dimensional image is illustrated by Figs. 6.11 and 6.12 for the upper limit $\tilde{r}_c = 0.4$ at $f_m = 0.5 f_r$ in the camera and for two values MTF_d of the display. The theoretical modulation limit ($f_m = 0.5 f_r$)

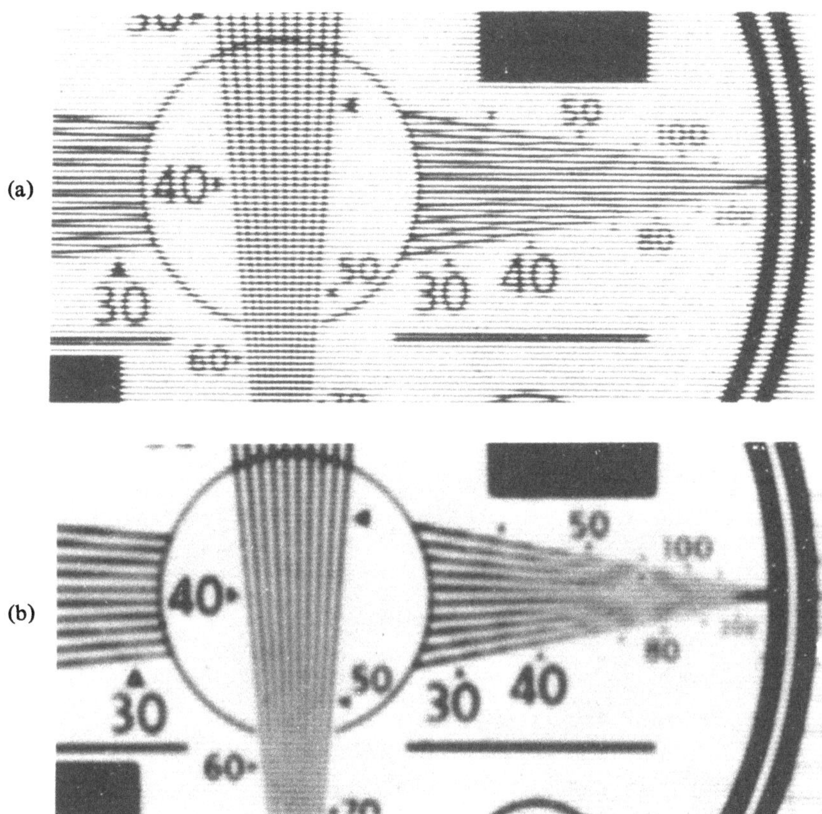

Fig. 6.11. Line raster process with camera MTF $\tilde{r}_c = 0.4$ at $f_m = 0.5 f_r$, $f_r = 70$ in. pattern. (a) High MTF_d generates interfering line structure; (b) MTF_d reduced to obtain a "flat field."

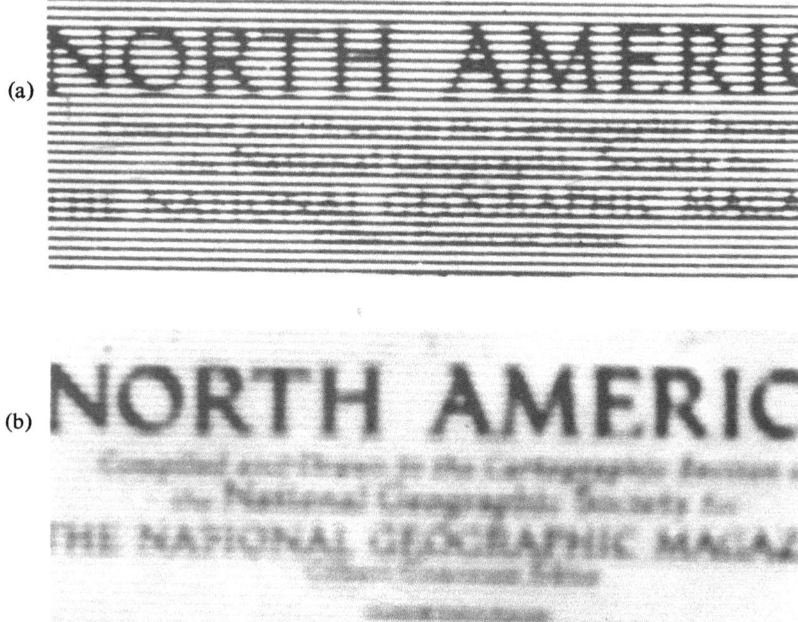

Fig. 6.12. Line raster process as in Fig. 6.11; (a) high MTF_d, (b) flat-field MTF_d.

occurs at the number 35 in the test patterns. Faint zero beat patterns are seen at 35 and at the carrier frequency (70). Note that the serrations of the slanting wedge lines increase with their angle α. The pronounced line structure caused by a high MTF_d in Figs. 6.11(a) and 6.12(a) is greatly attenuated by defocusing the display to obtain a flat field (lower MTF_d in Figs. 6.11b and 6.12b) which restores continuity (in y) and improves the detection of fine detail in spite of the reduced overall MTF.

A point raster is generated by two crossed line gratings. The raster plates used in printing processes have equal frequencies ($f_{r,x} = f_{r,y}$) at a 90 deg angle and generate two sets of spurious frequencies in perpendicular directions as illustrated by Fig. 6.13 for $f_{r,x} = f_{r,y} = 70$. Because of the preponderance of vertical and horizontal lines in most images taken on the ground, the point raster of printing processes is turned 45 deg to minimize beat patterns, which then have higher frequencies

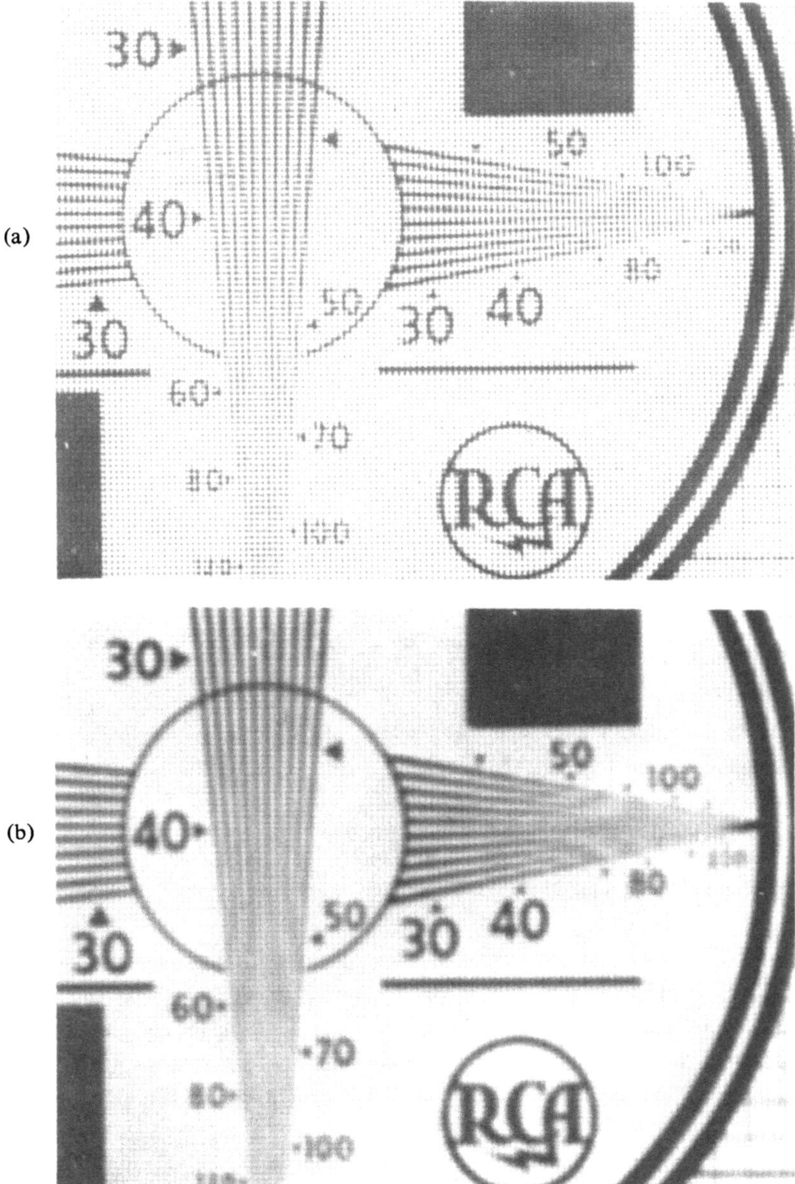

Fig. 6.13. Point raster process, $f_{r\,x} = f_{r\,y} = 70$, $\tilde{r}_c = 0.4$ at $f_m = 35$; (a) high MTF_d shows δ of structure, (b) MTF_d reduced to produce a "flat field."

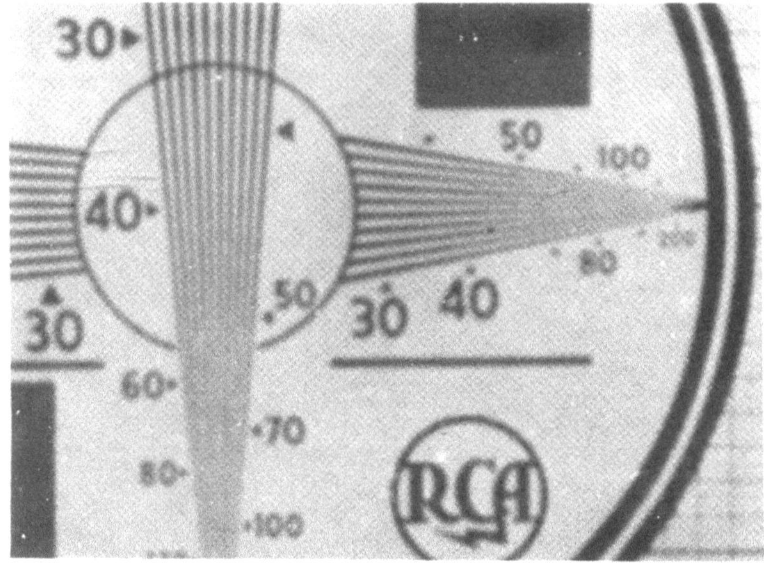

Fig. 6.14. Point raster as in Fig. 6.13(b) rotated 45 deg.

as demonstrated by Fig. 6.14. Figure 6.15 illustrates that a flat field is again essential to establish continuity and improve the detection of fine detail.

6.5. SYSTEM DESIGN

The MTF_c of the camera (in the y direction) determines the constants of the television system or vice versa. Equation (4) states that the raster frequency should be in the range $f_r = 2f_{0.4}$ to $2f_{0.05}$, where $f_{0.4}$ and $f_{0.05}$ are the spatial frequencies at which the camera response is 40% and 5%, respectively. A design for best utilization of the electrical frequency channel would select the lower raster frequency $(2f_{0.4})$, whereas a design for maximum resolving power requires the upper limit. In terms of television line numbers N and the raster line number n_r, Eq. (4) specifies the range $n_r = N_{0.4}$ to $N_{0.05}$. A commercial 525-line system, for example, has an active raster line number $n_r = 490$ lines and the

Fig. 6.15. (a) Point structure caused by high MTF_d interferes with detection of fine detail; (b) flat-field MTF_d improves detection of detail.

vertical camera response at this line number N is generally less than 40%.

A structureless field and low spurious response dictate a display system designed for an MTF_d of 2.5% or less at the raster frequency f_r. The MTF of a *good* commercial CRT is in the order of 27% at $f_r = 490$ *cycles* and, at a luminance $L = 40$ ft-L, the MTF of the eye (Schade, 1956) is 5% at a relative viewing distance $d/V = 4$ and about 0% for $d/V = 6$. The MTF_d of the display system (including the eye) is thus 1.35–0%. The peak-to-peak ripple is four times higher and is still visible at the shorter viewing distance. The line structure is very pronounced at close viewing distances and should be eliminated by vertical "spot wobble." The spot-wobble frequency should be outside the frequency spectrum of the system; about 20 MHz for standard CRT's, 50 MHz for a 20-MHz video system, and 140 MHz for 100-MHz systems and very high-resolution CRT's. Spot wobble is particularly recommended when the CRT image is magnified by overscanning the normal format. A laser-beam image recorder designed for a substantially rectangular frequency spectrum and a flat field increases the MTF_d and overall MTF of the system. It does, however, also increase the amplitude of the spurious response products. Figure 6.5 illustrates by rectangular broken lines that portions of the sidebands (D) are reproduced with unity response by a rectangular MTF_d. The modulation products shown in Fig. 6.9a are zero for $f/f_r > 0.5$ and have amplitudes equal to r_c for $f/f_r < 0.5$, which are the portions of curves 1–5 in Fig. 6.8 inside the rectangular first sideband. This condition recommends the use of higher raster frequencies $f_r \simeq 2f_{m(0.10)}$ to reduce spurious low frequencies.

The MTF's of the camera and display system are products of a number of components. It may thus occur that the MTF of the scanning aperture (beam) in the camera is much higher than the product, for example, when a high-resolution beam is used in combination with a light intensifier and a high-aperture lens. Calculation of the raster frequency with Eq. (4) may indicate a relatively low raster frequency at which the scanning beam leaves unscanned interline spaces. Although sufficient integration of the image flux occurs in the stages preceding scanning, the efficiency of signal conversion is reduced by interline charges not contacted by the beam and can result in undesirable secondary effects recommending the use of a larger beam or a higher raster frequency. A similar situation may occur in a display system containing several "copying" stages which "diffuse" the image of the actual scanning spot to provide a flat field. The current or light density in the scanning spot may then become excessive, which can result in saturation effects.

A "perfect" television system having equal rectangular MTF's in x and y, producing a structureless field, is anisotropic because the effective apertures δ_c and δ_d have a square base, causing an increase of the spatial frequency spectrum by $\sqrt{2}$ in the diagonal directions. The same anisotropy occurs in optical images formed with coherent monochromatic light by a lens having a square lens stop. A practical television system in which the MTF is bandwidth limited in x by the video system is similarly anisotropic, as is readily confirmed by observation.

For an analysis of the MTF in the displayed TV image as a function of direction it can be assumed that the apertures δ_c and δ_d have circular symmetry; their MTF's are isotropic. The electrical system, however, is anisotropic. The normal sine wave response $\tilde{r}_s(f, \theta)$ in any radial direction of a television system is the product

$$\tilde{r}_{s,\theta}(f) = (\tilde{r}_c\tilde{r}_d)_f \tilde{r}_{e(f/\cos\theta)} \tag{5}$$

where $\tilde{r}_e$ is the response of the electrical video system. The index $(f/\cos\theta)$ indicates that *the electrical response is a function of* θ; it is the normal response of the video system in the scanning direction (x) when the angle θ to the raster lines is zero. For angles $\theta > 0$ the frequency scale is stretched in first approximation by the factor $1/\cos\theta$.* Thus for the y direction $(\theta = 90$ deg$)$ the electrical response $\tilde{r}_e(90) = \tilde{r}_e(y)$ is unity; the video frequency scale is stretched to infinity and does not limit the "vertical" response of the system. A typical electrical response $\tilde{r}_e(\theta)$ with aperture correction (high-frequency boost) is shown by broken curves in Fig. 6.16. The aperture correction is 1.5 times for high frequencies in the x direction $(\theta = 0$ deg$)$ and decreases to unity for the y direction $(\theta = 90$ deg$)$. The solid curves are the products $\tilde{r}_{c(f)}\tilde{r}_{e(f/\cos\theta)}$ of the MTF$_c$ of the camera and the video system. The response for $\theta = 0$ is the "horizontal" video response limited by a sharp cutoff filter in the video system to $f/f_r = 0.52$ and made constant by the aperture correction up to the frequency $f/f_r = 0.3$. For $\theta = 90$ deg the response is the MTF$_c$ of the camera in the y direction.

The complete response of the system including a display is shown in Fig. 6.17 for an MTF$_d$ equal to MTF$_c$. Both have a 40% response at $f/f_r = 0.5$ to obtain a substantially flat field and conform to the upper limit for a maximum spurious response of 15%. The overall MTF has become much more isotropic. The rolloff of the response indicates a maximum overshoot-plus-underswing ripple of the order of 4% for edge transients at $\theta = 0$, decreasing to zero for $\theta = 90$ deg.

This system design is typical of commercial television systems except for a wider video passband. The reader can verify this by substituting a

---

* This is a reasonable approximation because only a small frequency band is needed for transmission of the electrical raster line-frequency and its harmonics.

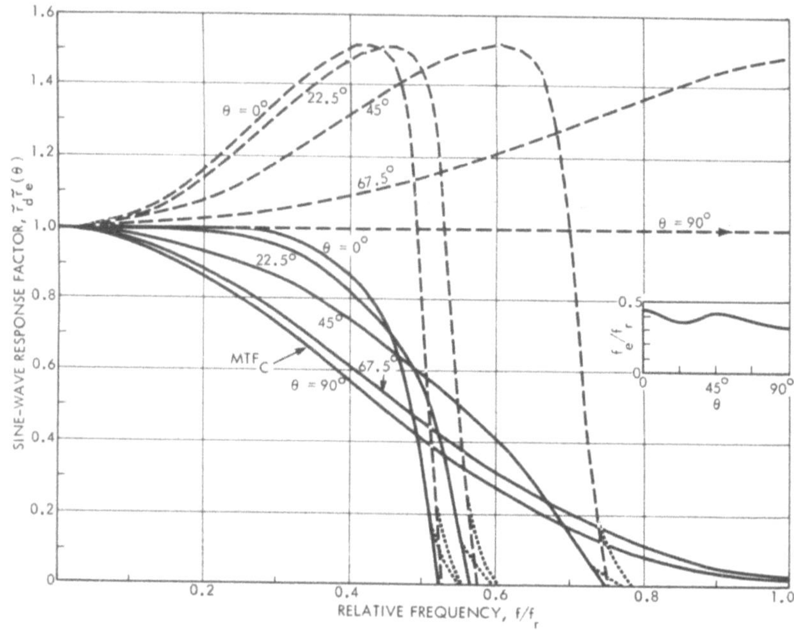

Fig. 6.16. Electrical response $\tilde{r}_e(\theta)$ of video system alone (broken lines) and response $\tilde{r}_e\tilde{r}_c(\theta)$ including the camera MTF_c (solid lines) of a well-designed TV system as function of direction θ.

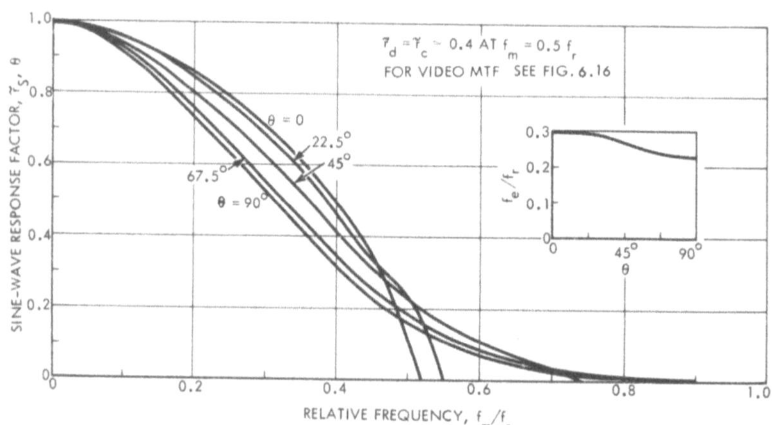

Fig. 6.17. Overall MTF and equivalent passbands (f_e/f_r) of TV system approaching optimum design as function of direction θ.

line number scale $N = 1000$ at $f/f_r = 1$ and decreasing the cutoff frequency of the video channel to $f/f_r = 0.34$ to obtain U.S. standards. The infrequent appearance of beat patterns in commercial black and white television images confirms that a spurious response of 15% is an acceptable upper limit. The system becomes isotropic when the raster frequency has twice the value at which the camera response is 5%, the MTF of the display system is 2% or less at the raster frequency, and the video system has unity response up to the resolution limit of the overall MTF product because the MTF of the system is then limited in all radial directions by the isotropic response of its two-dimensional circular apertures. This is the preferred system design for high resolving power. It should be pointed out that the MTF of a charge storage camera can become anisotropic because of "self-sharpening" of a low-velocity beam in x or y, which depends on a low or excessively high raster line density, respectively. The MTF_c in the y direction is readily measured with a horizontal pulse-gating circuit, and isotropy in the reproduced image can be tested visually by comparing the contrast of vertical and horizontal resolution bars in a standard Air Force test object, which can be made equal by adjusting the MTF of the video system by aperture correction circuits.

6.6. NOISE IN A RASTER PROCESS

The large-area signal-to-noise ratio (SNR_0) in an image can be determined by sampling or scanning an area of uniform intensity with a physical aperture of area a. The mean value of the light flux passing through the aperture is the signal and the rms value of the deviations in the aperture flux is the noise. For random deviations the signal-to-noise ratio is given by

$$\mathrm{SNR}_0 = (n_{(1)}a_e)^{1/2} \tag{6}$$

where $n_{(1)}$ is the photon, electron, or particle density per unit area. The *noise-equivalent area a_e of* the sampling aperture of a system is determined from the noise-equivalent passband* $\bar{N}_e = 2\bar{f}_e$ of the system com-

---

* The noise-equivalent passband is expressed in the following discussion by N_e in total TV lines (half-cycles/frame dimension), whereas f_e may be preferable when the spatial frequencies are expressed in cycles/mm.

ponents following the noise source, which is the rms value of the integrals $N_{e(\theta)} = \int \tilde{r}^2_{s(\theta)} \, dN$ evaluated for all values θ. The approximation

$$\bar{N}_e = (N_{e(0)}N_{e(90)})^{1/2} = (N_{e(x)}N_{e(y)})^{1/2} \qquad (7)$$

is adequate for most purposes. For round apertures the noise-equivalent area is given by $a_e = 1/1.09\bar{N}_e^2$. For rectangular areas $a_e = 1/\bar{N}_e^2$.

The SNR_0 obtained by scanning a photographic grain structure is independent of the direction of scanning, because the MTF is isotropic. This is no longer true when a grain or particle structure has been imaged by an anisotropic aperture process; the noise is anisotropic when the image is anisotropic.

A complete evaluation of the noise from various sources in a TV camera is beyond the scope of this section, which treats only the changes in the noise-equivalent passband of the complete system produced by a line raster process. It is evident that noise from a preamplifier, for example, does not involve the MTF_c of the camera, which is then unity. The results of the analysis cover a wide range of variations in the MTF_c of the camera for this purpose. Photoelectron noise in the video channel of a television system is the noise measured in the x direction ($\theta = 0$) with the aperture δ_c of the complete camera. The MTF_c is modified by the electrical response as shown by the solid curves $\theta = 0$ in Fig. 6.16. The noise in the y direction, however, is not determined by the MTF for $\theta = 90$ deg because of the spurious modulation frequencies from the raster process which increase the value of the integral $N_{e(y)}$. Similarly the integral $N_{e(s)x}$ for the overall system response including the display can be determined from Fig. 6.17 ($\theta = 0$), whereas the equivalent passband $N_{e(s)y}$ may be increased by the spurious frequencies generated by the raster process. It can be concluded from the preceding discussion that the increase of $N_{e(s)y}$ will be small for large apertures (low spurious signals) and can be quite large when the apertures δ_c and δ_d are small compared to $1/n_r$.

For an analysis it is unnecessary to examine waveform or phase distortion because the distribution of frequency components in a noise source is random. The sine wave spectrum for noise in the y coordinate is hence obtained by arranging all sine wave components in order of their frequencies (or television line numbers $N = 2f$) and combining response factors at equal frequencies by a quadrature addition (square root of the sum of the squares). This process has been carried out for several aperture combinations having Gaussian MTF's which are good approximations for two-dimensional apertures in television and photographic systems. The noise spectra are shown in Fig. 6.18. The aperture sizes are specified by their equivalent passbands $\bar{N}_e$ which correspond to the television line number $\bar{N}_e = 1.6 N_\delta$ where the response of a Gaussian spot has the value $\tilde{r} = 0.674$ (see Section 6.7). When both apertures δ_c and δ_d are large, i.e., when $\bar{N}_e$ is smaller than the raster constant (curve 1 in Fig. 6.18a), the sine wave spectrum is substantially the same as without raster; when $\bar{N}_{e(d)}$ is increased the high-"frequency" components increase considerably faster than without raster and show periodic

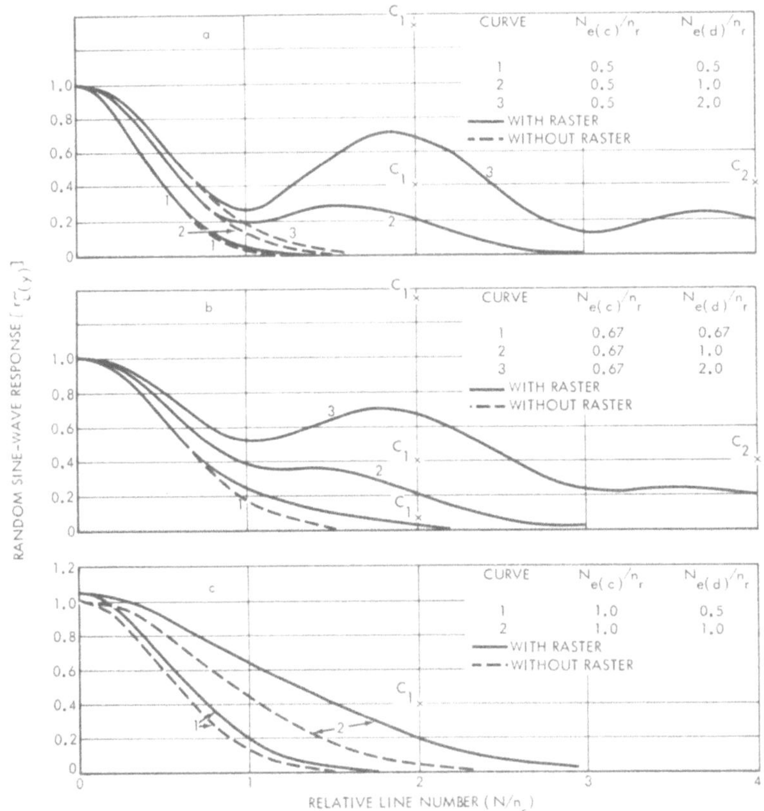

Fig. 6.18. Overall sine wave spectra of raster processes for various apertures, sizes δ_c and δ_d.

maxima and minima. These variations decrease when $\bar{N}_{e(c)}$ is increased (Fig. 6.18b) and disappear substantially for values $\bar{N}_{e(c)} = 1$ (Fig. 6.18c). It is concluded from a comparison without raster (broken curves) that the addition of a raster process may increase the sine wave response and the total aperture passband even for the "flat-field" condition $\bar{N}_{e(c)} = \bar{N}_{e(d)} = 0.67 n_r$ (Fig. 6.18b). The raster can therefore have a substantial negative aperture effect which increases the intensity and edge sharpness of the reproduced grain structure in the y coordinate.

The equivalent passbands $N_{e(s)y}$ of the system are plotted in Fig. 6.19 as a function of the relative passband $\bar{N}_{e(c)}/n_r$ of the camera aperture δ_c with $\bar{N}_{e(d)}/n_r$ as a parameter. Examination of these functions reveals the following facts.

(a) When both $\bar{N}_{e(c)}$ and $\bar{N}_{e(d)}$ are smaller than $0.7 n_r$ the aperture flux at successive raster points is correlated sufficiently (overlapping) to eliminate the effect of the raster. The noise-equivalent passband $N_{e(s)y}$ of the process can then be computed from the normal aperture response without raster or may be approximated with good accuracy by the cascade formula:

$$N_{e(s)y} = [(1/\bar{N}_{e(c)}^2) + (1/\bar{N}_{e(d)}^2)]^{-1/2} \qquad (8)$$

(b) When one or both values $\bar{N}_{e(c)}$ or $\bar{N}_{e(d)}$ are greater than n_r, the aperture flux is no longer correlated by at least one aperture and the noise-equivalent passband can be computed with good accuracy from the product

$$N_{e(s)y} \simeq (\bar{N}_{e(c)}\bar{N}_{e(d)}) \qquad (9)$$

(c) For all other values the aperture flux is partially correlated and the value $N_{e(s)y}$ should be computed as outlined above or may be approximated by the values computed for Gaussian apertures (Fig. 6.19).

For preamplifier or beam noise of the camera $N_{e(c)} \equiv N_{e(e)}$ of the video channel because the noise source is located after the raster process in the camera or not affected by it. It should be mentioned that a square aperture presents a special case because of its strongly periodic aperture

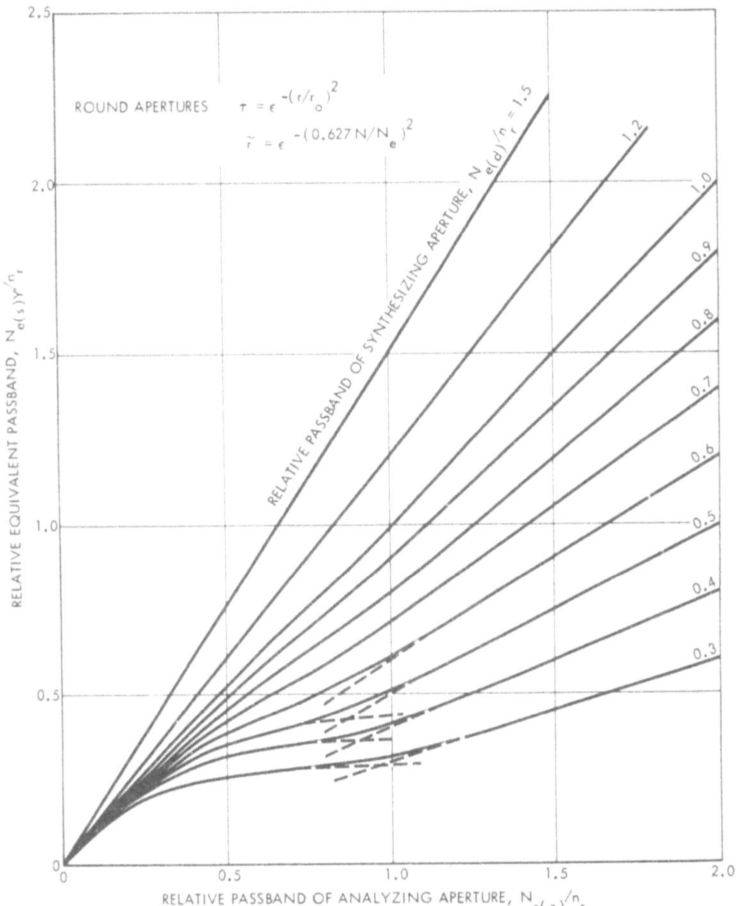

Fig. 6.19. Equivalent relative passband $N_{e(s)}y/n_r$ of systems containing a raster process as a function of the relative passband $N_{e(c)}/n_r$ of the analyzing aperture δ_c for various relative passbands $N_{e(d)}/n_r$ of the synthesizing aperture δ_d.

response and large number of terms, which cause periodic deviations from the characteristics shown in Fig. 6.19. The square aperture is of interest as a mathematical equivalent but its characteristics are in many cases undesirable for practical processes.

The equivalent passband $\bar{N}_e$ for determining the noise-equivalent sampling area a_e of the system can now be computed with Eq. (7).

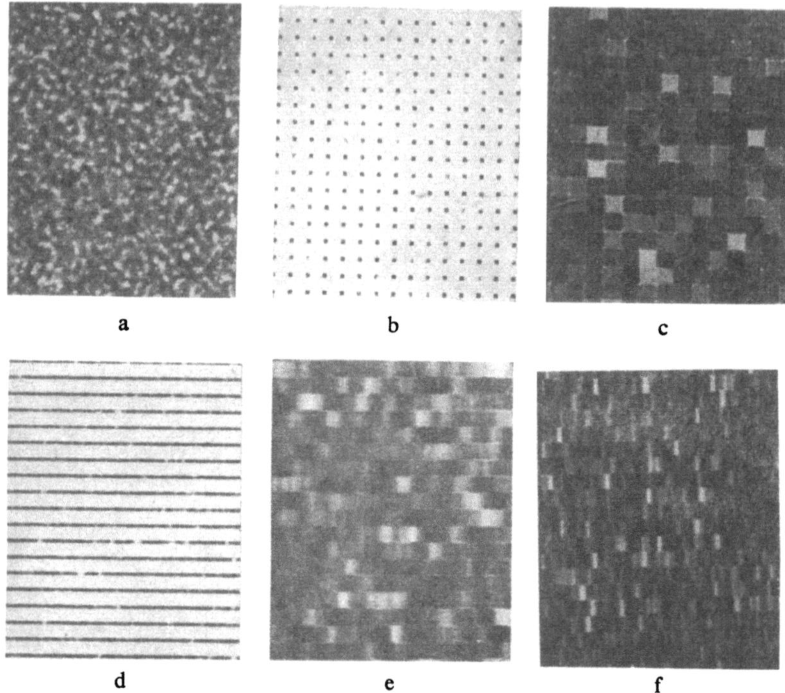

Fig. 6.20. Reproduction of photographic grain structure by point- and line-raster processes with rectangular apertures (highly magnified).

The greatly enlarged reproduction of a photographic grain structure by point- and line-raster processes is illustrated in Fig. 6.20. The original grain structure is shown in Fig. 6.20(a). The samples "seen" through a fine point-raster plate (δ_c small) are shown in Fig. 6.20(b); their reproduction by a square aperture providing a "flat" field is shown in Fig. 6.20(c). Reproduction of the same grain structure by a line-raster process using a square reproducing aperture is shown in Fig. 6.20(d, e). The anisotropic reproduction obtained with a vertical slit aperture is illustrated by Fig. 6.20(f).

A comparison of a line-raster process (a) using a round cosine-squared aperture δ_d with a continuous process (b) using the same apertures is shown with a lower magnification in Fig. 6.21. The slight increase in vertical sharpness by the raster process (a) observed in the originals may be lost in the printing process.

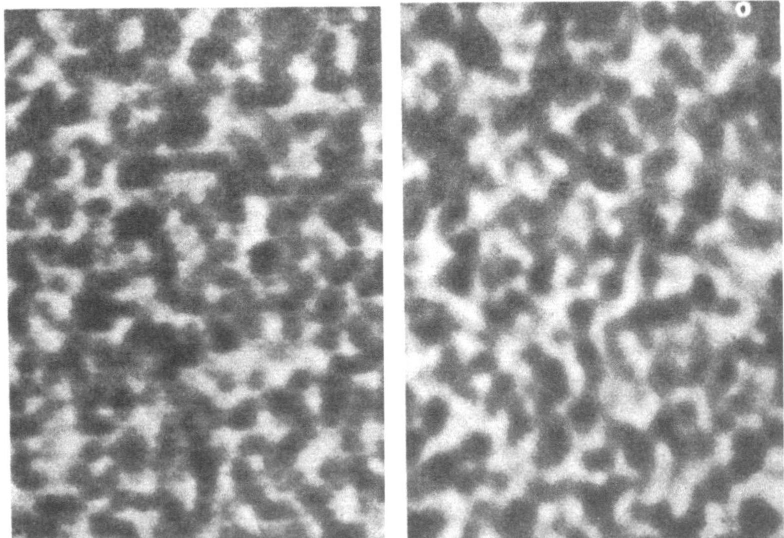

Fig. 6.21. Grain structure reproduced with (at left) and without (at right) line raster process by a round cosine-squared aperture.

6.7. CATHODE RAY TUBES FOR VISUAL DISPLAY OF TV IMAGES

6.7.1. The Gaussian Spot

The "spot size" of a cathode ray tube (CRT) is often defined as the half-amplitude width $W_{s(0.5)}$ of the spot profile, assuming that the half-amplitude width $W_{L(0.5)}$ of the line image generated by the spot has the same value. This is true only for a Gaussian spot having the intensity spread function

$$i/i_0 = \exp[-(r/r_0)^2] \tag{10}$$

and the half-widths

$$W_{s(0.5)} = W_{L(0.5)} = 1.665 r_0 \tag{11}$$

The diameter of a Gaussian spot may be defined by

$$\delta_e = 2r = 4r_0 \tag{12}$$

where the intensity limit is $0.0183i_0$. This choice results in a noise-equivalent passband $\bar{N}_e$ equal to that of a cosine-squared spot having the same diameter.

The normalized sine wave modulation transfer function (MTF) of a Gaussian spot is given by

$$\tilde{r} = \exp\left[-\left(\frac{\pi}{8}\ \frac{N}{N_\delta}\right)^2\right] \tag{13}$$

where the spatial frequency f is expressed in half-cycles by the television line number $N = 2f$ and related to the spot diameter by the reference line number

$$N_\delta = 1/\delta_e = 1/4r_0 \tag{14}$$

The normalized frequency spectrum [Eq. (13)] is shown in Fig. 6.22. The -3 dB and -6 dB points occur at

$$\tilde{r}_3 = 0.707 \quad \text{at} \quad N/N_\delta = 1.5 \qquad (-3 \text{ dB point})$$
$$\tilde{r}_6 = 0.502 \quad \text{at} \quad N/N_\delta = 2.12 \qquad (-6 \text{ dB point})$$

The equivalent passband has the value

$$\bar{N}_e = 1.6N_\delta, \qquad \text{where} \quad \tilde{r} = 0.674 \tag{15}$$

The MTF of the cosine-squared spot $(N_\delta = 1/\delta_{\cos^2} = 1/2r)$ is shown for comparison.

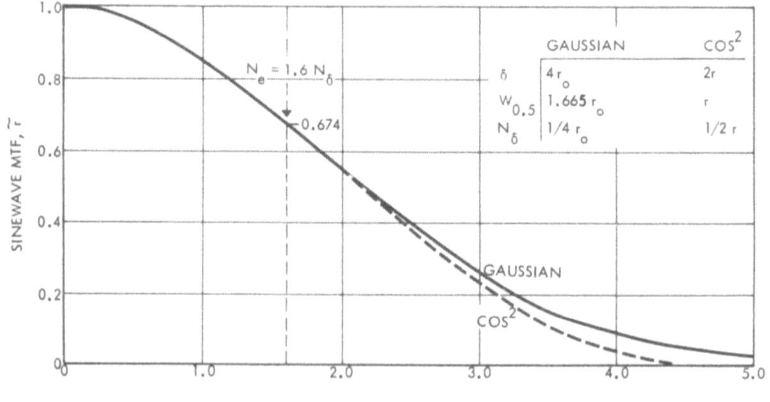

Fig. 6.22. Normalized MTF of Gaussian and cosine-squared spots.

6.7.2. The Composite Spot of CRT's

The spread function of the electron beam is similar to that of a camera tube, especially in high-resolution CRT's employing a limiting aperture in the electron gun. The spread function of the beam can theoretically be Gaussian, cosine-squared, or nearly uniform, depending on the aberrations in the main focusing lens (magnetic or electrostatic), the g_1 lens of the electron gun, and the radial energy selection by the limiting aperture, if used. The spread function of the light spot on the phosphor screen is larger because of light diffusion in the finite thickness of the phosphor and optical reflections on the faceplate surfaces which may spread a substantial fraction of the light into a flare or haze disk surrounding the small "nucleus" spot.

For the purposes of analysis the complex CRT light spot is decomposed into a sum of Gaussian spots having different diameters and intensities. Each of the spot components has a frequency spectrum inversely proportional to its diameter [Eq. (14)]. The relative amplitude scales of the MTF's are directly proportional to the component flux; i.e., to the *volume* of the component spot. Thus the intensity of a component spot having ten times the diameter and 1/100 the peak intensity of the spot nucleus is hardly visible, although it contains the same flux as the nucleus and reduces the high-frequency response factors of the total MTF to one-half.

The relations between a composite spot and its MTF are illustrated by Fig. 6.23 for a diameter ratio $\delta_2/\delta_1 = 3$ and equal flux values $\psi_1 = \psi_2$ in the component Gaussian spots. The example demonstrates that the half-width of the composite spot $W_{s(0.5)}$ is practically the same as the half-width of the nucleus alone without the flare spot, which has a 10% amplitude. The line width $W_{L(0.5)}$ is slightly larger than $W_{s(0.5)}$, but not at all in proportion to the flux ratio ψ_1/ψ_2 which determines the MTF of the composite spot.

The broken curve MTF_0 in Fig. 6.23 is the MTF for the nucleus spot alone, neglecting the flare spot. Two-to-one errors (in favor of a higher MTF) are not uncommon when the frequency spectrum is computed from a measured line profile, because the diameter of the CRT flare spots are often larger than ten times the nucleus spot diameter, requiring that the line profile be measured precisely over a distance ten times larger than where the amplitude has dropped to a few per cent!

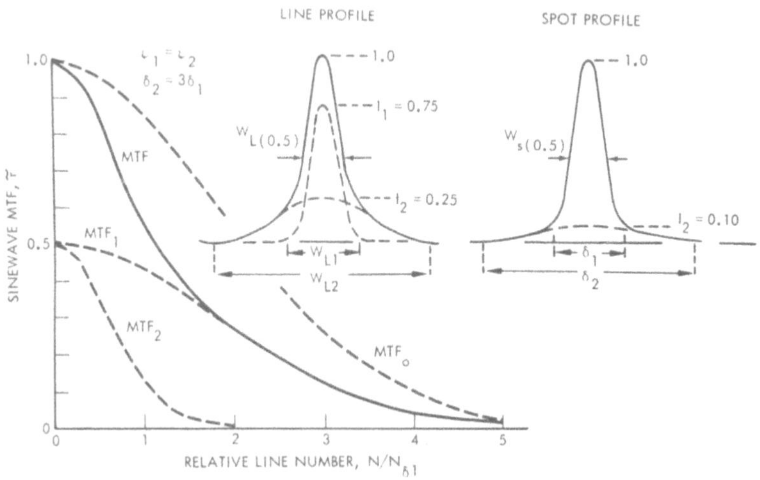

Fig. 6.23. Composite CRT spot, its line profile, and MTF.

This is quite difficult and requires sophisticated instrumentation because of noise and other instabilities in the pulse signals from a photomultiplier sampling the line profile of a CRT. Thus the line image of the nucleus is often assumed to be the true line image, because the low flare intensity above zero intensity is not detected on a linear scale.

It follows that the half-amplitude spots—or linewidths—are generally inadequate as a specification of CRT performance for half-tone images. The values $W_{s(0.5)}$ or $W_{L(0.5)}$ are of some use in digital recording, where the background base from the flare spot flux can be clipped out by underexposure of a high-contrast film.

6.7.3. Measured MTF's of High-Resolution CRT's

The direct measurement of the MTF by sine wave modulation of a continuous field is free of interpretation errors. The instrumentation is relatively simple (Schade, 1958) and measurements can be made by observing the sampled sine waves from the photomultiplier directly on a cathode ray oscilloscope.

The MTF of a 17-in. rectangular high-resolution monitor tube with magnetic focus is shown in Fig. 6.24.

The MTF is for a total HV supply current of 10 μA, resulting in a luminance $L = 10$ ft-L in a 10×10 in. raster (1000–2000 lines). The

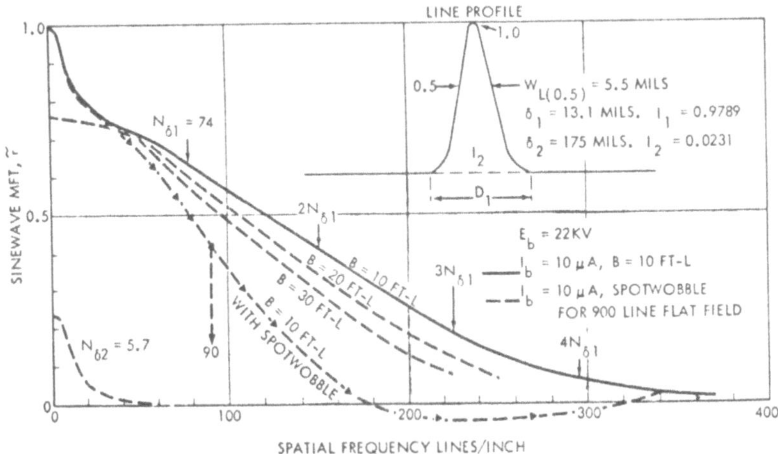

Fig. 6.24. Measured MTF's of RCA high-definition 17-in. rectangular picture tube.

white all-sulfide screen is operated at a potential of 22 kV. The frequency range of the MTF decreases 12% for L = 20 ft-L and 20% at L = 30 ft-L as shown by broken curves.

Many points are measured in the low-frequency region to obtain the MTF component of the flare spot which causes the initial drop in response. The component MTF's are obtained by curve fitting, beginning with the subtraction of a Gaussian curve fitting the high-frequency "tail" of the measured function.* This first component is the MTF_1 of the nucleus. The residual (MTF_2) is still a sum of Gaussian curves but is approximated by a single Gaussian curve representing the MTF_2 of the flare spot which contains 24% of the light flux because $\tilde{r}_{0,2}$ = 0.24. The index line numbers $N_{\delta 1}$ and $N_{\delta 2}$ are easily determined from the response factors. Their reciprocals furnish the spot diameters and the reference radius r_0 of the spread functions [Eq. (14)].

The half-width is sometimes measured by compressing a line raster until a flat field is obtained. The flat-field value W_{FF} is calculated by dividing the vertical dimension by the number of raster lines. This value is smaller than $W_{L(0.5)}$ because a Gaussian function will give a flat field when crossing over at the 54% intensity value. For composite spots the integration of the large flare component is complete and raises the 54%

* This is easily done in log-log coordinates.

amplitude crossover point to $i \simeq \psi_2 + 0.54\psi_1 = 0.65$ for the example. The width W_{FF} of the composite line at this intensity is approximately $1.35r_0 = 4.45$ mils. A flat field test required a line density of 225 lines/in. which gives $W_{FF} = 4.45$ mils, in exact agreement with the computed value.

6.7.4. Specifications for Display CRT's

Good utilization of a given television frequency channel requires that the overall system response remain above -12 dB (25%) in both x and y directions. A response of -6 dB at the spatial bandwidth limit N_c is desirable for the MTF of the CRT; however, this specification can be relaxed to -9 dB when aperture correction is provided and the signals from the camera remain above -6 dB.

The MTF shown in Fig. 6.24 is adequate for a 1000–1800-line display in a 10×10 in. format.

The raster line spacing in the y direction may be larger than the half-amplitude width $W_{L(0.5)}$ of the line profile; the lines are visible at close viewing distances. The line structure interferes with the observation of fine detail and must be eliminated. The ideal solution for establishing perfect continuity is a $(\sin x)/x$ type of line profile with first zeros on adjacent line centers, which requires very elaborate equipment. Simpler approximate solutions are obtained by widening the spot profile in the y direction to produce a "flat" field without affecting the MTF in the x direction. This is readily achieved by high-frequency spot wobble in the y direction. The spot-wobble frequency should be 1.5–2 times higher than the video passband to prevent interference. Spot wobble is even more essential when the normal format of the CRT is overscanned. Provision of a 2:1 magnification of the image by overscanning the format is very useful in direct-view, high-definition displays because it increases the effective MTF's of the CRT and the eye by a factor of two. The spot-wobble amplitude is adjusted to decrease the MTF to zero at the raster frequency $f_r = 2n_r$, where n_r is the raster line number. Thus for $n_r = 900$ active scanning lines in a 10×10 in. format the zero response point occurs at $N = 180$ lines/in. The MTF for this particular case was measured and is shown in Fig. 6.24, demonstrating the effective filtering of the carrier frequency f_r. The response at the line number $N = n_r = 90$ lines/in. has decreased from 60% to 43%, which is higher than obtainable by using a larger Gaussian spot giving a "flat" field.

6.7.5. Flicker

The "critical luminance" L_c of an unmodulated noninterlaced field at which *field flicker* is just perceptible is a sensitive function of the field frequency for a given persistence of the screen phosphor. The function measured with a P4 screen is shown in Fig. 6.25. Modulated fields are substantially free of flicker when their *average* brightness does not exceed the critical flicker brightness L_c of unmodulated fields.

When interlacing is used the lines of each field repeat at the frame rate. The field luminance contains a 30-cycle component for 60 fields 2:1 interlaced or a 20-cycle component for 60 fields 3:1 interlaced. The lines of the component fields are spaced closely together and integration occurs in the eye at the spatial frequency of the raster; i.e., at a line number N equaling twice the raster line number. Thus *interline flicker* is imperceptible at a normal viewing distance in an unmodulated 1000-line raster 2:1 interlaced because of the very low MTF of the eye at 2000 lines. Interline flicker becomes visible when the interlaced raster is observed with sufficient magnification, but a more prominent effect is an apparent drift or "crawl" of the lines, especially for interlacing factors

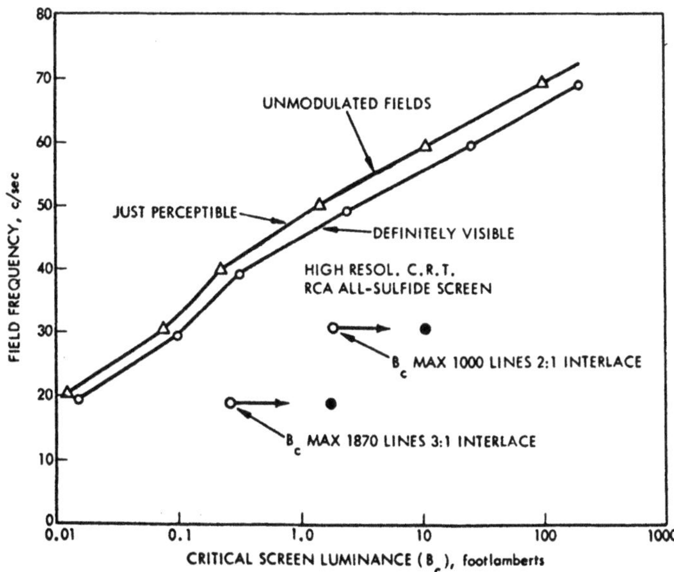

Fig. 6.25. Critical flicker frequency versus screen luminance (P4 phosphor).

greater than two, because the eye tends to follow the sequence of the lines.

Detail flicker is much more pronounced than interline flicker. It occurs when the camera tube beam provides a good signal from objects equaling one linewidth. The worst condition for flicker occurs when the beam is centered on a horizontal white line in a dark field, because the line signal will then repeat at the *frame* rate, causing a pronounced flicker. Detail flicker occurs generally when reproducing test patterns. The flicker disappears completely when the subject does not contain fine detail or when the camera tube beam is defocused or large enough to spread over two or more raster lines. This is often the case in commercial television systems in which detail flicker is rarely observed. Detail flicker cannot be eliminated by defocusing the picture tube. It can be reduced or eliminated by reducing the screen luminance. Its amplitude depends very much on the type of test object.

For a raster of 1000 lines, 60 fields 2:1 interlaced, the threshold *highlight* luminance $L_{c\mathrm{max}}$ of the CRT for extinguishing flicker is $L_{c\mathrm{max}} \simeq 2$ ft-L; flicker is visible at 4 ft-L as indicated on the 30-cycle coordinate in Fig. 6.25 by the small circle and arrowhead. For an 1870-line raster, 60 fields 3:1 interlaced, $L_{c\mathrm{max}}$ decreases to approximately 0.25 ft-L and flicker is visible at $L_{c\mathrm{max}} = 0.6$ ft-L. The black circles in Fig. 6.25 indicate values considered not objectionable for flicker and line "crawl." Higher values L_c can be tolerated by a trained observer.

6.7.6. System Requirements for Long-Persistence Picture Displays

The long storage time of a P38 phosphor* (0.47 sec time constant) can provide a substantial gain in signal-to-noise ratio by integration of signals and noise in a multiple-frame readout of a steady image. The phosphor requires a time of 2–2.5 sec to reach 97.5% of the steady-state luminance value. It follows that the phase stability of image signals and line raster must be very high to prevent loss of resolution by integration.

---

* This phospor has an exponential decay. It loses some efficiency in the scanned format area with use but does not retain a "burned-in" image. A faint afterimage is retained for a short time when a pattern has been displayed for a long time. Image carryover beyond the normal decay time does not occur when the display time per image is of the order of 1 min. The new image "overrides" the otherwise faintly visible (negative) afterimage.

When a line pattern is suddenly displaced by one half-cycle, the positive and negative amplitudes are reversed in polarity and a buildup or decay time of 2 sec is required for half-waves to approach steady-state values within 2.5%; i.e., the peak-to-peak amplitude is 5% below the steady-state value. It follows that phase drifts in the picture tube raster or the signal-generating system should not exceed 180 deg (one linewidth) at the limiting resolution in 2 sec to prevent a modulation loss greater than 5%. Sudden jumps (delta functions) in raster position or phase of video signals should not exceed 18 deg or 10% of one linewidth, and the peak-to-peak noise superimposed on deflection currents should not exceed 10% of one linewidth at the resolution limit.

A 2000-line definition, for example, requires that the peak-to-peak noise in the deflection current does not exceed 1/20,000 of the peak-to-peak deflection amplitude.

6.7.7. MTF Measurements of Long-Persistence Phosphors

The MTF of a display tube is determined by measuring the amplitude in the displayed optical image of sine wave test patterns with a microphotometer for a series of spatial frequencies. Long-persistence phosphor screens permit electrical scanning of the image (Schade, 1958), and a fixed slit position of the microphotometer, only at extremely slow scan rates, whereas the slit can be moved at relatively fast rates over a stationary image. The slow image scan method can be used to measure the low-frequency response of the tube but has been found impractical to measure the high-frequency response because of insufficient phase stability in available signal generators. To avoid inaccuracies of measurements caused by variations of light output over the phosphor screen and defocusing of the microphotometer image by curved screens, mechanical motion of the slit aperture must be limited to small displacements. Thus the positive or negative amplitude of very low-frequency patterns is moved by electrical displacement (centering controls) under the microphotometer slit and the phosphor is given sufficient time to allow buildup to steady-state conditions. The microphotometer slit can then be moved fairly fast to search out the peak amplitude. At spatial frequencies higher than 50 cycles/in. the slit displacement is small enough to measure positive and negative peaks with one position of the image. High stability of the image is absolutely necessary to obtain true maximum readings which cannot build up if there is the slightest motion of the pattern on the screen, because the crest values in a sinusoidal intensity distribution integrate rapidly. Thus, the microphotometer observation of a crest value in a high-resolution pattern reproduced by a camera chain on a long-persistence monitor screen is a very sensitive method for determining the phase stability of the system.

THE ALIASING PROBLEMS IN TWO-DIMENSIONAL SAMPLED IMAGERY

Richard Legault

7.1. INTRODUCTION

Most electrooptical imaging devices involve sampling at both the camera and display. The electron beam scanning of television and the optical–mechanical scanning of infrared imaging devices are examples of one-dimensional sampling in the imaging sensor. The excitation of cathode ray tube phosphors by a scanned electron beam is an example of sampling in the displayed image. We are all familiar with the pictures produced by imaging devices with one-dimensional sampling but less familiar with pictures produced by electrooptical devices sampling in two dimensions. Recently two-dimensional arrays of photoconductive or photovoltaic detectors have been used to sense a scene, and two-dimensional arrays of photoemitting diodes, etc. have been used to display imagery.

Chapter 6 by Schade presents an analysis of imaging devices using one-dimensional sampling. There is, in fact, very little difference in the analysis of one- and two-dimensional sampling. There is more freedom in the selection of sampling lattices and image reconstruction functions in the two-dimensional case, but most of the differences are of secondary importance.

When we produce an image for evaluation and interpretation by a human observer the effects of sampling on the interpretability of the image are of prime importance. For example, the effects of resolvable periodic display rasters are well known and discussed in Chapter 6. Arrays of photoemitting diodes present two-dimensional periodic patterns to the eye, similar to the one-dimensional scan rasters. As we shall see later, one consequence of sampling is the creation of spurious periodic patterns in the sampled image.

The current methods of analysis for electrooptical imaging devices make extensive use of the frequency analysis techniques of communication theory. The use of the modulation transfer function (MTF) as a measure of performance for imaging systems is the current fashion. Lavin (1971) proposes, with considerable justification, that we consider the phase transfer function (PTF) as well as the amplitude (MTF) characteristics of the optical transfer function when evaluating an electrooptical imaging device. Linfoot's (1966) monograph provides an excellent introduction to the use of spatial frequency analysis techniques for optical systems. Lavin's (1971) chapter develops the techniques for electrooptical systems. The analysis of two-dimensional sampling techniques makes extensive use of spatial frequency representations of scenes. The analytic simplicity of spatial frequency representations is counterbalanced by the difficulty of interpreting results expressed in terms of spatial frequencies. The human interpreter is capable of perceiving discrete frequencies (periodic patterns) with ease, but the perception of subtle differences in continuous spatial frequency content of scenes is difficult for the observer. Test charts for electrooptical imaging devices are generally periodic bar patterns or circularly periodic TV test patterns; however, natural scenes contain few periodic patterns, and natural scenes are of primary importance. We shall make every effort to interpret results derived in the spatial frequency domain into the more familiar spatial domain. The reader should be prepared to supply his own interpretations.

This chapter will develop a description of the image displayed by a sampled image electrooptical sensor. The aliasing of frequencies at the display will be analyzed. Finally, we will consider some design consideration for electrooptical sampled image systems which eliminate or minimize the effects of aliasing.

7.2. A BRIEF REVIEW OF ONE-DIMENSIONAL SAMPLING

The consequences of sampling a one-dimensional electrical signal are familiar. Sampling a voltage signal $v(t)$ every T seconds provides us with a set of values $v(mT)$, $m = \pm 0, 1, 2, 3, \ldots$. The problem is how

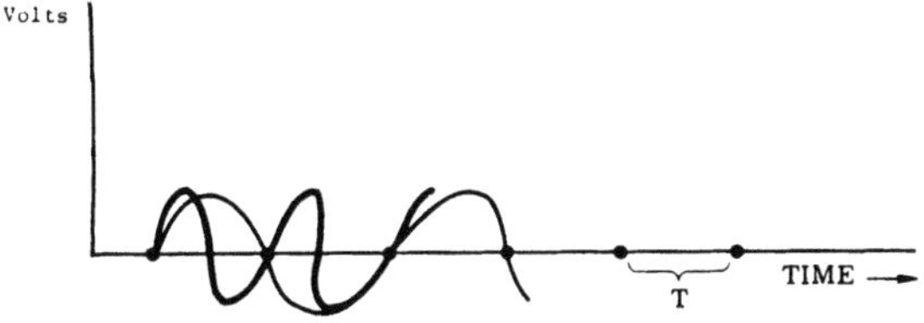

Fig. 7.1. Sampled values of frequencies $1/2T$ and $1/T$.

much do we know about $v(t)$ on the basis of the sampled values $v(mT)$. Figure 7.1 shows that periodic signals of frequencies $1/2T$ and $1/T$ give identical sampled values. If the signals $v(t)$ are sinusoidal, then for values of $f \leq \pi/T$ the set of sampled values for the functions $\sin\{2\pi t[f + (m/T)]\}$ is the same for all positive and negative integer values of m. Thus on the basis of samples spaced T seconds apart we are unable to distinguish between sinusoidal frequences $f + (m/T)$ for integer m. The set of frequencies $\{f + (m/T)\}$ is called the set of aliased frequencies. We shall have more to say about the set of aliased frequencies later. The next question is how to reconstruct the signal $v(t)$ from the sampled values $v(mT)$.

We shall use the notion of a reconstruction or interpolation function, used by Middleton (1960), to construct an estimate of the signal $v(t)$ from the sampled values. We introduce a reconstruction function $g(t)$ to be applied to the sample values. The estimated signal $\hat{v}(t)$ is given by

$$\hat{v}(t) = \sum_{m=-\infty}^{\infty} v(mT)g(t - mT) \tag{1}$$

The reconstruction formula applies the same reconstruction function at all sample values. The frequency representation of the estimated signal is revealing. The Fourier transform of $v(t)$ is given by

$$\hat{V}(f) = \int_{-\infty}^{\infty} v(t)e^{-2\pi ift}\, dt \tag{2}$$

After some manipulation of Eq. (1), we obtain

$$\hat{V}(f) = \frac{G(f)}{T} \sum_{m=-\infty}^{\infty} V\left(f + \frac{m}{T}\right) \tag{3}$$

Equation (3) shows that at every frequency f, $\hat{V}(f)$ is the sum of the aliased frequencies, $f + (m/T)$. The spectrum of $\hat{V}(f)$ is repetitive in

$$\sum_{m=-\infty}^{\infty} V\left(f + \frac{m}{T}\right)$$

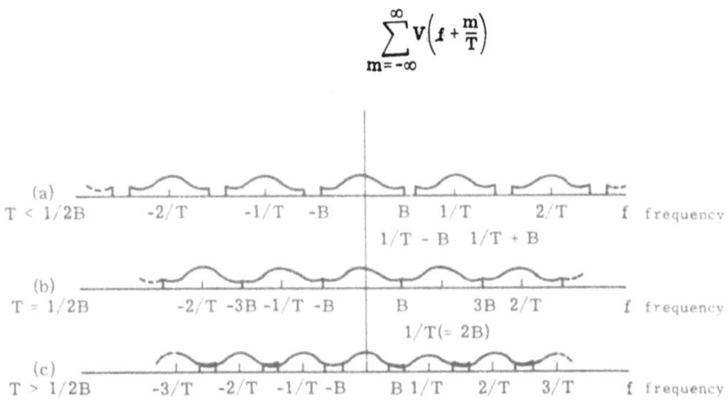

Fig. 7.2. Illustration of repetitive spectra.

the sense that $\sum_{m=-\infty}^{\infty} V[f + (m/T)]$ is periodic with period $1/T$ (Fig. 7.2). Clearly if $V(f)$ is nonzero only in an interval $|f| \leq 1/2T$, *and* we set $G(f) = T$ for $|f| \leq 1/2T$ and $G(f) = 0$ elsewhere, then

$$\begin{aligned} \hat{V}(f) &= V(f) & |f| &\leq 1/2T \\ &= 0 & &\text{elsewhere} \end{aligned} \tag{3a}$$

Thus if we sample the signal every T seconds and $V(f)$ is band limited with bandwidth $B = 1/2T$, then $V(f)$ can be exactly reproduced from the sampled values by passing the sample values through a bandpass filter of bandwidth $1/2T$. The transform of $G(f)$ is $g(t) = T[\sin(\pi t/T)]/\pi t$. An example of the reconstruction is given in Fig. 7.3. There is a tendency to interpret a sampling procedure as a bandpass filtering operation. It is not. The analysis represented by Eq. (3) shows that higher frequencies are not filtered but are added to lower frequencies. Even if $G(f)$, the

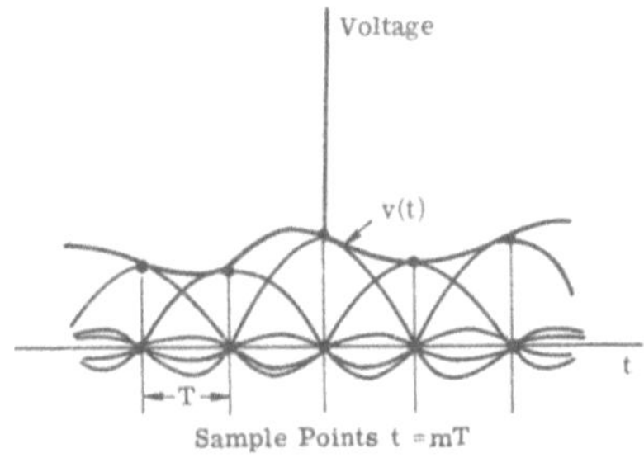

Fig. 7.3. Reconstruction of sampled function.

reconstruction function, is a bandpass filter with zero response outside the passband $|f| \leq 1/2T$, higher frequencies are aliased or added to the frequencies in the passband. Only if the original signal $V(f)$ is band limited, $|f| < 1/2T$, can we ensure, in principle, perfect reproduction of $V(f)$ from its sampled values. Thus sampling does not band limit the output signal, but if the original signal is band limited, then it can be reproduced from its sampled values.

It is apparent from Eq. (3) that if $V(f)$ is not band limited, there is nothing that we can do with the reconstruction function $G(f)$ to remove the aliasing. We must modify $V(f)$. Suppose we allow ourselves the option of prefiltering the signal $V(f)$ and obtain as an input to the sample the signal $M(f)V(f)$, where $M(f)$ is the prefilter. Suppose further that we want to minimize the average mean-square difference between the original and reconstructed signal. This criterion is formalized as

$$\min_{g(t),m(t)} \int_{-\infty}^{\infty} [v(t) - \hat{v}_m(t)]^2 \, dt \tag{4}$$

where

$$v_m(t) = \sum_{m=-\infty}^{\infty} \int_{-\infty}^{\infty} v(\tau)m(\tau - t) \, d\tau \, g\left(t - \frac{m}{T}\right)$$

Brown (1961) shows that if the power spectrum of $v(t)$ is monotone decreasing, a not unrealistic assumption, then the $M(f)$ and $G(f)$ which minimize Eq. (4) are bandpass filters of width $1/2T$. *It pays to prefilter the signal before sampling.*

The analysis of two-dimensional sampling proceeds along the same lines with some small but important differences. The theory of sampling in two or higher dimensions was developed by Peterson and Middleton (1962) in a classic paper. The reader who wants formal proofs for the results given in this chapter should read the Petersen and Middleton paper (1962). The main objective of this chapter is to provide the reader with a guide to the analysis of electrooptical sampled image systems and some of the problem areas of sampled image systems. Montgomery (1969) used the Petersen and Middleton results as a starting point for his analysis of electrooptical imaging systems. We have relied on these two references in preparing the analytic portions of this chapter. Correct results should be attributed to Petersen and Middleton (1962) and Montgomery (1969). The errors belong to the present author.

7.3. ELECTROOPTICAL SAMPLED IMAGE SYSTEMS

Electrooptical imaging systems generally involve some form of sampling: television involves electron beam scanning; infrared down- and forward-looking scanners involve optical–mechanical scanning. Such imaging devices can be considered as one-dimensional image sampling devices, which were discussed in Chapter 6. The introduction of silicon mosaic targets in low-light-level television tubes and rectangular detector arrays for infrared imaging devices provides clear examples of two-dimensional scene sampling. Modern displays using mosaics of light-emitting diodes are examples of two-dimensional sample displays. Plasma, liquid crystal, and electroluminescent displays are also discretely addressed. Figure 7.4 is a sketch of an electrooptical imaging system which provides two-dimensional sampling of the scene and a discrete display. The scene is represented by an intensity pattern $s(x, y)$. The optical system, with point spread function $m(x, y)$, focuses an image on a discrete two-dimensional array of detectors. Each detector integrates the incident radiation. The detector area weighting response is given by $a(x, y)$. The response is, of course, nonzero only over the area of the detector. Scanned arrays, such as those used in forward-looking infrared sensors, are frequently sampled. Such systems are properly considered as two-dimensional sampling systems.

The relationship of the detector output signal to the distribution of intensity in the scene is a function of the optical system and the de-

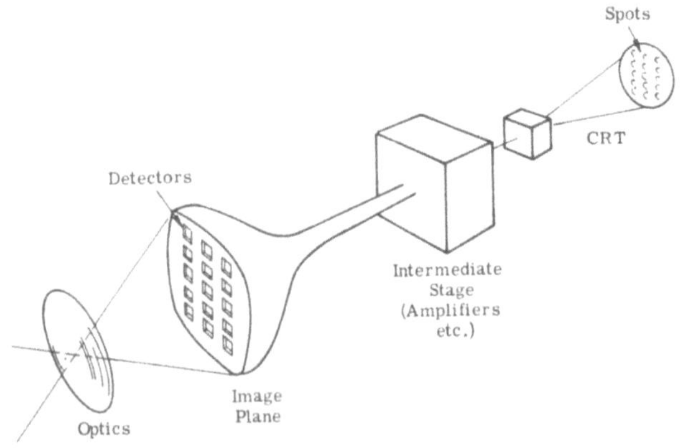

Fig. 7.4. Electrooptical sampled image system.

tector response function. Suppose the linear array is scanned in the x direction at c units of x per second. At any time t the signal from the y_j element of the array is given by

$$v'(t) = \iint s(x, y)a(x - ct, y - y_j) \, dx \, dy \tag{5}$$

Generally, there is a weighted integration of the detector's signal before sampling at a time t_i. If $h(t)$ is the response of the integration filter, then the output signal is given by

$$v(t) = \int v'(\tau)h(\tau - t) \, d\tau \tag{6}$$

Substituting Eq. (5) in Eq. (6) and rearranging the order of integration, we obtain

$$v(t) = \iint s(x, y) \int a(x - c\tau, y - y_j)h(\tau - t) \, d\tau \, dx \, dy \tag{7}$$

The effective detector area response function which reflects detector area and response as well as electronic integration and sampling is given by $a'(x, y)$:

$$a'(x - x_i, y - y_i) = \int a(x - c\tau, y - y_j)h(\tau - t_i) \, d\tau \tag{8}$$

where $x_i = ct_i$ and t_i is the sample time. If the detector array is discrete or we have a sampling of the electronic signals for display, we have a two-dimensional image sampling scheme. In principle, we still have a sampling system whether we sample by a discrete mosaic of detectors or an electronic sampling scheme.

The detector array or electronic sampling implies a sampling lattice in the focal plane of the optical system. The sampling lattice points are denoted by their coordinates (a_i, a_j). The signal from each detector after suitable amplification drives a discrete light source at the display. The light source has a spread function $b(x, y)$. This spread may be a defocused electron beam on a phosphor or a photoemitting diode. The coordinates of these light sources correspond to the coordinates of the detectors and are (a_i, a_j), although generally we should allow for a scale change, since the display arrays are considerably larger than the detector arrays. The notation describing some characteristics of components of the electrooptical imaging systems is given in Table 7.1.

TABLE 7.1

Notation for a Two-Dimensional Sampled Image System

$s(x, y)$	Distribution of intensity in the scene
$m(x, y)$	Point spread function of the optical system
$a(x, y)$	Response over the area of the detector
$b(x, y)$	Distribution of luminescence over the area of each discrete light source
(a_i, a_j)	Sample points
$o(x, y)$	Display scene

7.4. ANALYTIC REPRESENTATION OF TWO-DIMENSIONAL IMAGE SAMPLING

The usual sampling lattice in one dimension has a uniformly spaced sampling interval. The most easily visualized two-dimensional sampling lattice is a rectangular grid with equal spacings in both the x and y directions. We can generate a periodic lattice by defining two primary vectors $\mathbf{a}_1$ and $\mathbf{a}_2$. The points of the lattice are given by the linear combinations with integer coefficients

$$\mathbf{a}_n = n_1\mathbf{a}_1 + n_2\mathbf{a}_2, \qquad n_1, n_2 = 0, \pm 1, \pm 2, \pm 3, \ldots \tag{9}$$

There is a corresponding lattice in the spatial frequency domain (reciprocal lattice) which is generated by the vectors $\mathbf{b}_1$, $\mathbf{b}_2$ which are defined by

$$\mathbf{a}_1 \cdot \mathbf{b}_2 = 0, \qquad \mathbf{a}_2 \cdot \mathbf{b}_1 = 0 \tag{10a}$$

and

$$\mathbf{a}_1 \cdot \mathbf{b}_1 = 1, \qquad \mathbf{a}_2 \cdot \mathbf{b}_2 = 1 \tag{10b}$$

where in (10a) the primary and reciprocal lattice vectors are perpendicular. As an example, the square primary sampling lattice has the same appearance in the frequency (reciprocal) space but $\mathbf{a}_1 = \mathbf{b}_2$ and $\mathbf{a}_2 = \mathbf{b}_1$.

As we shall see, the reciprocal lattice plays a crucial role in the analysis of two-dimensional sampling. The intensity distribution at the detector plane is given by $\int s(\mathbf{u})m(\mathbf{u} - \mathbf{x})\,d\mathbf{u}$. From now on we shall use vector notation to denote a point in the plane $\mathbf{u} = (u_1, u_2)$ and a single integral sign to denote integration over the plane. The distribution of luminescence at the display is given by

$$o(\mathbf{x}) = \sum_{n_1=-\infty}^{\infty} \sum_{n_2=-\infty}^{\infty} \int s'(\mathbf{u})a(\mathbf{u} - \mathbf{x} - \mathbf{a}_1 n_1 - \mathbf{a}_2 n_2) \, d\mathbf{u} \, b(\mathbf{x} - n_1\mathbf{a}_1 - n_2\mathbf{a}_2) \tag{11}$$

where

$$s'(\mathbf{x}) = \int s(\mathbf{u})m(\mathbf{u} - \mathbf{x}) \, d\mathbf{u}$$

The spatial frequency representation of the display luminescence is given by the Fourier transform of $o(\mathbf{x})$,

$$O(r) = \int o(\mathbf{x}) \exp(-2\pi i \mathbf{r} \cdot \mathbf{x}) \, d\mathbf{x}, \qquad \mathbf{r} \cdot \mathbf{x} = r_1 x_1 + r_2 x_2 \tag{12}$$

A brief mathematical development of the Fourier transform of $o(\mathbf{x})$ is given in the appendix. The Fourier transform of $o(x)$ is

$$O(\mathbf{r}) = [B(\mathbf{r})/A] \sum_{[n]} A^*(\mathbf{r} + \mathbf{b}_n)S(\mathbf{r} + \mathbf{b}_n)M^*(\mathbf{r} + \mathbf{b}_n) \tag{13}$$

The summation $\sum_{[n]}$ is to be interpreted as $\sum_{n_1=-\infty}^{\infty} \sum_{n_2=-\infty}^{\infty}$ and $A(\mathbf{r})$, $S(\mathbf{r})$, and $M(\mathbf{r})$ denote the Fourier transforms of $a(\mathbf{x})$, $s(\mathbf{x})$, and $m(\mathbf{x})$, respectively. It is clear that the expression $\sum_{[n]} A^*(\mathbf{r} + \mathbf{b}_n)M^*(\mathbf{r} + \mathbf{b}_n)$ $S(\mathbf{r} + \mathbf{b}_n)$ contained in Eq. (13) is periodic with period $\mathbf{b}_n$. As we shall see, $B(\mathbf{r})$ denotes the reconstruction function role and $A^*(\mathbf{r})M^*(\mathbf{r})$ acts as a prefilter. In order to interpret Eq. (13), we should have a better picture of the sampling lattice and its impact on the sampled image.

We should consider first a simple interpretation of the spatial frequency representation $S(r_1, r_2)$. The value of $S(\mathbf{r})$ at the point (r_1, r_2) in the frequency plane is the amplitude of a sinusoidal wave with period $1/r_1$ in the x direction and period $1/r_2$ in the y direction. Figure 7.5 depicts such a wave with $r_1 = r_2 = k$. Clearly negative spatial frequencies have a definite physical interpretation. We say that a function $s(\mathbf{x})$ is periodic if there is a pair of vectors $\mathbf{a}_1$, $\mathbf{a}_2$ such that if $a_{[n]} = n_1 a_1 + n_2 a_2$, then $s(\mathbf{x}) = s(\mathbf{x} + \mathbf{a}_n)$ for all n. In short, the pattern repeats itself over parallelograms formed by the vectors $\mathbf{a}_1$ and $\mathbf{a}_2$. The checkerboard pattern of Fig. 7.6 is an example of a two-dimensional periodic pattern. The vectors $\mathbf{a}_1$ and $\mathbf{a}_2$ are given by $(d, 0)$ and $(0, d)$, respectively. It can be seen that the pattern repeats itself over squares of side d, which is the parallelogram specified by the vectors $(d, 0)$ and $(0, d)$.

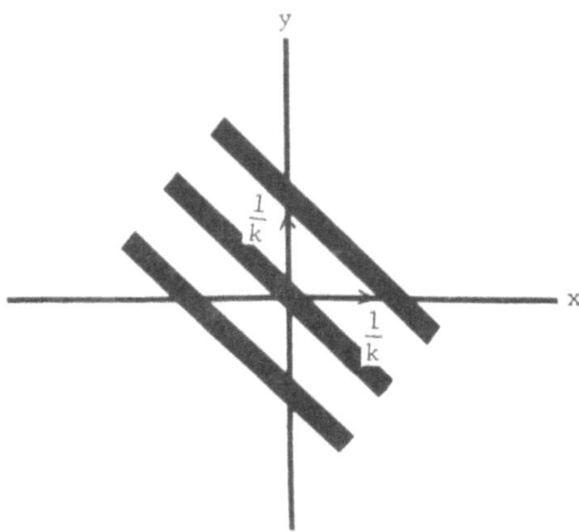

Fig. 7.5. Bar approximation to $\cos[2\pi(k_x x + k_y y)]$.

The sampling lattice that comes to mind almost immediately is the square lattice depicted in Fig. 7.7. It has been observed that the resolution for patterns at 45 deg to the lattice orientation is greater than for patterns oriented parallel to the rows and columns of the array. With a spacing of $1/2B$ in the x and y directions we have a spatial frequency limit of B. The spatial frequency limit in the 45 deg direction (see Fig. 7.7) is $\sqrt{2}B$. It has been observed that resolution in TV displays in a direction 45 deg to the scan direction is better than in the horizontal and vertical directions.

Fig. 7.6. Periodic two-dimensional pattern.

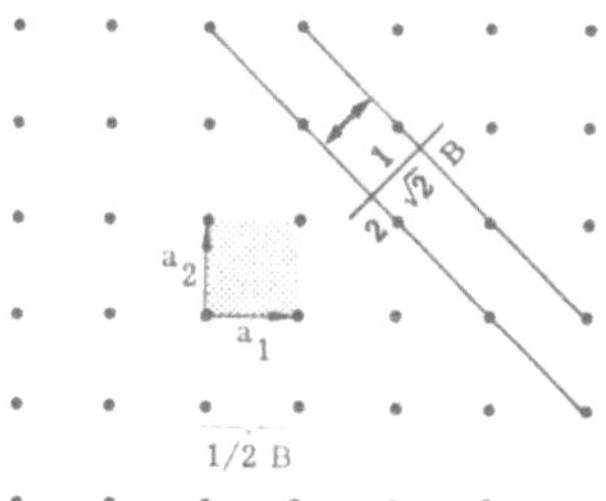

Fig. 7.7. Rectangular sampling lattice.

This effect is more evident in two-dimensional sampling arrays. The number of detectable line pairs is a factor of $\sqrt{2}$ greater in the 45 deg direction. This must appear strange to those familiar with photographic systems. The resolution in photographic systems is the same (or nearly so) in all directions, since the optical frequency response of the system is symmetric. Of course, the frequency response of the optical system for the two-dimensional sampling array is symmetric. The square sampling array introduces the asymmetry. I can find no justification for asymmetry in an electrooptical imaging system.

The asymmetry in frequency response can be explained by considering the reciprocal lattice in the frequency plane. Suppose that the frequency response of the optical system and detector-area weighting is unity, i.e., $M(\mathbf{r})$ and $A(\mathbf{r})$ equal one; then Eq. (13) can be rewritten as

$$O(\mathbf{r}) = [B(\mathbf{r})/A] \sum_{[n]} S(\mathbf{r} + \mathbf{b}_n) \qquad (14)$$

As we noted before, $\sum_{[n]} S(\mathbf{r} + \mathbf{b}_n)$ has period $\mathbf{b}_n$. This means that the pattern specified by $\sum_{[n]} S(\mathbf{r} + \mathbf{b}_n)$ repeats on parallelograms β defined by the vectors $\mathbf{b}_1$, $\mathbf{b}_2$ located at the lattice points $\mathbf{b}_n$. Thus, in a sense, the displayed information $O(\mathbf{r})$ conveys only that information about the scene $S(\mathbf{r})$ contained in the parallelogram β defined by $\mathbf{b}_1$ and $\mathbf{b}_2$. The parallelogram β defines the frequency reproduction capability of the sampling lattice defined by $\mathbf{a}_1$ and $\mathbf{a}_2$ in the same sense that $1/2T$ is the frequency reproduction limit (Nyquist limit) for a one-dimensional sampling lattice of spacing T. The display reproduction spatial frequency limits are given by the parallelogram β defined by $\mathbf{b}_1$ and $\mathbf{b}_2$, the reciprocal lattice specified by the sampling lattice.

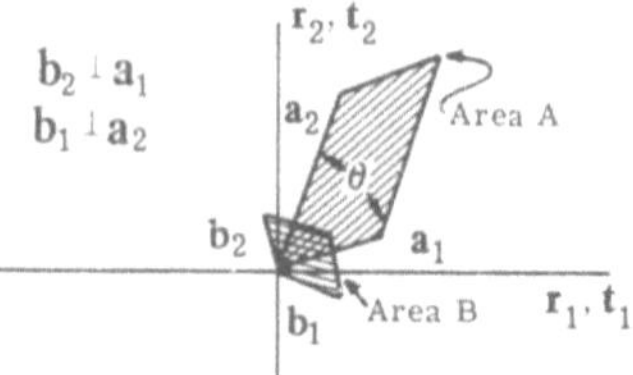

Fig. 7.8. Geometry of basis vector $\mathbf{b}_n$.

Figure 7.8 shows the geometry of the sampling lattice $\mathbf{a}_1$, $\mathbf{a}_2$ and the reciprocal lattice $\mathbf{b}_1$, $\mathbf{b}_2$. We see that the sampling lattice $\mathbf{a}_1$, $\mathbf{a}_2$ produces reciprocal lattice vectors $\mathbf{b}_1$, $\mathbf{b}_2$ whose associated parallelogram β (area B) is asymmetric. The reciprocal lattice vectors for a square lattice of unit spacing, $\mathbf{a}_1$ and $\mathbf{a}_2$ equaling $(1, 0)$ and $(0, 1)$, respectively, are $\mathbf{b}_1$ equals $(0, 1)$ and $\mathbf{b}_2$ equals $(1, 0)$. The parallelogram is a square of side one and the spatial frequency limit in the $(1, 1)$ direction implied by the square sampling lattice is a factor $\sqrt{2}$ higher than in either the $\mathbf{a}_1$ or $\mathbf{a}_2$ direction. The resolution in the $\mathbf{a}_1$ or $\mathbf{a}_2$ direction is the smallest resolution limit of the sampling lattice.

The requirement for $O(\mathbf{r})$, the Fourier transform of the displayed scene, to be identical to the real scene $S(\mathbf{r})$ is now apparent. If the spatial frequency representation of the scene is nonzero only on the parallelogram formed by $\mathbf{b}_1$ and $\mathbf{b}_2$, then $S(\mathbf{r} + \mathbf{b}_n)$ equals zero for $n = \pm 1$, $\pm 2, \ldots$ and is nonzero for r contained in β. Then Eq. (13) can be written as

$$O(\mathbf{r}) = [B(\mathbf{r})/A]S(\mathbf{r}) \tag{15}$$

for $M(\mathbf{r}) = A(\mathbf{r}) = 1$. If we let the spatial frequency distribution of luminance at a discrete light element of the display be a constant A over the spatial frequency parallelogram β, then $O(\mathbf{r}) = S(\mathbf{r})$ and we have perfect reproduction.

If the scene's spatial frequency representation $S(\mathbf{r})$ is nonzero over a bounded region in the frequency plane, we can call the scene spatial frequency limited. Clearly, for a spatial-frequency-limited scene we can find a pair of vectors $\mathbf{b}_1$ and $\mathbf{b}_2$ such that the associated parallelogram covers the spectrum of the scene $S(\mathbf{r})$. The reciprocal lattice specifies the sampling lattice $\mathbf{a}_1$, $\mathbf{a}_2$ and, in principle, perfect reproduction can be achieved. The spatial frequency representations of real world scenes are in practice limited in the way described above. Atmospheric paths and realistic optical systems, as well as nature, all affect the spatial frequencies in real scenes in the same way. Spatial frequency amplitude distributions of real scenes are almost always decreasing functions of frequency. Thus, we can find a spatial frequency limit above which we find very little power.

If the spatial frequency distribution of the scene lies partly outside the parallelogram β defined by $\mathbf{b}_1$ and $\mathbf{b}_2$, then perfect reproduction of the scene is not possible with the sampling lattice $\mathbf{a}_1$ and $\mathbf{a}_2$. The display frequency distribution $O(\mathbf{r})$ at a frequency point $\mathbf{r}$ is the sum of the generally complex values $\sum_n S(\mathbf{r} + \mathbf{b}_n)$. The set of spatial frequencies $\{\mathbf{r} + \mathbf{b}_n\}$ is called *the set of aliased spatial frequencies*.

7.5. EFFECTS OF ALIASING ON SAMPLED IMAGES

Equation (13) shows us how an image constructed from samples differs from an image which is not from a sampling system:

$$O(\mathbf{r}) = [B(\mathbf{r})/A] \sum_{[n]} A^*(\mathbf{r} + \mathbf{b}_n)S(\mathbf{r} + \mathbf{b}_n)M^*(\mathbf{r} + \mathbf{b}_n)$$

It is the disturbing summation $\sum_{[n]}$, which makes the output display from a sampled image system differ from an unsampled, linear, electrooptical imaging system. What are the consequences of aliasing on a sampled image? Equation (13) is a precise statement about the consequences of aliasing in the spatial frequency domain. Unfortunately, our eye–brain combination does not see a spatial frequency picture; we see an image in the spatial domain. Let us translate the effects of aliasing as given by Eq. (13) into effects on an image.

Two quite different aliasing effects can be noted. The first is a result of periodic patterns and the second is a result of aliased continuous frequency spectra. The best-known effects arise when a scene containing periodic components is sampled, causing Moiré patterns in the displayed image. If the scene contains a periodic pattern, then $S(\mathbf{r})$ takes on a nonzero value only at or near the frequency of the periodic pattern $\mathbf{r}_0$. However, we see from Eq. (13) that the periodic pattern of frequency $\mathbf{r}_0$ will appear on the displays at frequencies $\{\mathbf{r}_0 + \mathbf{b}_n\}$. This is particularly disturbing when $\mathbf{r}_0$ is near the limiting frequency of the sampling lattice, and the angle between a reciprocal lattice point and $\mathbf{r}_0$ is much greater

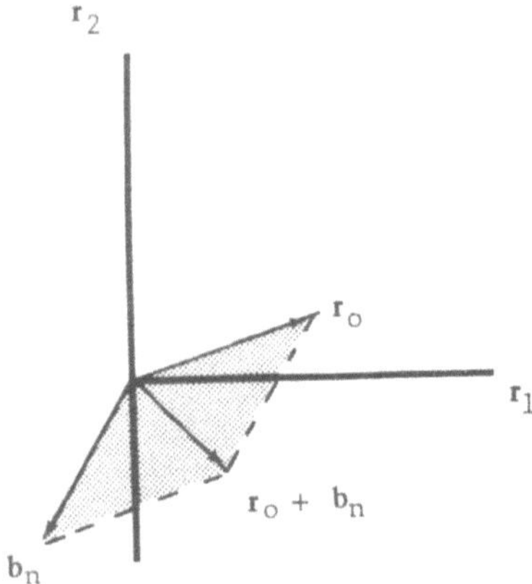

Fig. 7.9. Calculation of Moiré frequencies
from sampled image system.

than 90 deg. This occurs when the pattern is slightly rotated from one of the vectors forming the sampling lattice, causing a low-frequency pattern to appear. Figure 7.9 shows the computation of the Moiré pattern frequency. We see that r_0 now appears at a much lower frequency and the pattern has been rotated.

This can be simulated using conventional Moiré pattern transparencies. Figure 7.10a is a rectangular grid pattern. Figure 7.10b is a bar pattern. The two transparencies were laid one atop the other with the bar pattern slightly rotated in a clockwise direction with respect to the vertical direction of the grid pattern. Figure 7.11 is a photograph of the transparencies on the light table. Here the low-frequency rotated Moiré pattern is quite evident. The image in Fig. 7.12 is from an airborne scanner (sampling system) operating in the visible; the Moiré patterns in the plowed fields are quite evident. If a small object such as a tractor were present in the plowed field, its detection would be difficult because of this Moiré pattern.

When Moiré patterns appear in sampled images they interfere with the detection of objects. However, the geometry of the sampling lattice and the scene pattern must be just right to cause trouble from the Moiré pattern. We have seen a great many sampled images of real scenes and, although we have never measured the incidence of Moiré patterns in real-world sampled images, our impression is that they are few in number. We looked at 109 airborne images before finding the example in Fig. 7.12.

Fig. 7.10a. Image of rectangular grid transparency.

Fig. 7.10b. Image of bar pattern transparency.

Fig. 7.11. Image of superposition of two patterns—Moiré pattern.

The reader is invited to think of all the natural or man-made periodic patterns he can and assess for himself the magnitude of this problem. Common cases are ocean wave patterns and wind patterns in sand.

Aliasing which produces Moiré patterns depends on rather unusual properties of the scene. There is another kind of scene property which can occur more often but whose effects are less dramatic and more difficult to interpret. Suppose that for reasons of economy we are forced to make our sampling lattice rather coarse to save on the number of detectors used and display elements. Let us call the area A, the area of the parallelogram formed by the sampling lattice vectors $\mathbf{a}_1$ and $\mathbf{a}_2$. Then we can show that the area B covered by the reciprocal lattice parallelogram formed by $\mathbf{b}_1$ and $\mathbf{b}_2$ equals $1/A$. Thus, the coarser the sampling lattice, the smaller the area in the spatial frequency domain covered by the reciprocal lattice. There may then be large regions in the spatial frequency domain which are not covered by the reciprocal lattice parallelogram β. Figure 7.13 illustrates this situation. From Eq. (13) we see that the set of aliased frequencies is $\{\mathbf{r} + \mathbf{b}_n\}$, $\mathbf{r}$ contained in β. Thus, the frequencies in the cross-hatched area of Fig. 7.13 are

Fig. 7.12. Moiré patterns from plowed fields (scanner image).

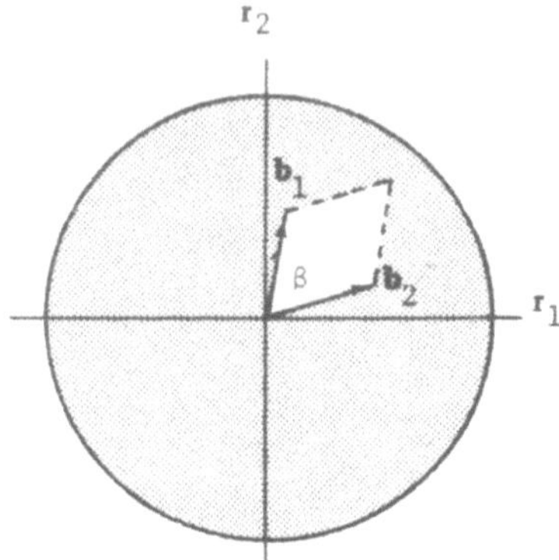

Fig. 7.13. Region of aliased frequencies.

aliased with frequencies in β. This is all rather obvious when we look in the spatial frequency domain. We have already shown what happens to the image when an aliased frequency $\mathbf{r}_0$ not in β happens to have a line spectrum of amplitude $S(\mathbf{r}_0)$, i.e., there is a periodic pattern of frequency $\mathbf{r}_0$ in the scene. This situation gives rise to the Moiré patterns. More often $S(\mathbf{r})$ is continuous over the region of aliased frequency, and the scene does not contain a periodic pattern for any aliased $\mathbf{r}$ since there is no scene amplitude at that line.

The following example gives us some insight. Figure 7.14(a) shows an array of square detectors of side d whose centers are spaced a distance d apart. Suppose the scene is composed of square resolution elements of side d. This gives us an image of the scene which looks like the screened images used by newspapers. Suppose that such a scene is incident on the detector array of Fig. 14(a) and that two of the scene elements coincide in the vertical direction with the detector array and the sides of the scene elements are parallel to the sides of the detector. Further suppose that the distance between the centers of the bright scene elements to the centers of the middle detector is d' and that an equal amount of light comes from each scene element. If the display is a discrete array of square light-emitting diodes, and if we define the display contrast as the ratio of the difference between the target and background brightness to the sum of the target and background brightness, we see that for $d' = 2/3d$, the two distinct scene objects will appear as a single object. For $d' > 2/3d$ two distinct objects appear on the display. For $d' > d$ the contrast between the scene elements and background is 100%. For $d' < d$ the contrast decreases as $(3d' - 2d)/(2d - d')$. Of course, for $d' < d$ the scene has spatial frequencies higher than the limiting frequency of the sampling lattice. These higher frequencies have reduced the scene contrast. Figure 7.14(b, c) illustrate the effect of shifting a sampling lattice (phase shift) with respect to the signal. We are assuming in Fig. 7.14(b, c) that we sample the signal only at the lattice points and do not integrate the signal over a detector area. Figure 7.14(b), sampling grid 1 sees only one point in the center of the signal, grid 2 sees the left-hand rise portion of the signal, and grid 3 sees two points in the center of the signal. Figure 7.14(b), grid 1 sees the left-hand edge, grid 2 sees the right-hand edge, and grid 3 sees both edges of the signal. Shifts in the sampling lattice with respect to the scene can shift or eliminate edges, and in some cases drop edges. These edges contain frequencies that fall well outside the spatial frequency limits of the sampling lattice.

Let us turn to Eq. (13) and the definition of the display spatial frequency contrast. We simplify the problem by assuming both the detector area weighting and optical frequency response are unity. Then

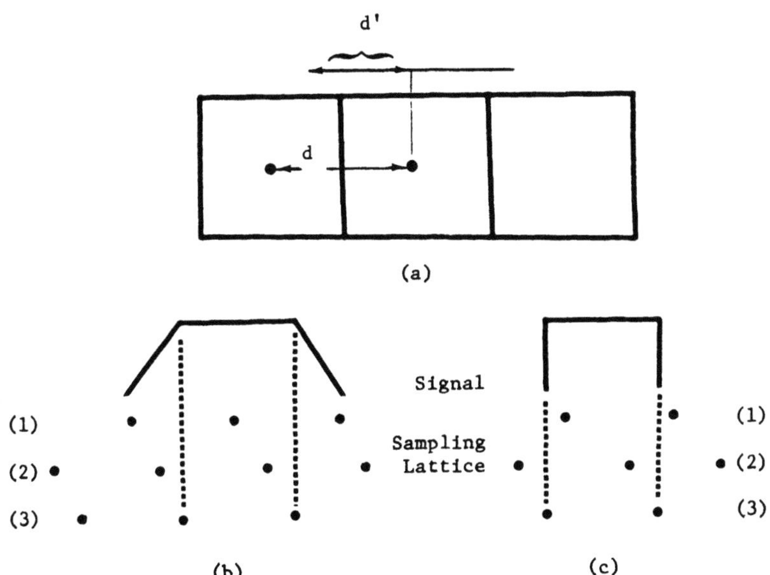

Fig. 7.14. Aliasing effects. (a) Contrast reduction as a function of distance between objects. (b) Phase shifts in sampling grid. (c) Edge effects.

Eq. (13) becomes

$$O(\mathbf{r}) = [B(\mathbf{r})/A] \sum_{[n]} S(\mathbf{r} + \mathbf{b}_n) \qquad (16)$$

There is no reason for the display element frequency response $B(\mathbf{r})$ to be other than zero over the region not covered by the limiting parallelogram. The displayed scene $o(\mathbf{x})$ is the Fourier transform of $O(\mathbf{r})$ given by

$$o(\mathbf{x}) = (1/A) \int_{\beta} [\exp(2\pi i \mathbf{x} \cdot \mathbf{r})] \sum_{[n]} S(\mathbf{r} + \mathbf{b}_n) \qquad (17)$$

Interchanging the summation and integration, we obtain

$$o(\mathbf{x}) = \frac{1}{A} s_\beta(\mathbf{x}) + \frac{1}{A} \sum_{[n] \neq 0} [\exp(-2\pi i \mathbf{x} \cdot \mathbf{b}_n)] \int_{\beta + \mathbf{b}_n} S(\mathbf{r})$$
$$\times \exp(2\pi i \mathbf{r} \cdot \mathbf{x}) \, d\mathbf{r} \qquad (18)$$

$s_\beta(\mathbf{x})$ is the image we would have obtained if we had bandpassed the original scene image $s(\mathbf{x})$. It is a "smeared" version of the original picture. The bandpass is, of course, determined by the sampling lattice. The remaining term is a weighted sum of bandpassed images where the bandpass is given by $\beta + \mathbf{b}_n$ (Fig. 7.15). The display image is then given as a bandpassed image plus the weighted sum of images with higher

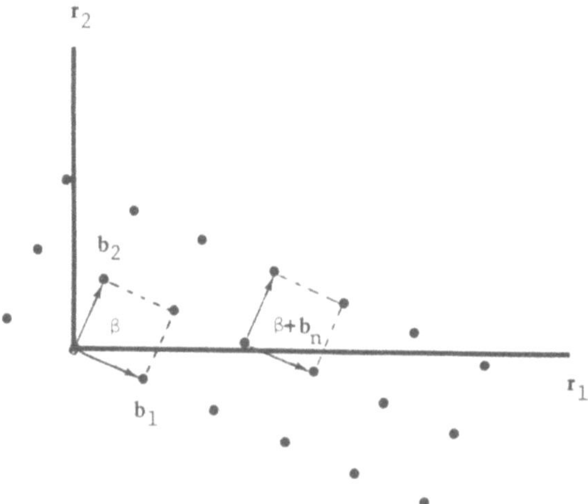

Fig. 7.15. Bandpasses for the displayed image.

bandpasses,

$$o(\mathbf{x}) = \frac{1}{A}\, s_\beta(\mathbf{x}) + \frac{1}{A} \sum_{[n]\neq 0} [\exp(-2\pi i\mathbf{x}\cdot\mathbf{b}_n)]s_{\beta+\mathbf{b}_n}(\mathbf{x}) \qquad (19)$$

The high-bandpass images $s_{\beta+\mathbf{b}_n}(\mathbf{x}) + s^{*}_{\beta+\mathbf{b}_n}(\mathbf{x})$ are essentially outlines of the edges of the original scene. We might consider the images $s_{\beta+\mathbf{b}_n}(\mathbf{x}) + s^{*}_{\beta+\mathbf{b}_n}(\mathbf{x})$ as a set of pictures of edges in the scene. The weightings $\exp(-2\pi i\mathbf{x}\cdot\mathbf{b}_n)$ are periodic over the sampling lattice and play an interesting role. The expression

$$[\exp(-2\pi i\mathbf{x}\cdot\mathbf{b}_n)]s_{\beta+\mathbf{b}_n}(\mathbf{x}) + [\exp(2\pi i\mathbf{x}\cdot\mathbf{b}_n)]s^{*}_{\beta+\mathbf{b}_n}(\mathbf{x})$$

is the high-bandpass image frequency shifted down into the bandpass β. The terms $\sum_{[n]\neq 0}[\exp(-2\pi i\mathbf{x}\cdot\mathbf{b}_n)]s_{\beta+\mathbf{b}_n}(\mathbf{x})$ produce Moiré patterns for periodic $s_{\beta+\mathbf{b}_n}(\mathbf{x})$. If the high-bandpass scene has a constant spectrum, $s_{\beta+\mathbf{b}_n} = \text{const}$, then the frequency-shifted version adds a sinc function, $[(\sin x)/x][(\sin y)/y]$, in the sampling lattice cell. If the high-bandpass spectrum is a high-frequency carrier modulated by a low frequency, $s_{\beta+\mathbf{b}_n}(\mathbf{x}) = [\exp(2\pi i\mathbf{x}\cdot\mathbf{b}_n)]t_\beta(\mathbf{x})$, then the frequency-shifted spectrum is $t_\beta(\mathbf{x})$, which is added to the bandpassed image $s_\beta(\mathbf{x})$. This type of high-bandpass image has a noiselike effect on the bandpass image $s_\beta(\mathbf{x})$. If $s_{\beta+\mathbf{b}_n}(\mathbf{x})$ contains an edge, then the frequency-shift term $\exp(-2\pi i\mathbf{x}\cdot\mathbf{b}_n)$ may shift, broaden, or eliminate the edge in the frequency-shifted image added to $s_\beta(\mathbf{x})$. The occurrence of any of these effects depends on where the edge is located, as specified by $\mathbf{x}$, in the sampling cell specified by $n_1\mathbf{a}_1$ and $n_2\mathbf{a}_2$. Shifts of the sampling lattice which are small relative to the dimensions of the sampling cell specified by $n_1\mathbf{a}_1$ and $n_2\mathbf{a}_2$ can have rather dramatic effects on the appearance of edges.

If $S(\mathbf{r})$ is symmetric and decreasing, the sampling grid is rectangular; then the big contributors are $s(\mathbf{x})_{\beta+\mathbf{b}_n}$, where $n = (1,1)$, $(-1,-1)$, $(1,-1)$, $(-1,1)$, $(0,1)$, $(1,0)$, $(-1,0)$, and $(0,-1)$. The edges at 45 deg to the sampling lattice should receive more emphasis than those parallel to the sampling lattice.

Equation (18) makes one point very clear. If most of the spatial frequency distribution is concentrated in β, say 95%, then the aliased frequencies probably will not cause much distortion of the band-limited image $s_\beta(\mathbf{x})$. *The output display of a sampled image system is essentially a spatially band-limited image plus the sum of frequency-shifted scene-edge images.*

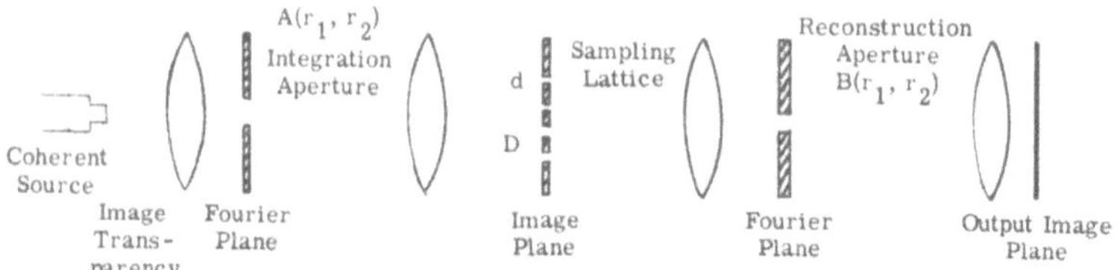

Fig. 7.16. Coherent simulation of sampling system.

Let us now look at a real scene band-limited image $s_\beta(\mathbf{x})$ and an edge image $s_{\beta+\mathbf{b}_n}(\mathbf{x})$. We shall also simulate a sampled image system with this experimental coherent optical setup (Fig. 7.16). Figure 7.17 is an aerial photograph of the streets in Baltimore, Maryland, and Fig. 7.18 is the same scene taken in the first image plane of the optical setup. The rectangular aperture $A(r_1, r_2)$ was set up to pass frequencies in the band -1.7 cycles/mm to 1.7 cycles/mm in both directions. The contrast of this bandpassed image is reduced. In general, the picture is a "smeared" version of the original. The image of Fig. 7.19 was made by placing a square annulus $a(r_1, r_2)$ which passes frequencies from 1.7 to 3.4 cycles/mm in both directions. In the Fourier plane this corresponds to the set of bandpasses $\beta + \mathbf{b}_n$ where $n_1, n_2 = (\pm 1, \pm 1), (0, \pm 1), (\pm 1, 0)$. If one looks closely, one can see that the street edges are bright while the street is dark. This is one justification for calling $s_{\beta+\mathbf{b}_n}(\mathbf{x})$ for $n \neq 0$ an edge image. Approximately 10% of the spatial frequency spectrum amplitude was concentrated in the square annulus of Fig. 7.19.

The sample spacing D was 0.3 mm and the sampling aperture d was 0.075 mm. One would like d much smaller; then the integration role played by the detector would be represented solely by the first Fourier plane aperture. The integration aperture $A(r_1, r_2)$ was a rectangular hole rather than the desired sinc function. Consequently, the integration function $a(x, y)$ is $(\sin 2\pi\omega x \sin 2\pi\omega y)/\pi^2 xy$ rather than the usual detector integration function $a(x, y) = 1$ for $1/-2\omega \leq x \leq 1/2\omega$ and $1/-2\omega \leq y \leq 1/2\omega$ and zero elsewhere. High-contrast scene elements can produce an irrelevant ringing in this simulation but for most portions of the usual scene the sinc function will fall off fast enough. This effect decreases as we make the $A(r_1, r_2)$ larger.

Figure 7.20 used a setting of $A(r_1, r_2)$ where $|\omega| \leq 3.4$ cycles/mm. This corresponds to the case of adjoining detectors. The image was taken in the final output plane of the experimental setup. The circular

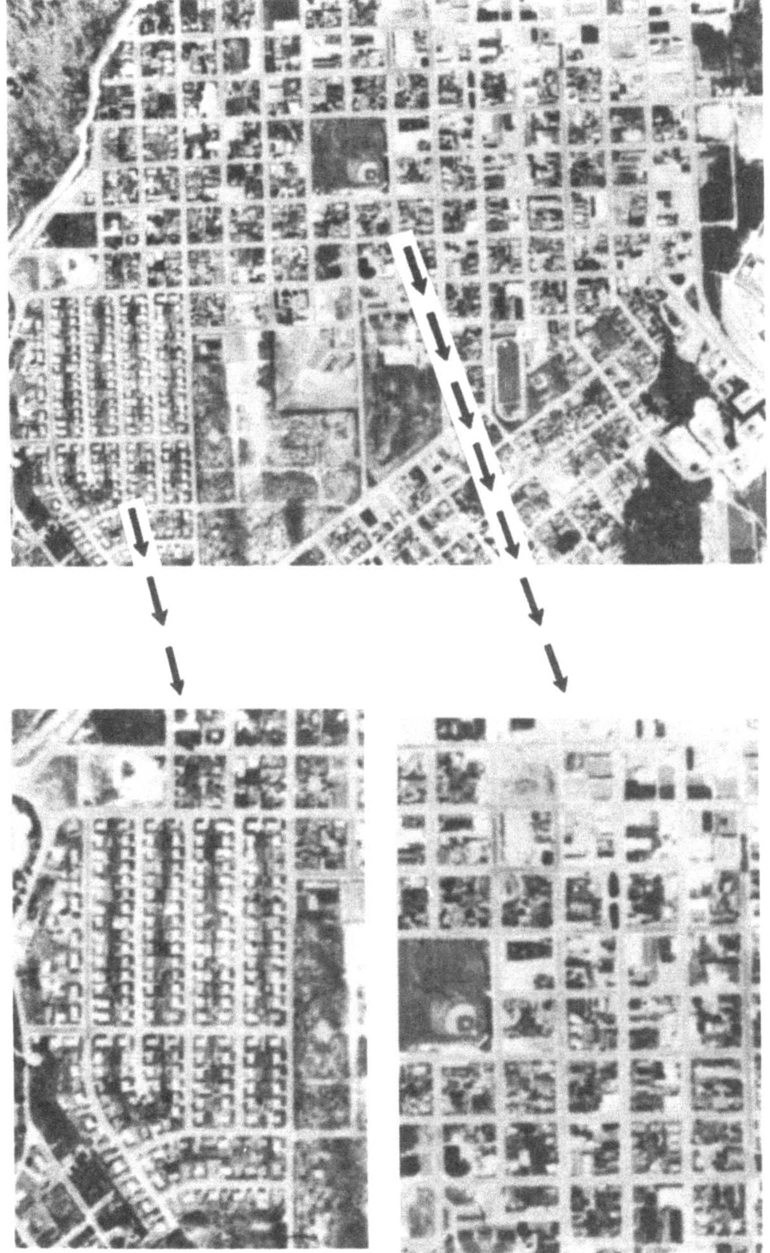

Fig. 7.17. Aerial photograph.

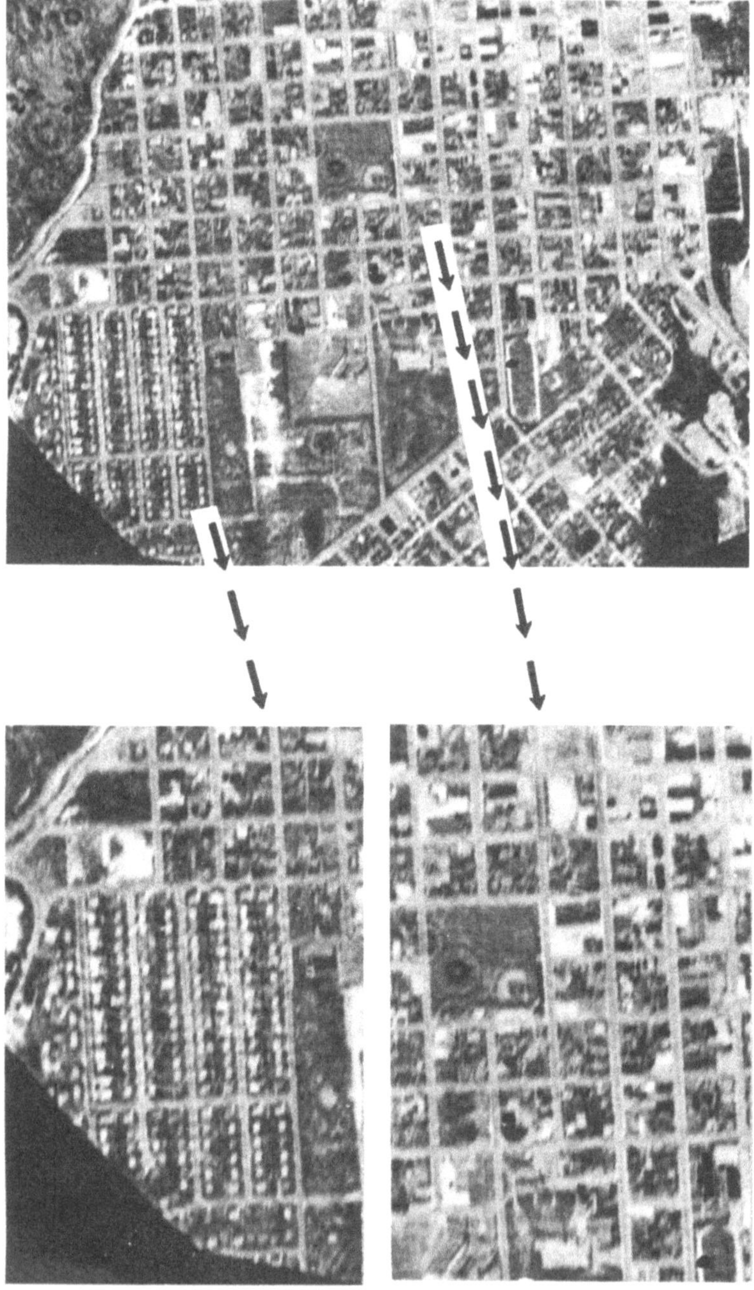

Fig. 7.18. Bandpassed image $s_\beta(\mathbf{x})$, 1.7 cycles/mm, rectangular bandpass.

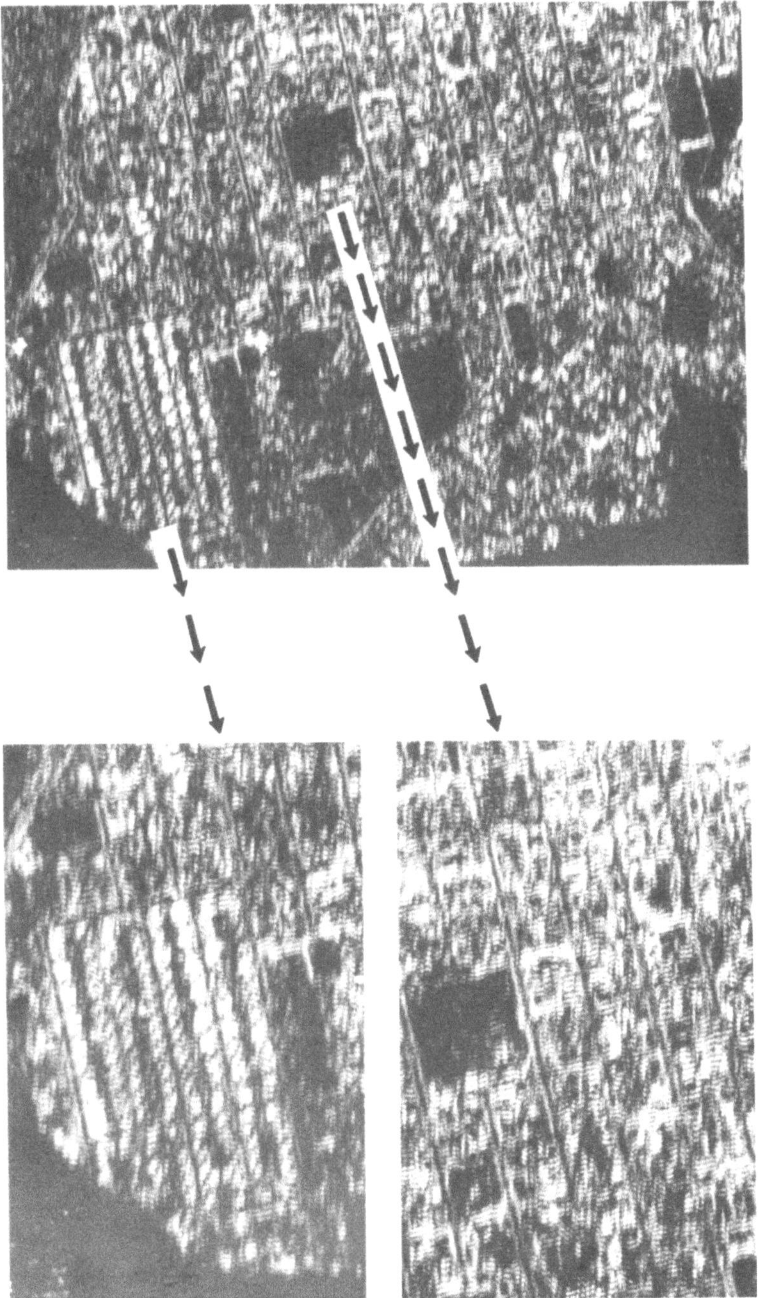

Fig. 7.19. Edge image $s_{\beta+\mathbf{b}_n}(\mathbf{x})$ annular bandpass, 1.7–3.4 cycles/mm.

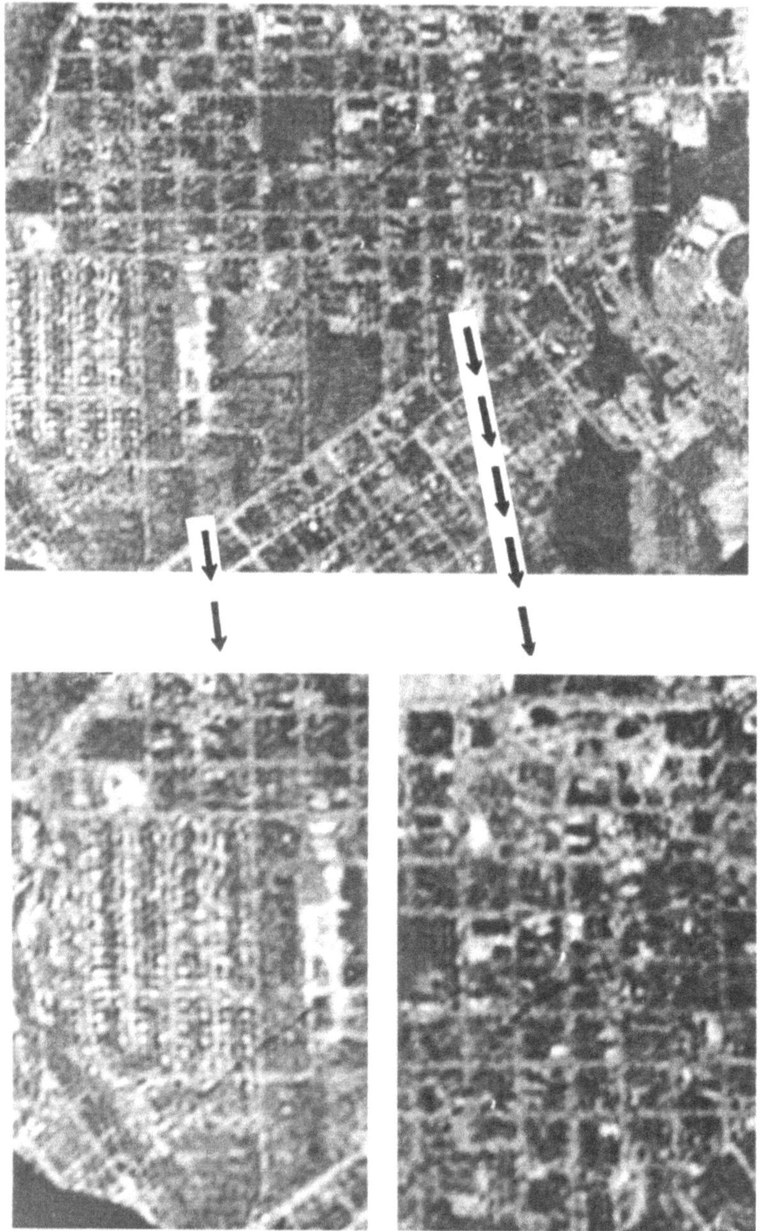

Fig. 7.20. Sampled image (simulated). Rectangular sampling grid spacing 0.3 mm.

ringing patterns in some portions of the scene is due to dust spots. The street patterns are clear in the bandpassed image $s_\beta(\mathbf{x})$ of Fig. 7.18 and in the sampled image of Fig. 7.20. The houses in Fig. 7.20 are not as clear. The reader's attention is directed to the street patterns. The sampling grid was oriented nearly parallel to the street pattern in the upper portion of the picture while the street pattern in the lower portion lies roughly 45 deg to the sampling grid. The contrast for the lower street pattern is higher than for the upper street pattern, as predicted earlier, despite the fact that the streets, at least some, in the upper portion are wider and the contrasts are about the same (see Fig. 7.17).

In summary, the effects of aliasing are clear for periodic patterns; The aliasing produces Moiré patterns. If the scene's spatial frequency is concentrated, say, 95% in the spatial frequency limit passband of the sampling lattice, then aliasing is probably not a problem. We really need more experimental evidence to establish the threshold; 95% is a very shaky threshold value. The sampled image is basically a bandpassed image plus some weighted edge images; the weight placed on an edge depends on the angles between the edge points and the sampling lattice. The reader is left to ponder the effects of aliasing further.

7.6. BEST SAMPLING LATTICES

The attachment to the square sampling lattice is a product of our one-dimensional experience. It is clear that we can select any noncollinear sampling lattice vectors $\mathbf{a}_1$ and $\mathbf{a}_2$. The only requirement is that the reciprocal lattice $\mathbf{b}_1$ and $\mathbf{b}_2$ provide a parallelogram which covers the spatial frequency distribution of the scene.

Atmospheric and optical spatial frequency responses are usually symmetric. They have no directional preference when they attenuate a spatial frequency. The asymmetric frequency response of a TV camera is due primarily to the sampling lattice. We are firmly convinced on the basis of experience that the average spatial frequency distribution of natural scenes is symmetric. This leads us to assert that, to a reasonable approximation, we may consider the spatial frequency distribution of a natural scene to be confined to a circle in the reciprocal (frequency) plane.

Generally, the repeated (repetitive) spectrum associated with a given reciprocal lattice which has no overlap will have areas not covered by

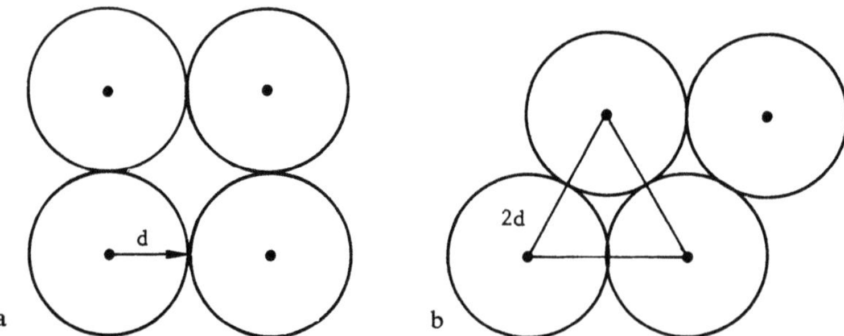

Fig. 7.21. Close packing of circles in the plane. (a) Square lattice; (b) 60° rhombic lattice.

the repeated spectrum. If we can minimize the area of noncoverage of a repeated spectrum, we are minimizing the amount of wasted frequency response. Figure 7.21 illustrates the repeated spectrum for two different reciprocal lattices. Geometers know that the pattern which minimizes the area not covered by circles of radius $d/2$ is formed by placing the centers of the circles at the vertices of an equilateral triangle of side d (Fig. 7.21b). Visual comparison of Fig. 7.21(a) and Fig. 7.21(b) should convince the reader that the area not covered is smaller in Fig. 7.21(b). The reciprocal lattice shown in Fig. 7.22 has $\mathbf{b}_1$ and $\mathbf{b}_2$ 120 deg apart and both vectors of length $1/d$. The sampling lattice (Fig. 7.23) shows

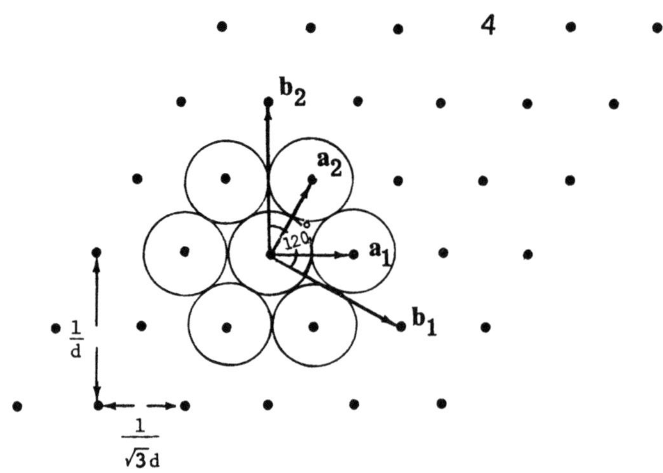

Fig. 7.22. Spectrum repetition for circular spectrum.

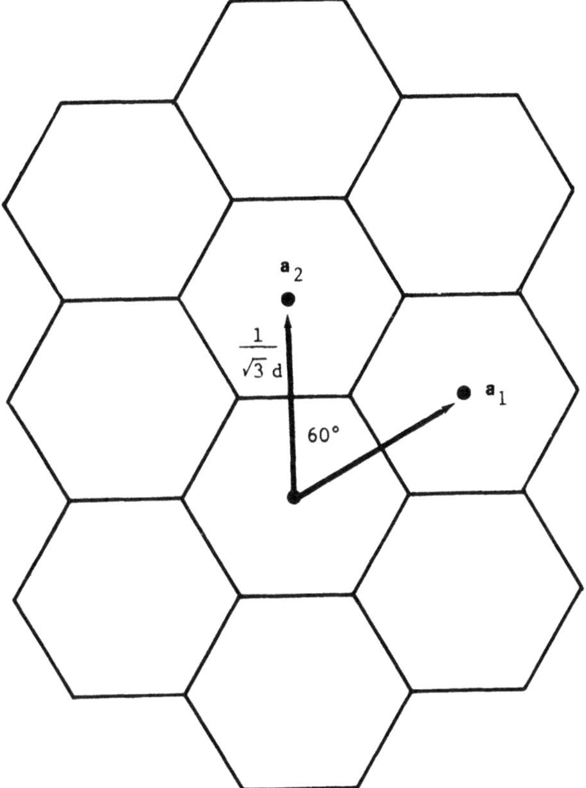

Fig. 7.23. "Best" detector pattern for circular spectrum.

that $\mathbf{a}_1$ and $\mathbf{a}_2$ are 60 deg apart and of length $1/\sqrt{3}\,d$. The detector array consists of rectangular hexagons whose sides are of length $1/3d$.

If we compare a square lattice and a 60 deg rhombic lattice with the same limiting spatial frequency response (both reciprocal lattices cover the circular spatial frequency spectrum), we see that, clearly, the hexagon detector is larger (in fact, the hexagon detector is 27% larger), and therefore more advantageous, since the larger detector collects more radiation. The 60 deg rhombic lattice requires 13% fewer detectors to cover the same area in the image plane, thereby reducing the cost of the electrooptical system. If one agrees that the spatial frequency distribution of nature is symmetric, then one must agree that the usual rectangular sampling lattice is rather inefficient.

7.7. SYSTEM DESIGN CONSIDERATIONS FOR SAMPLED IMAGE SYSTEMS

We must first consider the objectives in designing an electrooptical sampled image system. We would like the displayed image to be a faithful reproduction of the original scene or some modification of the scene, but we will probably have to be content with something less than perfection. One measure of faithfulness of reproduction is a small mean-square difference between the display image $o(\mathbf{x})$ and the scene $s(\mathbf{x})$. This mean-square error is measured by $\int [s(\mathbf{x}) - o(\mathbf{x})]^2 \, d\mathbf{x}$ and averaged over various scenes. There is another constraint, the cost of constructing the imaging device. There are four system-component specifications that should be chosen with some care for a two-dimensional sampled image system. They should be chosen with care for any imaging system, but the rather special characteristics of a sampled image system should be taken into account in selecting (1) the optical transfer function, (2) detector size and area responsivity, (3) the sampling lattice, and (4) spacing, area and light distribution of the sources at the display.

We have already discussed the design of a sampling lattice. A prime consideration is choosing the sampling lattice so that the limiting frequency parallelogram formed by $\mathbf{b}_1$ and $\mathbf{b}_2$ covers those frequencies where $S(\mathbf{r})$ is nonzero. We see from Eq. (13),

$$O(\mathbf{r}) = [B(\mathbf{r})/A] \sum_{[n]} A^*(\mathbf{r} + \mathbf{b}_n)S(\mathbf{r} + \mathbf{b}_n)M^*(\mathbf{r} + \mathbf{b}_n)$$

that either the detector or optical response can be used to prefilter the scene's spatial frequency distribution. As Eq. (13) shows, aliasing occurs before the display. There is no display response $B(\mathbf{r})$ which can eliminate aliasing. If we cannot afford the number of detector required to make $O(\mathbf{r})$ and $S(\mathbf{r})$ equal, then it can be shown (Petersen and Middleton, 1962) that the electrooptical system which minimizes the mean-square difference between the real and displayed images is specified as follows. The allotted number of detectors implies that only a limited portion of the spatial frequency can be covered by the parallelogram β. This determines the allowable lengths for $\mathbf{b}_1$ and $\mathbf{b}_2$. Select $\mathbf{b}_1$ and $\mathbf{b}_2$ such that β covers that portion of the spatial frequency spectrum where $S(\mathbf{r})$ is the largest. We then select that frequency $\mathbf{r} + \mathbf{b}_{n_0}$ from the set of frequencies

$\{\mathbf{r} + \mathbf{b}_n\}$ produced by letting $[n]$ vary, so that $\hat{O}(\mathbf{r})$

$$\hat{O}(\mathbf{r}) = \int \exp(-2\pi i \mathbf{x} \cdot \mathbf{r}) \text{ Average over scenes} \int s(\tau)s(\tau - \mathbf{x}) \, d\tau \, d\mathbf{x} \tag{20}$$

the scene's power spectrum, is a maximum. The best electrooptical imaging system sets $B(\mathbf{r})$ and $M^*(\mathbf{r})A^*(\mathbf{r})$ equal to one on the set of frequencies defined above and set equal to zero elsewhere. If the power spectrum $O(\mathbf{r})$ of the scene is decreasing, then the best system is obtained by setting $B(\mathbf{r})$ and $M^*(\mathbf{r})A^*(\mathbf{r})$ equal to one on the set of limiting frequencies β given by the sampling lattice.

We shall now turn our attention to the specifications of the optimal electrooptical sampled imaging system defined above. It is apparent that the response of the optical system $M(\mathbf{r})$ should be matched to the sampling lattice. If we have coarse sampling of the scene, we do not want a high-resolution optical system. We want an optical response $M(\mathbf{r})$ that passes only the limiting frequencies β implied by the sampling lattice.

The detector spatial frequency response $A^*(\mathbf{r})$ should also be selected so $A^*(\mathbf{r})$ has a flat response. If $M(\mathbf{r})$ bandpasses the image, then $A^*(\mathbf{r})$

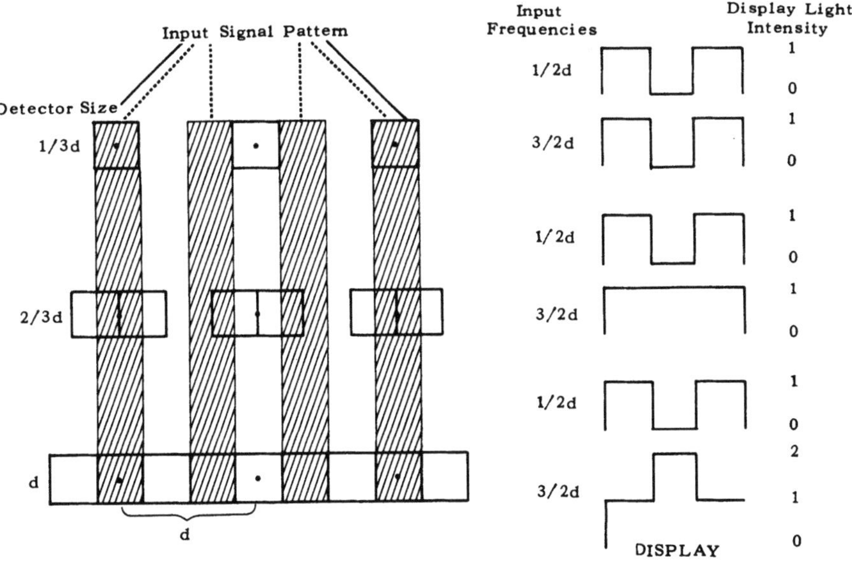

Fig. 7.24. Prefilter effects of detector size.

need not, but if $M(\mathbf{r})$ does not bandpass the scene, $A^*(\mathbf{r})$ should. The area responsivity of the detector is generally not controllable, but the shape and size of the detector are. Figure 7.24 illustrates the effects of various detector sizes. The signal is a bar pattern. If we use all the bars, the pattern has a frequency $3/2d$. If we use only the outer bars, the pattern period is $2d$ and the pattern frequency is $1/2d$. The spacing of the detectors is d. The scene signal is integrated over the detector area. We will show the display signals that arise when we use three different rectangular detectors in the focal plane of the imaging system. The signal from each detector drives a square light source at the display. The display element luminescence is proportional to the signal. The figure at the right depicts the light signals at the corresponding display elements. We see that for the small detectors the two frequencies give the same displayed signal. In short, the two frequencies are aliased as we expect. However, if we make the detectors larger, the two frequencies are not aliased at the display. Basically, we can assume the detector is of constant response over its area but the area and shape of the detector can be varied. A square detector of size $2d$ will be represented by

$$a(x, y) = 1 \qquad -d \le x \le d; \quad -d \le y \le d$$
$$ = 0 \qquad \text{elsewhere} \tag{21}$$
$$A(r_1, r_2) = (\sin 2\pi d r_1 \sin 2\pi d r_2)/\pi^2 r_1 r_2$$

The first zero contour in the frequency plane (r_1, r_2) occurs when r_1 or r_2 equals $1/2d$, and, to a rough approximation, we may consider the square of size $1/2d$ the bandpass of a square detector of size $2d$. The detector size which matches the detector bandpass to the limits of a sampling lattice of spacing $1/2B$ has $1/2d = B$ or $2d = 1/B$. Thus the "matched" detector is twice as large as the spacing between sampling centers and the edge of one detector will touch the edge of the center of the adjacent detector. This construction cannot be realized in a mosaic but can be realized in a scanning system by overscanning. The usual mosaic has adjacent detectors touching. The detector size is then $1/2B$ and the detector bandpass is $2B$. The aliased frequencies then lie in the band B to $2B$. The response $A(\mathbf{r})$ falls over the aliased frequency, but near the limits $A(\mathbf{r})$ is still fairly large. Fundamentally, the detector area carries part of the prefiltering burden. If the spatial frequency amplitudes for the sets of frequencies $\{\beta + \mathbf{b}_n\}$, $n = (0, 1)$, $(1, 0)$, $(0, -1)$,

$(-1, 0)$, $(-1, 1)$, $(1, -1)$, $(1, 1)$, and $(-1, -1)$, are large, we must depend on the optical response $M(\mathbf{r})$ for a bandpass prefiltering.

The specification of the spatial frequency distribution of a display light source can be easily obtained from Eq. (13). If we have prefiltered the image, then $O(\mathbf{r}) = S_\beta(\mathbf{r})$ if we have designed our display element with a specification of

$$B(\mathbf{r}) = 1/A^*(\mathbf{r})M^*(\mathbf{r}), \qquad r \in \beta$$
$$= 0 \qquad \qquad \text{elsewhere} \tag{22}$$

Generally $b(\mathbf{x})$ for each discrete light source should be an oscillating spatial distribution of light. The first zeros of the light intensity from a discrete light source fall on the center of the adjacent light sources. This is the motive for defocusing the spot on a cathode ray tube display so that the spot covers half of the adjacent scan line. The use of discrete light sources such as light-emitting diodes presents a problem similar to the adjacent detector problem. An element of a display mosaic of square light sources of side d will have a spatial frequency distribution given by

$$B(r_1 r_2) = (\sin \pi d r_1 \sin \pi d r_2)/\pi^2 r_1 r_2 \tag{23}$$

The first zeros occur at $-1/d \leq x, y \leq 1/d$. This is four times the area of the optimal bandpass of Eq. (22). The high and irrelevant spatial frequencies that appear on such a display can be disturbing to the viewer. One simple method of correcting for these high display frequencies is to make the distance between the display and viewer enough so that the eye response will eliminate the irrelevant high frequencies from the display. The sensible viewer will discover this for himself.

The well-designed electrooptical sampled imaging system eliminates aliasing. The displayed image will be a bandpassed version of the original scene.

APPENDIX

In this appendix the Fourier transform of $o(\mathbf{x})$ is developed using the Dirac delta properties. We shall not strive for complete rigor in this development. The analysis follows that of Petersen and Middleton

(1962). Let

$$f(\mathbf{a}_n) = \int s'(\mathbf{x})a(x - \mathbf{a}_n)\,d\mathbf{x}, \qquad o(\mathbf{x}) = \sum_{[n]} f(\mathbf{a}_n)b(\mathbf{x} - \mathbf{a}_n) \qquad (13a)$$

Using a two-dimensional delta function $\delta(u)$, we obtain

$$f(\mathbf{a}_n)b(\mathbf{x} - \mathbf{a}_n) = \int f(\mathbf{u})b(\mathbf{x} - \mathbf{u})\,\delta(\mathbf{u} - \mathbf{a}_n)\,d\mathbf{u} \qquad (13b)$$

Inserting Eq. (13b) in (13a) and interchanging the order of integration, we obtain

$$o(\mathbf{x}) = \int f(\mathbf{u})b(\mathbf{x} - \mathbf{u}) \sum_{[n]} \delta(\mathbf{u} - \mathbf{a}_n)\,d\mathbf{u} \qquad (13c)$$

The sum of delta functions can be considered as a two-dimensional periodic function on the lattice points $\mathbf{a}_n$ with a Fourier series expansion

$$\sum_{[n]} \delta(\mathbf{u} - \mathbf{a}_n) = (1/A) \sum_{[n]} \exp(-2\pi i \mathbf{u} \cdot \mathbf{b}_n) \qquad (13d)$$

where $A = |\mathbf{a}_1 \times \mathbf{a}_2|$, the area of the parallelogram formed by $\mathbf{a}_1$ and $\mathbf{a}_2$. Substituting Eq. (13d) in Eq. (13c), we obtain

$$o(\mathbf{x}) = (1/A) \sum_{[n]} \int f(\mathbf{u})b(\mathbf{x} - \mathbf{u}) \exp(-2\pi i \mathbf{u} \cdot \mathbf{b}_n)\,d\mathbf{u} \qquad (13e)$$

Equation (13e) is the sum of the convolution of $f(\mathbf{u})\exp(-2\pi i \mathbf{u} \cdot \mathbf{b}_n)$ and $b(\mathbf{u})$. Thus, we have

$$O(\mathbf{r}) = [B(\mathbf{r})/A] \sum_{[n]} F(\mathbf{r} + \mathbf{b}_n) \qquad (13f)$$

From the definition of $f(\mathbf{x})$ given above, we have $F(\mathbf{r}) = S(\mathbf{x})A^*(\mathbf{x})M^*(\mathbf{x})$ where the asterisk indicates the conjugate. Thus, we obtain Eq. (13).

Chapter 8

A SUMMARY

Lucien M. Biberman

8.1. AN OVERVIEW OF IMAGE QUALITY

Having spent many tens of pages saying more or less positive things, it becomes necessary to take a look from the other, less optimistic side. Possibly some of the most cogent remarks about the fallibility of some of the concepts we have put forth in this book were made by Brock (1965) in an invited paper before the SPSE. Before I quote from that paper, I should like to point out that Brock was addressing the topic of image quality in very good photography, a level of quality whose resolving power can be measured in hundreds of line pairs per millimeter as opposed to displays whose resolving power is usually measured in hundreds of lines across a major dimension. Brock therefore was talking about spatial frequencies perhaps two orders of magnitude greater than we normally consider for displays.

I have not quoted Brock's paper serially. I have taken things out of order but not out of context. I have done so in an effort to shorten the quotation and to cover a single topic in adjacent paragraphs even if they were not originally presented that way.

After Brock's remarks I shall look at the display problem and try and refocus his point of view upon our problems.

> "During the past thirty years great efforts have been expended in research on image evaluation, and considerable progress has been made in theory and instrumentation. From the standpoint of aerial photography, however, it is fair to say that no breakthrough of real practical

value has occurred. It must be twenty-five years since Selwyn first used the concept of the modulation transfer function (MTF) as a tool for the study of photographic image quality,* but it has taken a long time for the ideas to become widely accepted and applied. In the interim, the optical and mathematical aspects have been developed in great breadth and depth by many other workers, while the study of images by micro-densitometry and sinewave analysis has greatly increased. Nevertheless, non-technical observers might be forgiven for thinking that our advance has bogged down, that we are fanning out laterally instead of pressing forward to some clear objective and that the ordinary user of aerial photography is no better off for a simple standard of image quality than he was thirty years ago.

"Let us look squarely at the facts. MTF's are indeed calculated and measured to an ever increasing extent, but how are they used? By translation into resolving power. Edges are traced, converted with great labor and loss of accuracy into transfer functions, they are then interpreted (how else?) by turning them into resolving power. About twenty years ago, resolving power began to lose credit, yet it remains obstinately in use today as the basis of procurement specifications, the everyday test in the laboratory, and the universal standard of image quality without which photographic engineers cannot talk to each other, to their customers, or to the users of other imaging devices. The overwhelming majority of all image quality tests on photographic systems are still, let us face it, resolving power tests.

"The user may be forgiven for expecting a simple answer to his question 'How good is this system or this photograph'? but in truth his questions are not simple and there are no simple answers."

We must indeed ask the question in return, "Good for what purpose?" Though Brock does not say it directly, he infers quite correctly that we rarely specify which of Linfoot's three criteria we wish to pursue, but we are quick to specify and to test.

"It can be argued that we test systems or components with two different purposes in mind. One purpose is to evaluate the practical or operational value of a system. Three bar resolving power is used for this purpose; it tells us (or so we like to think), the least size of detail we can record. Its essential characteristic is that a permanent image is produced by the complete photographic system; in effect it is a picture, but made under controlled conditions and in such a way that we concentrate on fine detail. For the second purpose we desire to measure some physical property of the system or a component. Such properties

---

* In wartime reports to DSR Photographic Research Committee, Ministry of Aircraft Production, London.

may be of vital importance to the quality of the images that the system can produce;—they are required in design, specification and production control, but in measuring them we do not necessarily have to form a permanent image or to deduce anything about the practical performance of the system. The measurement of a lens MTF is an example of such a test. By itself the MTF is not 'image quality,' being merely an operator that we can apply to spatial frequency spectra.

"In relation to lenses the MTF or OTF is of a fundamental nature and provides an elegant link between theory and practice. MTF's can be found for other parts of the aerial photographic system, which can then be studied profitably by linear analysis. As applied to photographic emulsions the MTF is less satisfactory, since it must be inferred rather than observed and is affected by factors that have nothing to do with the spread function of which it is assumed to be the transform. The non-linearities make its use in image analysis clumsy, to say the least, yet no one, having become accustomed to its use, would agree to give it up.

"No doubt the rapid growth of interest in transfer functions was due in part to Schade's success in using them to link optics with electronics for television, and to a general desire to find analogies between photography and electronics, but television is naturally suited to a frequency approach. The whole network up to the final picture handles frequencies in time, and spatial frequencies easily transpose into the time domain. There is a definite cut-off frequency, and the system can be equalized and phase-corrected up to that point. Thus, there is an obvious requirement to know the transfer function of the lens and to keep it as high as possible up to the equivalent frequency in the time domain. It is not too difficult to achieve this, since the required bandwidth is small relative to that defined by the cut-off frequencies of the lenses. Nevertheless it is interesting to know that even television engineers sometimes find difficulty in translating transfer functions into practically useful terms, and have recourse to the transmission of pulses of standard shape.

"The value of the transfer function to the designer and research worker as a tool for image analysis and synthesis is not in question, but we are more concerned with it as an aid to the specification or measurement of practical image quality, which is not the same thing. When we see an MTF, how do we react to it, what physical meaning should we attach to it, and what does it tell us about the image quality the system will give?

"Judging by the general photographic literature, most people react as though they were seeing a graph of image quality versus diminishing detail size, the nature of the detail being conveniently unspecified and, hence, assumed to be anything one pleases. This is what people desire to have, and they naturally read it into what they are shown. But the only direct physical application of the MTF is to show how the system reduces modulation and sinusoidal inputs as a function of the spatial frequency in the image plane. Any deductions made by the expert about image quality for shapes other than sinusoidal will be found, on reflection, to be based on intuitive or intellectual translation of the object into its spectrum. In other words, our opinion is not based on the argument: 'the system is good because it has these frequency characteristics,' but 'because it has these frequency characteristics it will affect these spectra in this or that way.' The difference is subtle but must always be kept in mind. Thus we cannot compare systems just by looking at their MTF's; these must be considered in relation to the spectra of the objects that we want to image. Of course there are degrees of rigor in the application of this general statement. When a number of MTF's have the same shape the respective bandwidths will obviously have some significance—they can be graded for potential quality, as it were. But without reference to an object spectrum we cannot say that any one of them will actually be best.

"In general the concept of image quality does not emerge until we have defined some task for the system to perform; such tasks do not necessarily have to be performed by the components. Definition of the imaging task is a vital part of image evaluation. The more limited and precisely defined we can make the task, the more accurately can the performance of the system be evaluated; the converse is also true. Thus if the task is to image sinusoidal patterns on film at a definite modulation, we can evaluate systems quite accurately if we know their MTF's and other relevant data. But if the task is stated to be the imaging of the details likely to be found in aerial photography, then the evaluation becomes almost impossible and can hardly be said to have an accuracy at all, because the statement is so general and ill-defined. It does not help to know the MTF's unless we also know the Fourier spectra of the scenes. But these must vary so much that the general aerial scene could only be defined in a statistical way, which would inevitably be wrong for individual cases. This problem has always been approached by substituting some simple standard pattern, such as a three bar target, for the infinite complexity of aerial scenes, and judging the systems by the smallest size of the pattern that they will recognizably reproduce. This approach represents a backing off from the perfectly general problem in order to achieve some reasonably definite estimate of comparative performance by a practicable test. At the same time, to be realistic, the tests have to retain at least some of the numerous variables that enter into the practical situation, and because of this they cannot be highly precise or accurate.

"It is popular to draw analogies between photo optics and hi-fi audio, but the analog stops short at the testing and evaluation stage. We do not expect to buy a hi-fi amplifier on the strength of a single quality index. Its performance is expressed by many figures; thus bandwidth, power bandwidth, harmonic distortion, frequency response, phase shift, square wave response at high and low frequencies, signal-to-noise ratio, equalization, and tone compensation are the minimum expected in a specification, and even then we argue about the quality of the results and do not always agree that we have fully defined the performance. In contrast, when we buy an aerial camera system costing one hundred or one thousand times as much as the audio-amplifier, we expect its performance to be summed up in one number that will tell us exactly how any object will be imaged in any circumstances."

Brock, in concluding, says:

"Before we can make progress in the use of our new techniques it will be necessary to bypass two obstacles, the first of which is the existence and firm establishment of resolving power, and the second is the belief that science will give us a one number quality index that will supplant all previous evaluation techniques.

"Resolving power has been in use so long that it has come to be thought of as something fundamental which determines other aspects of image quality and has some very special significance. Whenever a new criterion of quality is proposed, we at once ask 'How does it relate to resolving power?' instead of considering it in more general terms. And because resolving power is used for so many different purposes, and gives a one number answer, it is assumed that any new technique must be inferior if it does not do the same. As we have already seen, resolving power can serve many purposes because it does not serve any of them very well."

Brock's conclusions, drawn from years of careful thought, must be heeded. We are quick to point out that in this book we do not elevate the transfer coefficient to mystical levels. Throughout our treatment we stress that the perceptual process is related to the ratio of desired signal to other forms of signal and noise. The variation of signal over the spatial frequency spectrum can well be described by the modulation transfer function. It can also be described by other functions in the spatial rather than the spatial frequency domain. The MTF is now a well-understood concept; it is reasonably measured with relatively few hidden traps in the measurement process, but the few traps are real and MTF measurement on a machine for that purpose should not be attempted before the method embodied in the machine is well understood.

The MTF is a function of signal and frequency—it does not include noise or noiselike functions. Noise definitely interferes with the perceptual process; for example, it is difficult to listen to a violin sonata when an aircraft overflies the outdoor concert bowl. Luminous noise and non-image-related luminous patterns interfere with visual perception; thus one must consider both the signal and its spectral power distribution, and the noise and its spectral power distribution, both in the time and space domains. The criteria for the level of signal-to-noise ratios required for perception is a matter that as yet we have not been able to construct from first principles, but rather have established them based partly upon the earlier empirical work of Johnson.

Thus to add our own summary after the conclusions of Brock, we must say that MTF is a powerful tool to aid in the calculation of the signal-to-noise ratio of information displayed to an observer. These signals and the displayed noise can be treated as a signal-to-noise ratio for a given frequency or frequency band. The signal-to-noise ratio at a frequency corresponding to the Johnson equivalent bar pattern can be used as an index to the probability of accomplishing a given visual task when proper corrections for bar length-to-width ratio are made, as outlined in Chapter 5. This index must be compared with a similar index computed on the basis not of periodic bar targets but on the basis of an aperiodic object, and a judicious weighting of the two results used as a more appropriate and more nearly correct index of the probability of accomplishing the visual task. In few cases the periodic model calculation and bar pattern models are highly appropriate and no weighting factors are needed. In more cases the aperiodic model is completely appropriate by itself. In the vast majority of cases the real answer is a judicious weighting of the two methods of computation. As yet the precise choice of weighting factors is more judgemental than scientific. However, a 50%–50% choice will produce quite acceptable results.

8.2. A FEW LAST REMARKS

In the foregoing material we have presented a positive approach toward the design and specification of display parameters. We have stressed that one can use the concept of an "equivalent bar pattern" and, by calculating the signal-to-noise ratio at the display output, we can indicate the probability of accomplishing, say, the recognition of the object corresponding to our "equivalent bar pattern."

We have shown that motion makes the task both more and less difficult depending on size and speed.

We have shown that rasters can cause trouble, but do not always cause serious trouble. Figures 8.1 and 8.2 show a photograph and a line-scanned version of that photograph. The research staff at Perkin-Elmer has examined many such pairs of photographs. They informally point out that perhaps one photograph in 500 gives rise to the very strange "staircase effect" shown in Figs. 8.1 and 8.2. The presence of such effects depends upon the proper occurrence of a sampling frequency or raster and object spatial frequencies sufficiently closely related to yield artifacts of a proper size to fool the observer. Ladders can look like staircases and fences can look like almost anything, but these same effects can make possible the counting of the rails in a one- or two-track railway that otherwise could not be resolved! A knowledge of the probable geometry and of the raster parameters can make possible an analysis of an image that is not easily done without the raster.

Thus there are good and bad characteristics of raster imagery.

We have drawn brief attention to three different but important factors in the earlier chapters of this book. Steedman and Baker (1960) have shown the increase in time and the increase in errors associated with small images. Crook *et al.* (1950) have shown that observers cannot sum numbers when vibrational amplitudes of only 0.02 in. are applied to otherwise acceptably large (1/8 in.) and bright images. Lastly Thompson (1957) has shown that observers prefer twice the distance for a display with a typical visible raster over that required for an apparently rasterless image.

These factors do not necessarily become incompatible, but may very well become so, especially as one attempts to maximize observer performance in close and cramped quarters.

Possibly the worst situation occurs in such equipment designs as are proposed for high-performance aircraft. In such installations instrument panel space is at a premium and there is a strong drive to make displays small. To make matters worse, such displays are often designed for low brightness, *so as not to impair the observer's normal night vision!* Further, in order to maximize the angular subtense of the image, the display–observer distance is made small.

Fig. 8.1. Photographic print before line scanning.

Fig. 8.2. **Photographic** print after line scanning. Note aliasing effects.

All of the above sounds reasonable if one does not bother to think the problems out beyond the point where a "sounds reasonable" verdict is reached.

Possibly the worst arguments involve the use of low illuminance or red phosphors to minimize the loss of dark adaptation. One look at Blackwell's data should suffice, even though these data do not include the time factor. In this situation, there does not seem to be a very good compromise. One either protects his dark-adapted seeing capability and does not see very well with his rod vision— or he sees with his cones and sacrifices his dark adaptation. At about 10^{-3} ft-L both the rods and cones are active to about an equal extent, but the sensation produced when *both* rods and cones are operating is highly disconcerting to say the least. One *knows* something is annoying, but does not know what. Personally, I find seeing at that level of luminance very disconcerting and *most irritating*.

For flight conditions encountered in high-performance maneuvering aircraft, vibration is a frequent but unwelcome guest. Not only do instruments vibrate, but the observer's head does a reasonable amount of jiggling. The problem is somewhat like trying to read when riding in a bus on a rough road. It probably does not matter if the head or the image moves; the relative motion probably is the trouble maker and the shorter the distance between observer and display, the worse the situation becomes. Twice the display size at twice the distance will materially improve viewing, by effectively halving the apparent vibration and maintaining the same image angular subtense.

The removal of raster structure not only increases comfort in these crowded conditions but does make structure of a size comparable with raster pitch reasonably visible when, with raster lines present, the detail was submerged in the line structure.

Finally, nothing compares with providing a bright, large, structureless display, driven by adequate signal with bandwidth matched to frequency content of the image.

REFERENCES

Attaya, W. L., and Brock, G. C. (1966). Study of Image-Evaluation Techniques. Itek Corp. Technical Report No. AFAL TR 66 343, AD 803-459.

Attaya, W. L., and Yachik, T. R. (1971). Photographic Quality Transfer Techniques. Technical Report No. Itek 71-9651-1, Optical Systems Division, Itek Corp., Lexington, Mass.

Allnatt, J. W. (1965). Subjective Quality of Colour-Television Pictures Impaired by Gain and Delay Inequalities between the Luminance and Chrominance Channels. *Proc. IEE* **112**(10):1819–1824.

Allnatt, J. W. (1966). Subjective Quality of Colour Television Pictures Impaired by Random Noise. *Proc. IEE* **113**(4):551–557.

Allnatt, J. W. (1968). Subjective Quality of Television Pictures Impaired by Sine-Wave Noise and Low-Frequency Random Noise. *Proc. IEE* **115**(3):371–375.

Anon (1952). Television Spot. Wobbler. *Electronics* **1952**(July):212.

Arguello, Roger J. (1971). Encoding Transmission and Decoding of Sampled Imagery. In *A Symposium on Sampled Images*. Perkin-Elmer Corp. Is 10763.

Attneave, F., and Olson, R. K. (1957). *J. Exp. Psychol.* **74**:149.

Bailey, H. H. (1970). Target Detection through Visual Recognition: A Quantitative Model. Rand Corp., Santa Monica, Calif., Memorandum RM-6158/1-PR.

Baldwin, Millard W., Jr. (1940). The Subjective Sharpness of Simulated Television Images. *Bell Syst. Tech. J.* **19**(4):563–586.

Baldwin, M. W., Jr., and Nielsen G., Jr. (1956). Subjective Sharpness of Simulated Color Television Pictures. *J. Opt. Soc. Am.* **46**(9):681–685.

Barnard, Thomas W. (1971). An Image Evaluation Method. In *A Symposium on Sampled Images*. Perkin-Elmer Corp. Is 10763.

Barnes, R., and M. Czerny (1932). *Z. Physik* **1932**:79.

Barrows, Robert S. (1957). Factors Affecting the Recognition of Small, Low-Contrast Photographic Images. *Photo. Sci. Engng.* 1, No. 1 (July).

Barstow, J. M., and Christopher, H. N. (1954). The Measurement of Random Monochrome Video Interference. *AIEE Trans.* (Part I) **1954**(Jan.):735.

Bennett, C. A., Winterstein, S. H., and Kent, R. E. (1966). Image Quality and Target Recognition. IBM, Federal Systems Div., Electronics Systems Center, Oswego, New York, July 1966.

Beurle, R. L., and Daniels, M. V. (1968). Visual Perception with Electronic Imaging Systems. Report No. J.8, Univ. of Nottingbean Dept. of Elec. Engng.

Beurle, R. L., and Hills, B. L. (1968). Visual Perception with Electronic Imaging Systems. Report No. J.5, Univ. of Nottingbean, Dept. of Elec. Engng. March 1968.

Biberman, L. M., Schnitzler, A. D., Rosell, F. A., Schade, O. H., Sr., Legault, R., Milton, A. F., and Snyder, H. L. (1971). Image Quality in Sampled Data Systems. Institute for Defense Analysis, Arlington, Va. IDA Research Paper p-741; IDA Log No. HQ 71-13139, Aug. 1971.

Blackman, F. S. (1968). Effects of Noise on the Determination of Photographic System Modulation Transfer Functions. *Photo. Sci. Engng.* Vol. 12, No. 5 (Sept.-Oct.).

Blackwell, H. Richard (1946a). Contrast Thresholds of the Human Eye. *J. Opt. Soc. Am.* **36**(11):624-714.

Blackwell, H. Richard (1946b). Visibility Studies and Some Applications in the Field of Camouflage. Summary Technical Report of Division 16, Vol. 2, National Defense Research Committee, Washington, D. C.; *J. Opt. Soc. Am.* **36**:624, Appendix A.

Blackwell, H. Richard, and D. W. McCready, Jr. (1958). Foveal Contrast Thresholds for Various Durations of Single Pulses. USN BuShips Contract NObs-72038, Index No. 2455-13-F, Univ. of Michigan Engng. Res. Inst., June 1958.

Born, M., and Wolf, E. (1965). *Principles of Optics.* 3rd ed. Pergamon, New York.

Borough, H. C., Fallis, R. F., Warnock, R. H., and Britt, J. H. (1967). Quantitative Determination of Image Quality. Boeing Company Report D2-114058-1, May 1967.

Bouman, M. A., Vos, J. J., and Walraven, P. L. (1963). Fluctuation Theory of Luminance and Chromaticity Discrimination. *J. Opt. Soc. Am.* **53**:121.

Bowen, H. M., Andreassi, J. L., Truax, S., and Orlansky, J. (1960). Optimum Symbols for Radar Displays. *Human Factors* **2**(1):28-33.

Brainard, R. C. (1967). Low-Resolution TV: Subjective Effects of Noise Added to a Signal. *Bell Syst. Tech. J.* **46**(1):233-260.

Brainard, R. C., Kammerer, F. W., and Kimme, E. G. (1962). Estimation of the Subjective Effects of Noise in Low-Resolution Television Systems. *IRE Trans. Information Theory* IT-**8**:99.

Brainard, R. C., Mounts, F. W., and Prasada, B. (1967). Low-Resolution TV: Subjective Effects of Frame Repetition and Picture Replenishment. *Bell Syst. Tech. J.* **46**(1):261-271.

British Patent 535,905.

Brock, G. C. (1965). Paper presented before the Soc. Photo. Sci. Eng., May 1965.

Brock, G. C. (1970). *Image Evaluation for Aerial Photography.* Focal Press, London—New York.

Brown, Earl B. (1971). Parameters of the Sampled Image System. In *A Symposium on Sampled Images.* Perkin-Elmer Corp. Is 10763.

Brown, W. M. (1961). Optimum Prefiltering of Sampled Data. *IRE Trans. Information* IT-**7**:269-270.

Bryngdahl, O. (1966). Characteristics of the Visual System: Psychophysical Measurements of the Response to Spatial Sine-Wave Stimuli in the Photopic Region. *J. Opt. Soc. Am.* **56**(6):811-821.

Bryngdahl, O., and Risenberg, L. (1964). New Phenomena in the Visual Response to Sinusoidally Varying Spatial Stimuli. February 10, 1964.

Campbell, F. W. (1968). The Human Eye As an Optical Filter. *Proc. IEEE* **56**(6): 1009–1014.

Campbell, F. W., and Green, D. G. (1965). Optical and Retinal Factors Affecting Visual Resolution. *J. Physiol.* **181**:576–593.

Carman, P. D., and Charman, W. N. (1964). Detection, Recognition, and Resolution in Photographic Systems. *J. Opt. Soc. Am.* **54**(9):1121–1130.

Charman, W. N. (1965). Resolving Power and the Detection of Linear Detail. *The Canadian Surveyor Z* **XIX**(2):190–205.

Charman, W. N., and Olin, A. (1965). Tutorial—Image Quality Criteria for Aerial Camera Systems. *Photo. Sci. Engng.* **9**(6):385–397.

Coltman, J. W. (1954a). Scintillation Limitations to Resolving Power in Imaging Devices. *J. Opt. Soc. Am.* **44**(3):234–237.

Coltman, J. W. (1954b). The Specification of Imaging Properties by Response to a Sine Wave Input. *J. Opt. Soc. Am.* **44**(6):468–472.

Coltman, J. W., and Anderson, A. E. (1960). Noise Limitations to Resolving Power in Electronic Imaging. *Proc. IRE* **48**(5):858–865.

Crook, M. N., Harker, G. S., and Hoffman, A. C. (1950). Effect of Amplitude of Apparent Vibration, Brightness, and Type Size on Numeral Reading. AF Technical Report No. 6248, Sept. 1950.

Cunniff, J. F. (1971). Generation and Recording of Sampled Images. In *A Symposium on Sampled Images*. Perkin-Elmer Corp. Is 10763.

Damon, R. W., Eshback, J. R., and Redington, R. W. (1958). Present and Future Performance of Photoconductive Camera Tubes. General Electric Co., Schenectady, N. Y. Research Report 58-RL-1990, July 1958.

Davidson, Michael (1968). Perturbation Approach to Spatial Brightness Interaction in Human Vision. *J. Opt. Soc. Am.* **58**:1300.

Davis, Samuel. (1969). *Computer Data Displays*. Prentice Hall, Englewood Cliffs, N. J. Prentice-Hall EE Series.

De Loor, G. P., Jurriens, A. A., Levelt, W. J. M., and Van deGeer, J. P. (1968). Line-Scan Imagery Interpretation. *Photogrammetric Eng.* **34**(5):502–510.

De Palma, J. J. and Lowry, E. M. (1961), *J. Opt. Soc. Am.* **51**:474.

De Palma, J. J., and Lowry, E. M. (1962). Sine-Wave Response of the Visual System. II. Sine-Wave and Square-Wave Contrast Sensitivity. *J. Opt. Soc. Am.* **52**:328–335.

Deutsch, S. (1965). Pseudo-Random Dot Scan Television Systems. *IEEE Trans. Broadcasting* 11-21, July 1965.

De Velis, John B. (1965). Comparison of Methods for Image Evaluation. *J. Opt. Soc. Am.* **55**(2):165–174.

de Vries, Hessel (1943). The Quantum Character of Light and Its Bearing upon Threshold of Vision, The Differential Sensitivity and Visual Acuity of the Eye. *Physica* **10**(7):553–564.

Elias, P. (1953). Optics and Communication Theory. *J. Opt. Soc. Am.* **43**:229.

Enoch, J. M. (1959). Effect of the Size of a Complex Display upon Visual Search. *J. Opt. Soc. Am.* **49**(3):280–286.

Erickson, R. A. (1971). Human Factors Research Techniques with Television. Naval Weapons Center, China Lake, Calif. Report NWCTP5072, Jan. 1971.

Erickson, R. A., and Hemingway, J. C. (1970a). Visibility of Raster Lines in a Television Display. *J. Opt. Soc. Am.* **60**(5):700–701.

Erickson, R. A., and Hemingway, J. C. (1970b). Image Identification on Television. Naval Weapons Center, China Lake, Calif. Report NWCTP5025, Sept. 1970.

Erickson, R. A., Linton, P. M., and Hemingway, J. C. (1968). Human Factors Experiments with Television. Naval Weapons Center, China Lake, Calif. Report NWCTP-4573, Oct. 1968.

Fellgett, P. (1953). Concerning Photographic Grain, Signal-to-Noise Ratio, and Information. *J. Opt. Soc. Am.* **43**:271 (1953).

Fellgett, P. B., and Linfoot, E. H. (1955). On the Assessment of Optical Images. *Phil. Trans. Roy. Soc. London* (Ser. A) **247**:369–407.

Fink, D. G. (1955). Color Television vs. Color Motion Pictures. *SMPTE* **1955**(June): 282.

Fink, D. G. (1957). *Television Engineering Handbook.* McGraw-Hill, New York.

Flamant, F. (1955) *Rev. Optique* **34**:433.

Fry, Glenn A. (1961). Relation of Blur Functions to Resolving Power. *J. Opt. Soc. Am.* **51**(5):560–563.

Fry, Glenn A. (1970). The Optical Performance of the Human Eye. In *Progress in Optics*, Vol. 8, p. 51, ed. E. Wolf. North-Holland, Amsterdam.

Gaven, J. V., Jr., Tavitian, J., and Harabedian, A. (1970). The Informative Value of Sampled Images As a Function of the Number of Gray Levels Used in Encoding the Images. *Photo. Sci. Engng.* **14**(1):16–20.

Gaven, J. V., Jr., Tavitian J., and Hollanda, P. A. (1971). The Relative Effects of Gaussian and Poisson Noise on Subjective Image Quality. *Appl. Opt.* **10**(9): 2171–2178.

Gilinsky, Alberta S. (1967). *Psychon. Sci.* **8**:395.

Gilinsky, Alberta S. (1968). *J. Opt. Soc. Am.* **58**:13.

Goodman, M. (1968). *Introduction to Fourier Optics.* McGraw-Hill, New York.

Greening, Charles P., and Wyman, Melvin J. (1970). Experimental Evaluation of a Visual Detection Model. *Human Factors* **12**(5):435–445.

Hairfield, H. W. (1970). Night and All-Weather Target Acquisition: State-of-the-Art Review. Part III: Television and Low-Light-Level Television Systems. Boeing Company Report D162-10116-3, May 1970.

Hallows, R. W. (1950). Spot Wobble Improves TV. *Radio Electronics* **1950**(May):29.

Hammill, H. B., and Holladay, T. M. (1968). The Effects of Certain Approximations in Image Quality Evaluation from Edge Traces. SPIE 13th Ann. Tech. Symp. Proc., Aug. 19–23, 1968.

Hanbury-Brown, R., and Twiss, R. Q. (1956). Correlation between Photons and Two Coherent Beams of Light. *Nature* **178**:4541.

Hardy, Arthur C., and Perrin, Fred H. (1932). *The Principles of Optics.* McGraw-Hill, New York.

Harris, J. L. (1964). Resolving Power and Decision Theory. *J. Opt. Soc. Am.* **54**(5): 606–611.

Harris, James L., Sr. (1966). Image Evaluation and Restoration. *J. Opt. Soc. Am.* **56**(5):569–574.

Hecht, S., Shlaer, S., and Pirenne, M. H. (1942). Energy Quanta and Vision. *J. Gen. Physiol.* **25**:819.

Hemingway, John C., and Erickson, Ronald A. (1969). Relative Effects of Raster Scan Lines and Image Subtense on Symbol Legibility on Television. *Human Factors* **11**(4):331–338, Aug. 1969.

Higgins, George C., and Perrin, Fred H. (1958). The Evaluation of Optical Images. *Photo. Sci. Engng.* **2**(2):66.

Higgins, G. C., and Stultz, K. (1950). *J. Opt. Soc. Am.* **40**:135.

Higgins, George C., and Wolfe, Robert N. (1955). The Relation of Definition to Sharpness and Resolving Power in a Photographic Systems. *J. Opt. Soc. Am.* **45**(2):121–129.

Higgins, G. C., Wolfe, R. N., and Lamberts, R. L. (1956). Relationship between Definition and Resolving Power with Test Objects Differing in Contrast. *J. Opt. Soc. Am.* **46**(9):752–754.

Hollanda, Peter A., Scott, Frank, Harabedian, Albert. The Informative Value of Sampled Images As a Function of the Number of Scans per Scene Object and the Signal-To-Noise Ratio. *Photo. Sci. Engng.* **14**(6):1970.

Horgan, S. H. (1913). Half-Tone and Photo-Mechanical Processes. Inland Printing, Chicago.

Howell, William C., and Kraft, Conrad L. (1959). Size, Blur, and Contrast As Variables Affecting the Legibility of Alpha-Numeric Symbols on Radar-Type Displays. WADC Technical Report 59-536, Sept. 1959.

Huang, T. S. (1965). The Subjective Effect of Two-Dimensional Pictorial Noise. *IEEE Trans. Information Theory* **1965**(Jan.):43.

Hufnagel, Robert E. (1965). Random Wavefront Effects. In *The Practical Application of Modulation Transfer Functions* (symp. held at Perkin-Elmer Corp., March 6, 1963), pp. 244–247. Soc. Photo. Sci. and Eng.

Itek Corp., Lexington, Mass., Technical Note No. 1. Image Evaluation for Reconnaissance. Proc. of a Symp. April 3 and 4, 1963, Air Force Avionics Laboratory, Reconnaissance Division, Wright-Patterson Air Force Base, Ohio.

James, T. H., and Higgins, G. C. (1960). Fundamentals of Photographic Theory. In *Structure of the Developed Photographic Image*, 2nd ed., Chapter 13. Morgan and Morgan, New York.

Johnson, A. D., and Cowden, D. G. (1967). Considerations in Specifying Display System CRT Design Objectives. *Information Display* **4**(3):46–51.

Johnson, John. (1958). Image Intensifier Symposium, Fort Belvoir, Va., Oct. 6-7, 1958, AD 220160.

Johnston, Dorothy M. (1965). Search Performance As a Function of Peripheral Acuity. *Human Factors* **7**(6):521–526 (527–535).

Johnston, Dorothy M. (1968). Target Recognition on TV As a Function of Horizontal Resolution and Shades of Gray. *Human Factors* **10**(3):201–210.

Jones, R. Clark (1959). Quantum Efficiency of Human Vision. *J. Opt. Soc. Am.* **49**:645.

Jones, R. Clark (1955). New Method of Describing and Measuring the Granularity of Photographic Materials. *J. Opt. Soc. Am.* **45**:799.

Jones, Robert A. (1967). An Automated Technique for Deriving MTF's from Edge Traces. *Photo. Sci. Engng.* **11**(2):102–106.

Jones, Robert A. (1968). Accuracy Test Procedure for Image Evaluation Techniques. *Appl. Opt.* **7**(1):133–136.

Jones, Robert A., and Kelly, John T. (1969). Reliability of Image Evaluation Techniques. *Appl. Opt.* **8**(5):941–945.

Kaestner, Paul T. (1967). Visual Simulation. *Information Display* **4**(2):49–54.

Kazan, B., and Knoll, M. (1968). *Electronic Image Storage*. Academic, New York.

Ketchel, James (1969). The Effects of High-Intensity Light Adaption on Electronic Display Visibility. *Information Display* **6**(3):71–76.

King, W. (1968). Diffraction-Based Criteria of Image Quality in Optical Design. In *Recent Developments in Optical Design*. Perkin Elmer Corp.

Kinzly, R. E., Mazurowski, M. J., and Holladay, T. M. (1968). Image Evaluation and Its Application to Lunar Orbiter. *Appl. Opt.* **7**(8):1577–1586.

Kishner, S. J. (1970). Nonlinear Filtering of Image Noise. SPSE Conf. on Electronics Imaging Systems, April 1970, Palo Alto, Calif.

Kling, J. W., and Riggs, L. A. (1971). *Woodworth and Schlosberg's Experimental Psychology*, 3rd ed. Holt, Rinehart, and Winston, New York.

Klingberg, C. L. *et al.* (1963). A Study of Photointerpreter Performance in Change Discrimination. U. S. Dept. of Commerce, AD 431, 495, Nov. 1963.

Klingberg, C. S., Elworth, C. S., and Filleau, C. R. (1970). Image Quality and Detection Performance of Military Photointerpreters. Boeing Co. Report D162-10323-1, Sept. 1970.

Kornfeld, G. H., and Lawson, W. R. (1971). Visual Perception Models. *J. Opt. Soc. Am.* **61**:811.

Lachs, Gerald (1968). Effects of Photon Bunching on Shot Noise in Photoelectric Detection. *J. Appl. Phys.* **39**:4193.

Langner, Guenther O. (1970). Light Gating Brightens CRT Image for Large Projection Displays. *Electronics* **43**(25):78–83.

LaPorte, H. R., and Calhoun, R. L. (1966). Clue Utilization in Target Recognition: A Methodological Study. Autonetics Technical Report T6-2258/501, Sept. 1966.

Lavin, H. P. (1971). In *Photoelectronic Imaging Devices*, Vol. 1, Chapter 15. L. M. Biberman and S. Nudelman, eds. Plenum, New York.

Lawson, Walter R. (1968). Image Intensifier Evaluation. Thesis submitted to Univ. of Rochester.

Lawson, Walter R. (1971). Electrooptical System Evaluation. In *Photoelectronic Imaging Devices*, Vol. 1, p. 375. L. M. Biberman and S. Nudelman, eds. Plenum, New York.

Lee, Yuk Wing (1960). *Statistical Theory of Communication*. Wiley, New York.

Legault, R. L. (1971). Visual Detection Process for Electrooptical Images: Man—The Final Stage of an Electrooptical Imaging System. In *Photoelectronic Imaging Devices*, Vol. 1, Chapter 4, pp. 69–87. L. M. Biberman and S. Nudelman, eds. Plenum, New York.

Leibowitz, H. (1953). *J. Opt. Soc. Am.* **43**:902.

Levi, L. (1970). Vision in Communication. In *Progress in Optics*, Vol. 8, p. 343, E. Wolf, ed. North-Holland, Amsterdam.

Limansky, I. (1958). A New Resolution Chart for Imaging Systems, *The Electronic Engineer* **1958**(June).

Linfoot, E. H. (1956). Transmission Factors and Optical Design. *J. Opt. Soc. Am.* **46**(9):740-752.

Linfoot, E. H. (1958a). *Physica* **XXIV**(6):476.

Linfoot, E. H. (1958b). Quality Evaluations of Optical Systems. *Optical Acta* **5**(1-2): 1-14.

Linfoot, E. H. (1964). *Fourier Methods in Optical Image Evaluation.* Focal Press, New York.

Linfoot, E. H. (1966). *Fourier Methods in Optical Image Evaluation.* Focal Press, London and New York.

Lowry, E. M., and DePalma, J. J. (1961). Sine-Wave Response of the Visual System. I. The Mach Phenomenon. *J. Opt. Soc. Am.* **51**:740-746.

Ludvigh, E. (1953), U. S. Navy School of Aviation Medicine and Kresge Eye Institute, Bureau of Medicine and Surgery Project No. NM001 075.01.04.

Ludvigh, E., and McCarthy, E. F. (1938). Absorption of Visible Light by the Refractive Media of the Human Eye. *Arch. Ophthal.* **20**:37.

Luxenberg, H. R., and Kuehn, R. L. (1968). *Display System Engineering.* McGraw-Hill, New York.

MacDonald, D. E., and Watson, J. T. (1956). Detection and Recognition of Photographic Detail. I. Empirical Data Applicable to the Prediction of Performance of Diffraction Limited Systems. *J. Opt. Soc. Am.* **46**(9):715-720.

McGlamery, Benjamin L. (1966). An Image Restoration Experiment on Turbulence-Degraded Images. U. S. Dept. of Commerce Technical Report AD 656-124, July 1966.

McLean, Michael V. (1965). Brightness Contrast, Color Contrast, and Legibility. *Human Factors* **7**(6):521-526.

Meiron, Joseph (1968). The Use of Merit Functions Based on Wavefront Aberration in Automatic Lens Design. *Appl. Opt.* **7**(4):667-672.

Mertz, Pierre, and Gray, Frank (1934). A Theory of Scanning and Its Relation to the Characteristics of the Transmitted Signal in Telephotography and Television. *Bell Syst. Tech. J.* **XIII**:464-515.

Mertz, P., Fowler, A. D., and Christopher, H. N. (1950). Quality Rating of Television Images. *Proc. IRE* **38**(11):1269-1283.

Middleton, D. (1960). An Introduction to Statistical Communications, pp. 206-210. McGraw-Hill, New York.

Miller, J. W., and Ludvigh, E. (1960). Visual Search Techniques. National Academy of Sciences Publication 712, p. 170.

Montgomery, W. D. (1969). Some Consequences of Sampling in FLIR Systems. Institute for Defense Analyses, Research Paper P-543, Sept. 1969.

Moon, P. (1936). *The Scientific Basis of Illuminating Engineering*, 1st ed., pp. 422-423. McGraw-Hill, New York.

Moon, P., and Spencer, D. E. (1944). On the Stiles–Crawford Effect. *J. Opt. Soc. Am.* **34**:319 (1944).

Moss, Hilary (1968). Narrow-Angle Electron Guns and Cathode Ray Tubes. Academic, New York.

National Bureau of Standards (1954). Optical Image Evaluation. Circular 526.

Offner, Abe (1965). Optical Design and Modulation Transfer Function. In *The Practical Application of Modulation Transfer Functions* (symp. held at Perkin-Elmer Corp. March 6, 1963), pp. 240–243. Soc. of Photo. Sci. and Eng.

Ogle, Kenneth N. (1962). Blurring of Retinal Image and Foveal Contrast Thresholds of Separated Point Light Sources. *J. Opt. Soc. Am.* **52**(9):1035–1039.

Papoulis, Althanasios (1968). *Systems and Transforms with Applications in Optics.* McGraw-Hill, New York.

Parton, J. S., and Moody, J. C. (1961). Performance of Image Orthicon Type Intensifier Tubes. In *NASA Proc. Image Intensifier Symp., Ft. Belvoir, Va.,* 24–26 Oct. 1961. NASA SP 2.

Patel, A. S. (1966). Spatial Resolution by the Human Visual System. The Effect of Mean Retinal Illuminance. *J. Opt. Soc. Am.* **56**(5):689–694.

Perrin, F. H. (1960). Methods of Appraising Photographic Systems. Part 1. Historical Review. Part 2. Manipulation and Significance of the Sinewave Function. *J. SMPTE* **69**(3).

Perrin, F. H., and Altman, J. H. (1953). Studies in the Resolving Power of Photographic Emulsions. VI. The Effect of the Type of Pattern and the Luminance Ratio in the Test Object. *J. Opt. Soc. Am.* **43**(9):780–790.

Petersen, D. and Middleton, D. (1962). Sampling and Reconstruction of Wave-Number Limited Functions in N-Dimensional Euclidean Spaces. *Information and Control* **1962**:279–323.

Poole, H. H. (1966). *Fundamentals of Display Systems.* Spartan Books.

Prosser, R. D., and Allnatt, J. W. (1965). Subjective Quality of Television Pictures Impaired by Random Noise. *Proc. IEE* **112**(6):1099–1102.

Rabedeau, M. E., and Bates, A. D. (1965). Image Quality Requirements in Optical System for Reproducing Typewritten Documents. *Appl. Opt.* **4**(4):439–443.

Lord Raleigh (1881). *Phil. Mag.* **11**:214.

Reeves, Prentice (1920). The Response of the Average Pupil to Various Intensities of Light. *J. Opt. Soc. Am.* **4**:35 (1920).

Riesenfeld, James (1967). Relationships between Information Capacity and Resolving Power. *Photo. Sci. Engng.* **11**(6):415–418.

Roetling, P. G., Trabka, E. A., and Kinzly, R. E. (1968). Theoretical Prediction of Image Quality. *J. Opt. Soc. Am.* **58**(3):342–345.

Rose, Albert (1942). The Relative Sensitivities of Television Pickup Tubes, Photographic Film, and the Human Eye. *Proc. IRE* **30**:295.

Rose, Albert (1948a). The Sensitivity Performance of the Human Eye on a Absolute Scale. *J. Opt. Soc. Am.* **38**(2):196–208.

Rose, Albert (1948b). Television Pickup Tubes and the Problem of Vision. *Advan. Electron.* **1**:131.

Rose, Albert (1957). Quantum Effects in Human Vision. *Advan. Biol. and Med. Phys.* **5**:211.

Rosell, F. A. (1968). Noise Limitations to the Detection of Isolated Square Images on a TV Monitor. Westinghouse Aerospace, Baltimore, Oct. 1968.

Rosell, F. A. (1969). Limiting Resolution of Low-Light-Level Imaging Sensors. *J. Opt. Soc. Am.* **59**(5):539–547.

Rosell, F. A. (1971). In *Photoelectronic Imaging Devices*, Vol. 1, Chapter 14; Vol. 2, Chapter 22. L. M. Biberman and S. Nudelman, eds. Plenum, New York.

Rosell, F. A., Svensson, E. L., and Willson, R. H. (1972). *J. Appl. Opt.* **11**(5): 1058–1067.

Schade, Otto H., Sr. Patent 3,030,440. Vertical Aperture Correction.

Schade, Otto H., Sr. (1948a). Electro-Optical Characteristics of Television Systems: I. Characteristics of Vision and Visual Systems. *RCA Rev.* **9**(March):5–37.

Schade, Otto H., Sr. (1948b). Electro-Optical Characteristics of Television Systems: II. Electro-Optical Specifications for Television Systems. *RCA Rev.* **9**(June):247–286.

Schade, Otto H., Sr. (1948c). Electro-Optical Characteristics of Television Systems: III. Electro-Optical Characteristics of Camera Systems. *RCA Rev.* **9**(Sept.):491–531.

Schade, Otto H., Sr. (1948d). Electro-Optical Characteristics of Television Systems: IV. Correlation and Evaluation of Electro-Optical Characteristics of Imaging Systems. *RCA Rev.* **9**(Dec.):653–687.

Schade, Otto H., Sr. (1951). Image Gradation, Graininess and Sharpness in Television and Motion Picture Systems: I. Image Structure and Transfer Characteristics. *J. SMPTE* **56**(2):137–171.

Schade, Otto H., Sr. (1952). Image Gradation, Graininess and Sharpness in Television and Motion Picture Systems: II. The Grain Structure of Motion Picture Images—An Analysis of Deviations and Fluctuations of the Sample. *J. SMPTE* **58**(March):181–222.

Schade, Otto H., Sr. (1953). Image Gradation, Graininess and Sharpness in Television and Motion Picture Systems: III. The Grain Structure of Television Images. *J. SMPTE* **61**(2):97–164.

Schade, Otto H., Sr. (1955). Image Gradation, Graininess and Sharpness in Television and Motion Picture Systems: IV. A and B Image Analysis in Photographic and Television Systems (Definition and Sharpness). *J. SMPTE* **64**(11):593–617.

Schade, Otto H., Sr. (1956). Optical and Photoelectric Analog of the Eye. *J. Opt. Soc. Am.* **46**(9):721–739.

Schade, Otto H., Sr. (1958). A Method of Measuring the Optical Sine Wave Spatial Spectrum of Television Image Display Devices. *J. SMPTE* **67**(Sept.):561–566.

Schade, Otto H., Sr. (1964a). Modern Image Evaluation and Television. *Appl. Opt.* **3**(1):17–21.

Schade, Otto H., Sr. (1964b). An Evaluation of Photographic Image Quality and Resolving Power. *J. SMPTE* **73**(2):81–119.

Schade, Otto H., Sr. (1967). The Resolving-Power Functions and Quantum Processess of Television Cameras. *RCA Rev.* **28**(3):460–535.

Schade, Otto H., Sr. (1971). In *Photoelectronic Imaging Devices*, Vol. 2, Chapter 16 and 17. L. M. Biberman and S. Nudelman, eds. Plenum, New York.

Schnitzler, Alvin D. (1971a). Communication Theory of Night Vision—Human Eye Systems. Paper presented to 1971 Ann. Mtg. of Opt. Soc. Am. in Tucson, Ariz.

Schnitzler, Alvin D. (1971b). Low-Light-Level Performance of Visual Systems. Institute for Defense Analyses Paper P-743.

Schnitzler, Alvin D. (1973). Analysis of Electrooptical Imaging Systems. To be published.

Schwartz, Mischa (1959). *Information Transmission, Modulation, and Noise*. McGraw-Hill, New York.

Scott, Frank (1963). Image Evaluation for Reconnaissance. Proc. of a Symposium, Itek Report 9048-6, 9-17, April 1963.

Scott, F. (1964). Photographic Image Simulation. *J. Photo. Sci.* 12:139–142.

Scott, Frank (1965). Film Modulation Transfer Functions. In *The Practical Application of Modulation Transfer Functions* (symp. held at Perkin-Elmer Corp., March 6, 1963), pp. 248–251. Soc. Photo. Sci. and Eng.

Scott, F. (1966). Three-Bar Target Modulation Defectability. *Photo. Sci. Engng.* 10(1):49–52.

Scott, F. (1967a). A New Objective Characterization of Film Image-Structure. *Photo. Sci. Engng.* 11(1):7–10.

Scott, F. (1967b). A Line-Scan Image Generator. *Photo. Sci. Engng.* 11(5):348–351.

Scott, F. (1968). The Search for a Summary Measure of Image Quality—A Progress Report. *Photo. Sci. Engng.* 12(3):154–164.

Scott, F., Scott, Roderick M., and Shack, Roland V. (1963). The Use of Edge Gradients in Determining Modulation-Transfer Functions. *Photo. Sci. Engng.* 7(6): 345–349.

Scott, Frank, Hollanda, Peter A., and Harabedian, Albert (1970). The Informative Value of Sampled Images As a Function of the Number of Scans per Scene Object. *Photo. Sci. Engng.* 14(1):21–27.

Scott, Roderick M. (1959). Contrast Rendition As a Design Tool. *Photo. Sci. Engng.* 3(5):201.

Scott, Roderick (1965). Introduction. In *The Practical Application of Modulation Transfer Functions* (symp. held at Perkin-Elmer Corp. March 6, 1963), pp. 237–239. Soc. Photo. Sci. and Eng.

Scott, Roderick (1971). Sampled Images—An Introduction. In *A Symposium on Sampled Images*. Perkin-Elmer Corp. Is 10763.

Self, Herschel (1969). Image Evaluation for the Prediction of the Performance of a Human Observer. Presented at the NATO Symp. on Image Evaluation, Kunstlerhaus, Munich 18–22 Aug. 1969.

Sellner, Harvey (1971). Transfer Function Compensation. In *A Symposium on Sampled Images*. Perkin-Elmer Corp. Is 10763.

Selwyn, E. W. H. (1948) *Phot. J.* 88B:6.

Sendall, R. L., and Rosell, F. A. (1972). E/O Sensor Performance Analysis and Synthesis (TV/IR Comparison Study), Air Force Systems Command, 4950th Test Wing (Technical) PMMA, Wright–Patterson Air Force Base, Ohio, Final Report, November 1972.

Shannon, Robert R. (1969). Some Recent Advances in the Specification and Assessment of Optical Images. Optical Sciences Center, Technical Report 49, 15 Dec. 1969.

Shaw, R. (1962). The Application of Fourier Techniques and Information Theory to the Assessment of Photographic Image Quality. *Photo. Sci. Engng.* 6(5):281–286.

Sherr, Sol (1970). *Fundamentals of Display System Design.* Wiley—Interscience, New York.

Simonds, J. L., Kelch, J. R., and Higgins, G. C. (1964). Analysis of Fine-Detail Reproduction in Photographic Systems. *Appl. Opt.* 3(1):23–28.

Sipley, L. W. (1958). *The Photomechanical Halftone.* American Museum of Photography, Philadelphia, Penn.

Slocum, Gerald K., Hoffman, William C., and Heard, James L. (1967). Airborne Sensor Display Requirements and Approaches. *Information Display* 4(6):44–51.

Smith, F. Dow (1963). Optical Image Evaluation and the Transfer Function. *Appl. Opt.* 2(4):335–350.

Smith, Russell L., Garfinkle, David R., Groth, Hilde, and Lyman, John (1966). Effects of Display Magnification, Proprioceptive Cues, Control Dynamics and Trajectory Characteristics on Compensatory Tracking Performance. *Human Factors* 8(5):427–434.

Smith, Sidney L. (1963). Color-Coded Displays for Data-Processing Systems. *Electro-Technology* 1963(April):63–69.

Snyder, H. L. *et al.* (1967). Low-Light-Level TV Viewfinder Simulation Program. Phase A: State-of-the-Art Review. USAF Report AFAL-TR-67-293, Nov. 1967.

Steedman, William C., and Baker, Charles A. (1960). Target Size and Visual Recognition. *Human Factors* 2(3):121–127.

Stultz, K. F. and Zweig, H. J. (1959) *J. Opt. Soc. Am.,* **49**:693.

Stultz, K. F., and Zweig, H. J. (1962). Roles of Sharpness and Graininess in Photographic Quality and Definition. *J. Opt. Soc. Am.* **52**(1):45–50.

Thompson, F. T. (1957). Television Line Structure Suppression. Scientific Paper 8-1041-P14 Westinghouse Res. Labs., Pittsburgh, Pa., 27 June, 1957, and *J. SMPTE* **66**:602–606.

Tourshov, S. I. U. S. Patent 2,728,013.

Trott, Timothy (1966). Sine Wave Testing and its Practical Implications (Proc. SPIE Symp., El Paso). *Geometrical Optics* VIII(March):1–7.

Turnage, Rodger Elmo, Jr. (1966). The Perception of Flicker in Cathode Ray Tube Displays. *Information Display* 3(3):38–52.

Vallerie, Leroy L. (1966). Displays for Seeing without Looking. *Human Factors* 8(6):507–513.

Van Meeteren, A. (1969). Modulation Sensitivity and Spatial Bandwidth of the Eye at Low Luminances. Paper presented at Opt. Soc. Am. Mtg., Chicago, Ill., 21–24 Oct. 1969.

Wagenaar, W. A., and van Meeteren, A. (1968). Discrimination and Identification in Noisy Line-Scan Pictures. In *AGARD Conf. Proc.,* No. 29.

Wald, George, Brown, Paul K., and Gibbons, Ian R. (1963). The Problem of Visual Excitation. *J. Opt. Soc. Am.* **53**:20.

Walker, Roger S. (1968). Simplified Methods for Determining Display Screen Resolution Characteristics. *Information Display* **5**(1):28–31.

Weaver, L. E. (1968). The Quality Rating of Color Television Pictures. *J. SMPTE* **77**(6):610–612.

Weiss, Helmut (1969). Optimum Spot Size of a Scanned CRT Display. *Information Display* **6**(6):51–52.

Westheimer, G. (1970). Image Quality in the Human Eye. *Optica Acta* **17**(9):641–658.

Williams, Leon G. (1966). Target Conspicuity and Visual Search. *Human Factors* **8**(1):80–92.

Woehl, Walter E. (1968). Comparison of Image Degredation in Photographic and Image Orthicon Systems. *Information Display* **5**(1):32–36.

Wolfe, Robert N. (1962). Width of the Human Visual Spread Function As Determined Psychometrically. *J. Opt. Soc. Am.* **52**(4):460–469.

Wolfe, Robert N., and Eisen, Fred C. (1953a). Psychometric Evaluation of Sharpness of Photographic Reproductions. *J. Opt. Soc. Am.* **43**(10):914–922.

Wolfe, Robert N., and Eisen, Fred C. (1953b). *J. SMPTE* **61**(5):590–604.

Wurtz, Jim E. (1967). High-Resolution Cathode Ray Tubes for the System Designer. *Information Display* **4**(3):39–45.

Wyman, M. J., Gilmour, J. D., Snyder, H. L., Jahns, D. W., and McGrath, J. J. (1968). Report on Low-Altitude Test 4.1—Visual Target Acquisition. Vol. 7. Simulation Studies, Basic Validation. Joint Chiefs of Staff, JTF-2, Nov. 1968.

Zaitzeff, L. P. (1971). Target Background Scaling and Its Impact on the Prediction of Aircrew Target Acquisition Performance. Boeing Co. Report D180-14156-1, Dec. 1971.

Zweig, Hans (1956). Autocorrelation and Granularity. *J. Opt. Soc. Am.* **46**:805, 812.

Zworykin, V. K., and Morton, G. A. (1954). "The Kinescope" and "The Electron Gun," Chapters 11 and 12 in *Television*, 2nd ed., pp. 383–484. Wiley, New York.

INDEX

Index

P

Parameters
  display, 63
  sensitometric, 35
Passband
  equivalent, 266
  noise-equivalent, 263
Pattern
  bar, 181, 202
    criterion for recognition, 216
    detection, 174
      probability of, 218
    equivalent, 186, 224, 225, 319
  beat, 64
  Moiré, 292
  motion, 206
  periodic, 292
    test, 180
  sine wave, 174
  spurious periodic, 280
  targets, three-bar, 5
  unstandard test, 174
Perception flash data, 124
Performance, 76*ff*, 119
  human visual system, 142
  ideal photon counter, 142
  index to, 123
  observer, 63
Phosphors, persistence MTF measurement, 277
Photocurrent, 171
Photoelectron incremental signal, 171
Photographic image quality, 313
Photometric display characteristics, 89
Photon counter, 141
  ideal, 122, 128
  model, 135
Photoprocess sensor, primary, 170
Photosurface irradiance, 170
Point raster, 257
Point spread function, 149, 285
Poisson probability
  density, 129
  distribution, 171
  law, 121
Power
  noise, spectral analysis, 162
  resolving, 2, 13, 21, 317

Preamplifier, 173
  noise current, 180
Preferred magnification, 47
Prefiltering, 236, 283
Primary sensor photoprocess, 170
Probability
  bar pattern detection, 218
  correct, 95
  detection, 53, 129, 131, 133, 134, 141
  false alarm, 129, 132*ff*, 145
  Gaussian density, 129
  identification, 224
  Poisson
    density, 129
    distribution, 171
    law, 121
  recognition, 216, 221*ff*
  response, 133
    spurious, 129, 133
Procedures, experimental, 35, 174
Process
  electrooptical imaging, 169
  raster, 238
    line, 256
    noise in, 263
    spectra of, 265
  scanning, 239
  superposition, 147
Product
  cross, evaluation of, 252
  modulation, 248
Psychophysical experiments, 34, 63, 188, 214
Psychophysical factors, 1, 2, 24*ff*, 59*ff*, 95*ff*, 188*ff*, 214*ff*
Pulse
  one-dimensional rectangular, 176
  unit rectangular input, 176

Q

Q (quantum yield), 122
  real image detecting device, 123
Quality, 11, 13
  correlation, 27, 28
  criterion display, 93